physica status solidi (b)
basic research

Volumes 133 to 138

Subject and Author Index 1986

Volumes 93 to 98

physica status solidi (a)
applied research

AKADEMIE-VERLAG BERLIN

ISSN 0370-1972/0031-8965 phys. stat. sol. (b)/(a), Berlin **R19** (1986) 1—595

International Classification System for Physics*)

60. Condensed matter: structure, mechanical and thermal properties
    61. Structure of liquids and solids; crystallography
        (see also 68.20. Solid surface structure, 71. Electron states)
    62. Mechanical and acoustical properties of condensed matter
        (see also 61.70. Defects in crystals, 68.30. Surfaces and interfaces)
    63. Lattice dynamics and crystal statistics
        (see also 65. Thermal properties, 66.70. Thermal conduction, 68.30. Dynamics of surface and interface vibrations, 78.30. Infrared and Raman spectra)
    64. Equations of state, phase equilibria, and phase transitions
    65. Thermal properties of condensed matter
        (see also 63. Lattice dynamics; for thermodynamic properties of quantum fluids, see 67.40; for thermal properties of solid helium, see 67.80)
    66. Transport properties of condensed matter (nonelectronic)
    67. Quantum fluids and solids; liquid and solid helium
    68. Surfaces and interfaces; thin films and whiskers
        (for impact phenomena, see 79; for crystal growth, see 61.50)
70. Condensed matter: electronic structure; electrical, magnetic, and optical properties
    71. Electronic states
        (see also 63. Lattice dynamics, 73. Electronic structure and electrical properties of surfaces, interfaces, and thin films)
    72. Electronic transport in condensed matter
        (for surfaces, interfaces, and thin films, see 73)
    73. Electronic and electrical properties of surfaces, interfaces, and thin films
    74. Superconductivity
    75. Magnetic properties and materials
    76. Magnetic resonances and relaxation in condensed matter; Mössbauer effect
    77. Dielectric properties and materials
        (for conductivity phenomena, see 72.20 and 72.80)
    78. Optical properties and condensed-matter spectroscopy and other interactions of matter with particles and radiation
        (for phonon spectra, see 63)
    79. Electron and ion emission by liquids and solids; impact phenomena

---

*) Excerpt; reproduced with permission of International Council for Scientific and Technical Information (ICSTI).

(The Substance Classification is given on cover three)

physica status solidi (b)

basic research

physica status solidi (a)

applied research

SUBJECT AND AUTHOR INDEX

Volumes 133 to 138, 1986 Volumes 93 to 98, 1986

ISSN 0370-1972/0031-8965

Edited by
E. BORCHARDT

AKADEMIE-VERLAG BERLIN

Subscriptions and orders for this Subject and Author Index
or single copies of physica status solidi (b) and physica status solidi (a) should be directed
in the GDR:

to the Postzeitungsvertrieb or
to the Akademie-Verlag Berlin, Leipziger Str. 3—4, DDR-1086 Berlin;

in the other socialist countries:

to a book-shop for foreign language literature or
to the competent news-distributing agency;

in the FRG and BERLIN (WEST):

to a book-shop or
to the wholesale distributing agency Kunst und Wissen, Erich Bieber OHG, Wilhelmstr. 4—6, D-7000 Stuttgart 1;

in the other Western European countries:

to Kunst und Wissen Erich Bieber GmbH, General Wille-Str. 4, CH-8002 Zürich;

in USA and CANADA:

to VCH Publishers, Inc., 303 NW 12th Avenue, Deerfield Beach, FL 33442—1788, USA;

in other countries:

to the international book- and journal-selling trade,
to Buchexport, Volkseigener Außenhandelsbetrieb der Deutschen Demokratischen Republik, Postfach 160, DDR-7010 Leipzig, or
to the Akademie-Verlag Berlin, Leipziger Straße 3—4, DDR-1086 Berlin.

Schriftleiter und verantwortlich für den Inhalt der Zeitschriften:

Professor Dr. E. Gutsche, Leipziger Str. 3—4, Postfach 1233, DDR-1086 Berlin.

Herausgeber und verantwortlich für den Inhalt des Registers:

Dipl.-Phys. E. Borchardt, Leipziger Str. 3—4, Postfach 1233, DDR-1086 Berlin.

Verlag:

Akademie-Verlag Berlin, Leipziger Str. 3—4, DDR-1086 Berlin; Fernruf: 2236221 und 2236229; Telex-Nr.: 114420;
Bank: Staatsbank der DDR, Berlin, Kto.-Nr.: 6836-26-20712.

Veröffentlicht unter der Lizenznummer 1310 und 1620 des Presseamtes beim Vorsitzenden des Ministerrates der Deutschen Demokratischen Republik.

Gesamtherstellung:

VEB Druckerei „Thomas Müntzer", DDR-5820 Bad Langensalza.

Bestellnummer des Registers: 1068/85 R 19.

Printed in the German Democratic Republic.

AN (EDV) 20831/20735

Preface

The nineteenth index volume contains volumes 133 to 138 of "physica status solidi (b) — basic research" as well as volumes 93 to 98 of "physica status solidi (a) — applied research", i.e. all volumes having appeared in 1986 containing 1469 papers.

Beginning from January 1986 the International Classification System for Physics (Condensed Matter, items 60 and 70) of the International Council for Scientific and Technical Information, Abstracting Board completed by a Substance Classification will be used for quoting papers in pysica status solidi. Nevertheless, before taking over these subject classifications into the index volume they were once more carefully revised and in some cases improved and completed in order to get a most uniform and clear classification for all papers. The average number of quotations per paper in this index volume is about 2.8.

Again I have to express my thanks to the readers for their interest, and I will always readily take up any critical remarks and proposals for improvements. I am especially obliged to Prof. Dr. E. Gutsche, Dr. H.-J. Hänsch, Dr. H. Lange, Dr. S. Oberländer, and Dipl. Phys. K. Müller for giving me advice and unselfish help. Furthermore I have to thank Mrs. E. Schwan for her exact and reliable secretary's work at the manuscript.

Berlin, June 1987

Eva Borchardt

International Classification System for Physics*)

60. Condensed matter: structure, mechanical and thermal properties
61. Structure of liquids and solids; crystallography
    (see also 68.20. Solid surface structures, 71. Electron states)
    61.10. X-ray determination of structures
        (for specific determinations, see 61.55 to 61.80)
    61.12. Neutron determination of structures
        (for specific determinations, see 61.55 to 61.80)
    61.14. Electron determination of structures
        (for specific determinations, see 61.55 to 61.80)
    61.16. Other determination of structures
        (for specific determinations, see 61.55 to 61.80)
    61.20. Classical, semiclassical, and quantum theories of liquid structure
        (for electronic states, see 71; for liquid helium, see 67)
    61.25. Studies of specific liquid structures
    61.30. Liquid crystals
    61.40. Amorphous and polymeric materials
    61.50. Crystalline state (including molecular motions in solids)
        (for magnetic structure and spin systems, see 75.25)
    61.55. Specific structures of elements and alloys
    61.60. Specific structures of inorganic compounds
    61.65. Specific structures of organic compounds
    61.70. Defects in crystals
        (see also 61.80. Radiation damage, 62. Mechanical and acoustical properties, 71.55. Impurities and defect levels, 76.30. EPR of color centers and other defects, 78.50. Impurity and defect absorption in solids)
    61.80. Radiation damage and other irradition effects
        (for techniques of structure determination, see 61.10 to 61.16; for electron and ion impact phenomena, see 79.20)
    61.90. Other topics in structure of liquids and solids

62. Mechanical and acustical properties of condensed matter
    (see also 61.70. Defects in crystals, 68.30. Surfaces and interfaces)
    62.10. Mechanical properties of liquids
        (for viscosity of liquids, see also 66.20)
    62.20. Mechanical properties of solids (related to microscopic structure)
    62.30. Mechanical and elastic waves
    62.40. Anelasticity, internal friction, and mechanical resonances
        For thermomechanical effects, see 65.70
        For magnetomechanical effects, see 75.80
        For piezoelectric effects, see 77.60
        For elasto-optical effects, see 78.20
    62.50. High-pressure and shock-wave effects in solids
    62.60. Acoustical properties of liquids
        For lattice dynamics, phonons, see 63
        For second sound in quantum fluids, see 67.40
    62.65. Acoustical properties of solids
        For magnetoacoustic effects in metals, see 75.80
        For acoustoelectric effects, see 72.50
        For acousto-optical effects, see 78.20
    62.80. Ultrasonic relaxation
        (see also 74.30. Ultrasonic attenuation in superconductors)
    62.90. Other topics in mechanical and acoustical properties of condenced matter

63. Lattice dynamics and crystal statistics
(see also 65. Thermal properties, 66.70. Thermal conduction, 68.30. Dynamics of surface and interface vibrations, 78.30. Infrared and Raman spectra)
63.10. General theory
63.20. Phonons and vibrations in crystal lattices
63.50. Vibrational states in disordered systems
63.70. Statistical mechanics of lattice vibrations
(see also 65. Thermal properties of condensed matter, and 66.70. Thermal conduction)
63.75. Statistical mechanics of displacive phase-transitions
For order–disorder and statistical mechanics of model systems, see 64.60
For crystallographic aspects of polymorphic and order–disorder transformations, see 61.50
63.90. Other topics in lattice dynamics and crystal statistics

64. Equations of state, phase equilibria, and phase transitions
64.10. General theory of equations of state and phase equilibria
64.30. Equations of state of specific substances
(see also 65.70. Thermal expansion)
64.60. General studies of phase transitions
(for magnetic, superconducting, and quantum fluid critical phenomena, see 75, 74, and 67, respectively)
64.70. Phase equlibria, phase transitions, and critical points of specific substances
64.75. Solubility, segregation, and mixing
64.80. Other phase properties of systems
64.90. Other topics in equations of state, phase equilibria, and phase transitions

65. Thermal properties of condensed matter
(see also 63. Lattice dynamics; for thermodynamic properties of quantum fluids, see 67.40; for thermal properties of solid helium, see 67.80)
65.20. Heat capacities of liquids
65.40. Heat capacities of solids
(for specific heat of superconductors, see 74.30; for specific heat of magnetic systems, see 75.40)
65.50. Thermodynamic properties and entropy
65.70. Thermal expansion and thermomechanical effects
(see also 64.30. Equations of state)
For thermal conduction in nonmetallic liquids, see 66.60; for nonmetallic solids, see 66.70
For electronic thermal conduction, see 72.10 and 72.20
For thermal conductivity of superconductors, see 74.30
For pyroelectric and electrocaloric effects, see 77.70
65.90. Other topics in thermal properties of condensed matter

66. Transport properties of condensed matter (nonelectronic)
66.10. Diffusion and ionic conduction in liquids
66.20. Diffusive momentum transport
(for viscosity of liquids, see also 62.10)
66.30. Diffusion in solids
66.60. Thermal conduction in nonmetallic liquids
(for thermal conduction in liquid metals, see 72.15)
66.70. Nonelectronic thermal conduction and heat-pulse propagation in nonmetallic solids
(for thermal conduction in solid metals, see 72.15; for statistical mechanics of lattice vibrations, see 63.70)
66.90. Other topics in nonelectronic transport properties

67. Quantum fluids and solids; liquid and solid helium
    - 67.20. Quantum effects on the structure and dynamics of nondegenerate fluids
    - 67.40. Boson degeneracy and superfluidity of helium-4
    - 67.50. Fermi fluids; liquid helium-3
    - 67.60. Mixed systems; liquid helium-3, -4 mixtures
    - 67.70. Films (including physical adsorption)
    - 67.80. Solid helium and related quantum crystals
    - 67.90. Other topics in quantum fluids (e.g., neutron-star matter)

68. Surfaces and interfaces; thin films and whiskers (for impact phenomena, see 79; for crystals growth, see 61.50)
    - 68.10. Fluid surfaces and fluid–fluid interfaces
    - 68.15. Liquid thin films
    - 68.20. Solid surface structures
    - 68.25. Mechanical and acoustical properties of solid surfaces and interfaces
    - 68.30. Dynamics of solid surfaces and interface vibrations
    - 68.40. Surface energy of solids; thermodynamic properties
    - 68.45. Solid–fluid interface processes
    - 68.48. Solid–solid interfaces (including bicrystals) (for grain boundaries, see 61.70)
    - 68.55. Thin film growth, structure, and epitaxy
    - 68.60. Physical properties of thin films: nonelectronic
    - 68.70. Whiskers and dendrites: growth, structure, and nonelectronic properties
    - 68.90. Other topics in the structure and nonelectronic properties of surfaces and thin films

70. Condensed matter: electronic structure; electrical, magnetic, and optical properties

71. Electronic states (see also 63. Lattice dynamics, 73. Electronic structure and electrical properties of surfaces, interfaces, and thin films)
    - 71.10. General theories and computational techniques
    - 71.20. Electronic density of states determinations (including energy states of liquid semiconductors) (see also 65.40. Electronic heat capacity)
    - 71.25. Nonlocalized single-particle electronic states
    - 71.30. Metal–insulator transitions
    - 71.35. Excitons and related phenomena (including electron–hole drops)
    - 71.36. Polaritons (including photon–phonon and photon–magnon interactions)
    - 71.38. Polarons and electron–phonon interactions (see also 63.20. phonon–electron interactions in lattices)
    - 71.45. Collective effects
    - 71.50. Localized single-particle electronic states (excluding impurities)
    - 71.55. Impurity and defect levels
    - 71.60. Positron states (see also 78.70. Positron annihilation)
    - 71.70. Level splitting and interactions (see also 75.10. — in magnetic phenomena, 75.30. Exchange and superexchange interactions, 73.20. Electronic surface states)
    - 71.90. Other topics in electron states

72. Electronic transport in condensed matter
    (for surfaces, interfaces, and thin films, see 73)
    72.10. Theory of electronic transport; scattering mechanism
    72.15. Electronic conduction in metals and alloys
    72.20. Conductivity phenomena in semiconductors and insulators
        (for nonelectronic conduction, see 66.70)
    72.30. High-frequency effects; plasma effects
    72.40. Photoconduction and photovoltaic effects; photodielectric effects
    72.50. Acoustoelectric effects
    72.55. Magnetoacoustic effects
    72.60. Mixed conductivity and conductivity transitions
    72.70. Noise processes and phenomena
    72.80. Conductivity of specific semiconductors and insulators
    72.90. Other topics in electronic transport in condensed matter

73. Electronic structure and electrical properties of surfaces, interfaces, and thin films
    73.20. Electronic surface states
        (for emission and impact phenomena, see 79)
    73.25. Surface conductivity
    73.30. Surface double layers, Schottky barriers, and work functions
    73.40. Interfaces
    73.60. Electronic properties of thin films
    73.90. Other topics in electrical properties of surfaces, interfaces, and thin films

74. Superconductivity
    74.10. Occurrence, critical temperature
    74.20. Theory
    74.30. General properties
    74.40. Fluctuations and critical effects
    74.50. Tunneling phenomena, Josephson effect, and proximity effects
    74.55. Type-I superconductivity
    74.60. Type-II superconductivity
    74.70. Superconducting materials
    74.90. Other topics in superconductivity

75. Magnetic properties and materials
    75.10. General theory and models of magnetic ordering
        (see also 71.25. Nonlocalized single-particle electronic states, 71.70. Level splitting and interactions)
    75.20. Diamagnestism and paramagnetism
    75.25. Spin arrangements in magnetically ordered materials
        (neutron studies, etc.)
    75.30. Magnetically ordered materials, other intrinsic properties
        (for critical point effects, see 75.40)
    75.40. Critical-point effects, specific heats, short-range order
        (see also 65.40. Heat capacities)
    75.50. Studies of specific magnetic materials
    75.60. Domain effects, magnetization curves, and hysteresis
    75.70. Magnetic films and plates
    75.80. Magnetomechanical and magnetoelectric effects, magnetostriction
    For galvanomagnetic effects, see 72.15 and 72.20
    For magneto-optical effects, see 78.20
    75.90. Other topics in magnetic properties and materials

76. Magnetic resonances and relaxation in condensed matter; Mössbauer effect
    76.20. General theory of resonances and relaxation
    76.30. Electron paramagnetic resonance and relaxation
    76.40. Diamagnetic and cyclotron resonances
    76.50. Ferromagnetic, antiferromagnetic, and ferrimagnetic resonances; spin-wave resonance (see also 75.30. Spin waves)
    76.60. Nuclear magnetic resonance and relaxation
    76.70. Magnetic double resonances and cross effects
    76.80. Mössbauer effect; other γ-ray spectroscopy
    76.90. Other topics in magnetic resonances and relaxation

77. Dielectric properties and materials
    (for conductivity phenomena, see 72.20 and 72.80)
    77.20. Permittivity
    77.30. Polarization and depolarization effects
    77.40. Dielectric loss and relaxation
    77.50. Dielectric breakdown and space-charge effects
    77.55. Dielectric thin films
    77.60. Piezoelectricity and electrostriction (for piezo-optical effects, see 78.20)
    77.70. Pyroelectric and electrocaloric effects
    77.80. Ferroelectricity and antiferroelectricity
    77.85. Electrical resonances
    77.90. Other topics in dielectric properties and materials

78. Optical properties and condensed-matter spectroscopy and other interactions of matter with particles and radiation
    (for phonon spectra, see 63)
    78.20. Optical properties and materials
    78.30. Infrared and Raman spectra and scattering
    78.35. Brillouin and Rayleigh scattering
    78.40. Visible and ultraviolet spectra
    78.45. Stimulated emission
    78.50. Impurity and defect absorption in solids
    78.55. Photoluminescence
    78.60. Other luminescence and radiative recombination
    For photoconduction and photovoltaic effects, see 72.40
    78.65. Optical properties of thin films
    78.70. Other interactions of matter with particles and radiation
    78.90. Other topics in optical properties of condensed matter and other interactions of matter with particles and radiation

79. Electron and ion emission by liquids and solids; impact phenomena
    79.20. Impact phenomena, including electron spectra and sputtering
    79.40. Thermionic emission
    79.60. Photoemission and photoelectron spectra
    79.70. Field emission and field ionization
    79.75. Exoelectron emission
    79.80. Resonance tunneling
    79.90. Other topics in emission and impact phenomena in condensed matter

Substance Classification (according to physica status solidi)

S1. Transition metals and their alloys and compounds
(for oxides, see S10)
- S1.1. Fe
- S1.2. Other 3d metals
- S1.3. Noble metals
- S1.4. Lanthanides
- S1.5. Actinides
- S1.61. Carbides, silicides, germanides
- S1.62. Nitrides, phosphides, arsenides
- S1.63. Sulfides, selenides, tellurides

S2. Alkali metals and their alloys and compounds
(for alkali halides, see S9.11)

S3. Alkaline earth metals and their alloys and compounds
(for alkaline earth halides, see S9.12)

S4. Other metals and their alloys

S5. Nonmetallic elements
- S5.11. Silicon
- S5.12. Germanium
- S5.2. Solidified Gases

S6. Group IV compounds
(for transition metal compounds, see S1.61)

S7. Group V compounds
(for transition metal compounds, see S1.62)
- S7.1. III–V compounds
- S7.11. Phosphides
- S7.12. Arsenides
- S7.13. Antimonides
- S7.15. Mixed crystals
- S7.16. Higher compounds

S8. Chalcogenides
(for transition metal compounds, see S1.63)
- S8.1. II–VI compounds
- S8.11. Sulfides
- S8.12. Selenides
- S8.13. Tellurides
- S8.15. Mixed crystals
- S8.16. Higher compounds

S9. Halides
- S9.11. Alkali halides
- S9.12. Alkaline earth halides
- S9.15. Mixed crystals
- S9.16. Higher compounds

S10. Oxides
- S10.1. Simple oxides
- S10.15. Mixed crystals

S11. Other higher compounds
- S11.1. Salts of oxy-acids
- S11.2. Spinels, garnets, ferrites

S12. Organic compounds
- S12.1. Quasi-one-dimensional systems

For comments on the use of the classification scheme see p. 10.

Some Comments on the Use of the Subject and Author Index

The "International Classification System for Physics" (the condensed matter sections 60 and 70) completed by a "Substance Classification" (according to physica status solidi) is used. On cover two the second-level headings of the condensed matter sections and on pages 4 to 8 the third-level headings are reproduced. As a help for using the subject index on pages 11 and 12 in addition also a series of important fourth-level headings is given. The substance classification is published on cover three and on page 9.

All papers will be quoted under each relevant section. There is made a distinction between one or two principal entries (refering to the main subject matter of the paper) and secondary entries (refering to subsidiary subject matter and substance, respectively).In the principal entries the title of the paper is given in its original language and in English. In addition, every principal entry is marked by * in front of the title of the paper. In the secondary entries the title is only given in English.

On the line following the title the subject classification is given (in parantheses) in the same manner as in the volumes. Each entry gives the volume, the number of the issue, the initial and final page numbers, and the year in which the paper appeared. A paper of physica status solidi (b) or physica status solidi (a) is marked by (b) or (a) in front of the line, respectively. The letter "K" before the page numbers indicates a short note published in the offset part of the journal.

Review articles appear, at the head of the list of entries under a given subject and are given the subtitle "Review Articles". These articles are listed in the chronological order of their publication, irrespective of being a paper of physica status solidi (b) or physica status solidi (a). The subject classification for a review article consisting of several parts is based on the article as a whole rather than on the individual parts.

The entries for original papers, short notes, and errata are made together in a single list for each subject and this follows the entries for review articles. These are again ordered according to data of publication, irrespective of being a paper of physica status solidi (b) or physica status solidi (a). Original papers are sometimes published in several parts under the same principal title (in general with special subtitles). The various parts of papers of this type are treated separately and each part is classified according to its particular contents.

In quoting errata the following rule is obeyed. If the erratum refers to a paper quoted in the preceding index, it is listed in the chronological order of its publication otherwise the erratum is quoted together with the paper to which it refers.

The author index gives the names of the authors in alphabetical order. Each name is followed by the bibliographic details of the paper. A paper of physica status solidi (b) or physica status solidi (a) is again marked by (b) or (a) in front of the number of the volume, respectively. In the last column the subject classification of the paper is presented in the same manner as in the volumes. When several papers have been published by the same author they are arranged according to the dates of publication, irrespective of being a paper of physica status solidi (b) or physica status solidi (a). To find a paper of your interest please choose first the special entry of the subject index and then use the volume number and page.

In using the subject index attention should be given to the following details:

61.10 to 61.16 Theory and experimental techniques of determination of structures
- for specific determinations see 61.55 to 61.80

61.16 Other determination of structures
- electron microscopy and field-ion microscopy determinations
- EPR and NMR determinations

61.25 Studies of specific liquid structures
- macromolecular and polymer solutions
- molten salts
- liquid metals

61.40 Amorphous and polymeric materials
- glasses, polymers, elastomers, plastics

61.50 Crystalline state (including molecular motions in solids)
- physics of crystal growth
- crystal symmetry: models, space groups, crystalline systems and classes
- crystal morphology and orientation
- crystallographic aspects of polymorphic and order–disorder transformations

61.70 Defects in crystals
- grain and twin boundaries
- stacking faults and other planar or extended defects
- doping and implantation of impurities
- impurity concentration, distribution, and gradients

61.80 Radiation damage and other irradiation effects
- channeling, blocking, energy loss of particles
- X-rays, γ-rays, electrons and positrons, neutrons, atoms and molecules, ions
- for ion implantation see 61.70

63.20 Phonons and vibrations in crystal lattices
- phonon–phonon, phonon–electron, and phonon–defect interactions

64.60 General studies of phase transitions
- order–disorder and statistical mechanics of model systems
- equilibrium properties near single critical points, critical exponents
- dynamic critical phenomena
- multicritical points
- metastable phases

64.80 Other phase properties of systems
- stoichiometry and homogeneity
- microstructure

66.30 Diffusion in solids
- theory of diffusion and ionic conduction in solids
- self-diffusion and ionic conduction in non-metals
- diffusion, migration, and displacement of impurities and other defects

71.25 Nonlocalized single-particle electronic states
- techniques of band-structure calculation
- measurement of Fermi surface parameters (including dHvA, magnetoacoustic, positron annihilation, and cyclotron resonance studies, etc.)
- effective mass and g-factors
- band structure of crystalline metals, semiconductors, and insulators

71.45 Collective effects
- exchange, correlation, dielectric and magnetic functions, plasmons
- Fermi-Thomas models
- calculations of total electronic binding energy

71.70 Level splitting and interactions
- crystal and ligand fields
- spin–orbit coupling, Zeeman, Stark, and strain splitting
- exchange interactions
- nuclear states and interactions

72.10 Theory of electronic transport; scattering mechanism
- general formulation of transport theory
- scattering by phonons, magnons, and other nonlocalized excitations
- scattering by point defects, dislocations, surfaces, and other imperfections (including Kondo effect)

72.15 Electronic conduction in metals and alloys
- galvanomagnetic and other magnetotransport effects
- thermomagnetic and thermoelectric effects
- collective modes
- scattering mechanism and Kondo effect

72.20 Conductivity phenomena in semiconductors and insulators
- scattering mechanism
- piezoresistance
- galvanomagnetic and other magnetotransport effects
- thermomagnetic and thermoelectric effects

73.40 Interfaces
- contact resistance, contact potential, rectification
- tunneling: general
- p–n junctions
- MIS, MIM, MSM, SIS, and SMS structures

75.30 Magnetically ordered materials, other intrinsic properties
- saturation moments and magnetic susceptibility
- spin waves
- exchange and superexchange interactions
- anisotropy
- magnetic impurity interactions
- magnetic phase boundaries
- magnetocaloric effect

75.50 Studies of specific magnetic materials
- ferromagnetism
- antiferromagnetics, ferrimagnetics
- amorphous magnetic materials
- magnetic liquids

75.60 Domain effects, magnetization curves, and hysteresis
- high coercitivity materials
- fine-particle systems
- magnetic aftereffects
- magnetic annealing and temperature-hysteresis effects

78.20 Optical properties and materials
- general theory
- optical constants and parameters
- optical rotatory power
- birefringence
- piezo-, elasto-, acousto-, electro-, magneto-, and thermo-optical effects

78.60 Other luminescence and radiative recombination
- electroluminescence
- cathodo- and iono-luminescence
- thermoluminescence
- sono- and tribo-luminescence
- chemiluminescence

78.70 Other interactions of matter with particles and radiation
- positron annihilation
- X-ray scattering, absorption, and emission
- microwave and radiofrequency (including resonances)

79.20 Impact phenomena, including electron spectra and sputtering
- electron impact: Auger and secondary emission
- atom, molecule, and ion impact
- atomic and molecular beam interactions

Subject Index

60. Condensed matter: structure, mechanical and thermal properties

61. Structure of liquids and solids; crystallography (see also 68.20. Solid surface structures, 71. Electron states)

61.10. X-ray determination of structures (for specific determinations, see 61.55 to 61.80)

Original Papers and Short Notes

M. V. Kruglov, I. K. Solomin, and A. V. Lunev
Determination of the Photoemission Generation Depth with Use of Experiments on the Dynamic Scattering of X-Rays
(79.60; 61.10; S5.11; S5.12)
(b) Vol. **133**, No. 1, p. 47—55 (1986).

M. Derno, G. Herms, and H. Steil
X-Ray Study on the Boron Coordination in Sodium Borate Glasses at Higher Temperatures
(61.40; 61.10; S11.1)
(a) Vol. **93**, No. 1, p. K5—K8 (1986).

G. Bottoni, D. Candolfo, A. Cecchetti, F. Masoli, E. Mello Ceresa, and A. Papoian
Magnetic Parameters and Crystal Lattice Modifications of Doped CrO_2 Particles Subjected to Thermal Treatment
(61.70 and 75.50; 61.10; S10.1)
(a) Vol. **93**, No. 1, p. K41—K45 (1986).

J. D. Stephenson
*The Tetragonal–Cubic Phase Transition and Coalescence of Double Topographs in Top-Seeded (001) $BaTiO_3$ Single Crystals Using White Synchrotron X-Radiation Topography (WSRT)
(61.10 and 77.80; 61.80; S11.1)
(a) Vol. **93**, No. 2, p. 463—470 (1986).

H. Cerva and W. Graeff
*Contrast Investigations of Surface Acoustic Waves by Stroboscopic Topography. III. Contrast in Transmission Case
(61.10 and 68.25; S11.1)
(a) Vol. **93**, No. 2, p. K129—K132 (1986).

E. Ratner, A. Appelbaum, R. Brener, and M. Eizenberg
Interactions of Thin Al Films with Ni–Cr Alloy and Bilayer Films Deposited on Si
(68.55; 61.10; 79.20; S1.2; S4)
(a) Vol. **94**, No. 1, p. 61—69 (1986).

G. V. Smirnov, Yu. V. Shvydko, U. van Bürck, and R. L. Mössbauer
*Experimental Study of the Influence of Lattice Vibrations on the Suppression Effect at Room Temperature
(61.10 and 76.80; 63.20; S10.15)
(b) Vol. **134**, No. 2, p. 465—475 (1986).

A. Ya. Dzyublik
*Effect of Forced Vibrations on the Scattering of X-Rays and Mössbauer Radiation by a Crystal. II. Dynamical Effects
(61.10 and 76.80; 63.20; 78.30)
(b) Vol. **134**, No. 2, p. 503—513 (1986).

T. Ungár and G. Szenes
Investigation of N-Implanted Fe Single Crystals by X-Ray Line Broadening
(61.70; 61.10; S1.1)
(a) Vol. **94**, No. 2, p. 839—844 (1986).

J. Härtwig and V. Holý
The Absolute "Lattice Parameter" of Crystals with Defects Measured with X-Ray Methods
(61.50; 61.10)
(b) Vol. **135**, No. 1, p. 37—47 (1986).

V. Fritzsche and P. Rennert
A Reduced Angular Momentum Expansion in the Multiple-Scattering Theory
(79.20 and 79.60; 61.10; 61.14; 68.20)
(b) Vol. **135**, No. 1, p. 49—60 (1986).

V. V. Aristov, A. Yu. Nikulin, A. A. Snigirev, and P. Zaumseil
*Experimental Investigation of X-Ray Bragg Diffraction on the Periodic Surface Relief of a Perfect Crystal
(61.10 and 68.20; S5.11)
(a) Vol. **95**, No. 1, p. 81—86 (1986).

E. Wolska and J. Baszyński
Prereactional Transformations in the Topotactic Conversion γ-FeOOH $\rightarrow$ γ-Fe_2O_3
(64.70; 61.10; 76.50; 78.30; S11.2)
(a) Vol. **95**, No. 1, p. 87—92 (1986).

H. J. Juretschke
*Anomalous Asymmetries in X-Ray π-Modes near n-Beam Interactions
(61.10; S5.12)
(b) Vol. **135**, No. 2, p. 455—466 (1986).

R. Subramanian and V. C. Sahni
*Vector Spaces and the Labelling of Diffraction Peaks in Crystals and Quasi-crystals
(61.10)
(b) Vol. **135**, No. 2, p. K95—K97 (1986).

I. Tomov
*A New Diffraction Method for the Determination of the Effective Depth of X-Ray Penetration
(61.10)
(a) Vol. **95**, No. 2, p. 397—405 (1986).

K. T. Gabrielyan, J. Kuběna, and V. Holý
*A Study of the Nature of Dislocation Loops in Growth Striations in Silicon by X-Ray Double Crystal Topography
(61.10 and 61.70; S5.11)
(a) Vol. **95**, No. 2, p. 579—588 (1986).

U. Pietsch
*Erratum to: X-Ray Structure Amplitudes for GaAs and InP
(61.10; S7.11; S7.12)
(a) Vol. **95**, No. 2, p. 749 (1986).

P. W. Deutsch and E. S. Banas
*Computational Simulation of Ice Ih X-Ray Powder Spectrum at 20 K
(61.10; S10.1)
(b) Vol. **136**, No. 1, p. K1—K3 (1986).

W. A. Keller and H. S. Campos
*Self-Consistent Model of Electromagnetic Dynamical Scattering
(61.10)
(a) Vol. **96**, No. 1, p. 83—88 (1986).

F. Cembali, E. Gabilli, R. Lotti, M. Servidori, and A. Zani
*Ambiguity in the Strain Profile Determination by Simulation of Double Crystal X-Ray Rocking Curves
(61.10 and 61.80; S5.11)
(a) Vol. **96**, No. 1, p. K31—K34 (1986).
Erratum
(a) Vol. **97**, No. 2, p. K215 (1986).

S. Hashimoto and H. Iwasaki
*Microdomain Model Analysis for Diffuse X-Ray Scattering from Quenched Cu–40 and 60 at% Au Alloys
(61.10; S1.2; S1.3)
(a) Vol. **96**, No. 2, p. 435—444 (1986).

J. D. Stephenson and B. Strocka
*White Synchrotron Radiation Topography of $Ba_{1.03}Fe_{11.97}O_{19}$ and $Ba_{1.06}Fe_{10.92}Mn_{1.02}O_{19}$ Single Crystals
(61.10 and 75.50; 61.70; S11.2)
(a) Vol. **96**, No. 2, p. 597—602 (1986).

A. L. Golovin and U. Pietsch
*X-Ray Investigations of Submicrometer Layer Heterostructures
(61.10; 68.55; S7.12; S9.15)
(a) Vol. **96**, No. 2, p. K111—K115 (1986).

A. SAKAI, H. SAKA, and T. IMURA
Reduction of Dislocation Density in Si by Thermal Cyclic Annealing
(61.70; 61.10; S5.11)
(a) Vol. **97**, No. 1, p. 57—63 (1986).

U. PIETSCH, V. G. TSIRELSON, and R. P. OZEROV
*X-Ray Electron Charge Density Distribution in Silicon
(61.10; S5.11)
(b) Vol. **137**, No. 2, p. 441—447 (1986).

M. SARKAR
*An Approximate Approach to *ZAF* Factor in Electron-Probe Microanalysis
(61.10; S1.2)
(a) Vol. **97**, No. 2, p. 343—346 (1986).

C. JOURDAN and J. GASTALDI
Synchrotron X-Ray Topography Study of Stresses Appearing to Crystal Boundaries during the Aluminium Recrystallization Process
(61.50 and 61.70; 61.10; S4)
(a) Vol. **97**, No. 2, p. 361—369 (1986).

N. MATTERN, A. G. ILINSKII, H. HERMANN, and A. V. ROMANOVA
X-Ray Diffraction Study on Amorphous and Liquid Fe–B Alloys
(61.25 and 61.40; 61.10; S1.1)
(a) Vol. **97**, No. 2, p. 397—401 (1986).

P. YANG, S. S. JIANG, H. N. ZHOU, and D. FENG
*The X-Ray Dynamical Diffraction Contrast Images of Brazil Twin Boundaries in Natural Quartz
(61.10 and 61.70; S10.1)
(a) Vol. **97**, No. 2, p. 411—420 (1986).

U. PIETSCH, V. G. TSIRELSON, and R. P. OZEROV
*X-Ray Electron Density Distribution of GaAs
(61.10; S7.12)
(b) Vol. **138**, No. 1, p. 47—52 (1986).

I. TOMOV
*Orientation Dependence of the Effective Depth of X-Ray Penetration
(61.10; S1.2)
(a) Vol. **98**, No. 1, p. 43—50 (1986).

K. SUMIYAMA, N. OHSHIMA, and Y. NAKAMURA
Magnetic Properties of Metastable α-Mn-Type $Mn_{1-y}Fe_y$ Alloys Produced by Vapor Quenching
(75.50; 61.10; 76.80; S1.1)
(a) Vol. **98**, No. 1, p. 229—238 (1986).

A. M. AFANASEV, R. M. IMAMOV, E. KH. MUKHAMEDZHANOV, and V. N. PEREGUDOV
*Fluorescence Accompanying X-Ray Diffraction in the Grazing Incidence Bragg-Laue Geometry
(61.10; 78.70)
(a) Vol. **98**, No. 2, p. 367—375 (1986).

A. Tressaud, S. Khaïroun, J. P. Chaminade, and M. Couzi
Structural Phase Transitions in $Rb_2KM^{III}F_6$ Elpasolites. I. Crystal Chemistry and Calorimetric Studies
(64.70; 61.10; 65.50; S9.16)
(a) Vol. **98**, No. 2, p. 417—421 (1986).

F. Cembali, M. Servidori, S. Solmi, Z. Šourek, U. Winter, and P. Zaumseil
Structural and Electrical Characterization of Boron Inplanted in Preamorphized Silicon Layers
(61.70; 61.10; 72.80; S5.11)
(a) Vol. **98**, No. 2, p. 511—516 (1986).

61.12. Neutron determination of structures (for specific determinations, see 61.55 to 61.80)

Original Papers and Short Notes

B. V. Kuvaldin, E. V. Zolotoyabko, E. M. Iolin, and E. A. Raitman
*Neutron Topography Study of Large Single Crystals by the Local Rocking Curve Method
(61.12 and 61.50; S1.2; S5.11)
(a) Vol. **94**, No. 1, p. 45—52 (1986).

A. M. Balagurov, B. Mroz, N. C. Popa, and B. N. Savenko
Investigation of the Structural Phase Transitions in $LiKSO_4$ in a Wide Temperature Range
(64.70; 61.12; 77.80; S11.1)
(a) Vol. **96**, No. 1, p. 25—31 (1986).

U. Hoppe
*Correction of the Scattering Cross-Section of Neutron Diffraction Measurements for Inelastic Scattering in the Case of Amorphous Silica
(61.12; S10.1)
(b) Vol. **137**, No. 1, p. K5—K8 (1986).

B. Chalupa, R. Michalec, L. Horalik, and P. Mikula
*The Study of Neutron Acoustic Effect by Neutron Diffraction on InSb Single Crystal
(61.12; S7.13)
(a) Vol. **97**, No. 2, p. 403—409 (1986).

61.14. Electron determination of structures (for specific determinations, see 61.55 to 61.80)

Original Papers and Short Notes

G. RUDAKOFF and B. MÜLLER
RHEED Study on Surfaces of Amorphous Materials
(68.20; 61.14; S10.1)
(a) Vol. **93**, No. 2, p. 471—478 (1986).

A. E. SMITH and D. F. LYNCH
*Calculated Low Energy Electron Total Current Spectra for the Al(111) Surface
(61.14; 68.20; S4)
(b) Vol. **134**, No. 1, p. 11—15 (1986).

G. KÄSTNER
*A Comparison of Differential Equations that Control Transmission of Strained, Non-Centrosymmetric Crystals by Fast Electrons
(61.14)
(b) Vol. **134**, No. 2, p. 485—493 (1986).

V. FRITZSCHE and P. RENNERT
A Reduced Angular Momentum Expansion in the Multiple-Scattering Theory
(79.20 and 79.60; 61.10; 61.14; 68.20)
(b) Vol. **135**, No. 1, p. 49—60 (1986).

H. MENNINGER, M. BEER, and H. LANGE
Surface Imperfection Induced Electrical Leakage Paths in Silicide Schottky Diodes Investigated by SEM/EBIC and AES
(61.70 and 73.30; 61.14; 79.20; S1.61)
(a) Vol. **96**, No. 1, p. 353—362 (1986).

C. MANOLIKAS, J. VAN TENDELOO, and S. AMELINCKX
The "Devil's Staircase" in CdP_2 and ZnP_2
(61.70; 61.14; 61.16; S7)
(a) Vol. **97**, No. 1, p. 87—102 (1986).

K. HELMING and B. RAUSCHENBACH
*Determination of Implantation-Induced Textures by Electron Diffraction
(61.14 and 68.55; 61.70; S1.2; S1.62)
(b) Vol. **138**, No. 1, p. K1—K6 (1986).

E. CHRZANOWSKI, W. DOLIŃSKI, S. MRÓZ, and E. WOŹNIAK
Investigation of $Pb_{1-x}Sn_xTe$ Crystal Surfaces by Quantitative AES, LEED, and Kikuchi Pattern Observation
(68.20; 61.14; 79.20; S8.15)
(a) Vol. **98**, No. 2, p. 359—366 (1986).

K. WATANABE, Y. KIKUCHI, and H. YAMAGUCHI
*Screening Effect on Lattice Images of Silicon
(61.14; S5.11)
(a) Vol. **98**, No. 2, p. 409—416 (1986).

61.16. Other determination of structures (for specific determinations, see 61.55 to 61.80)

Original Papers and Short Notes

J. P. Zhang, H. Q. Ye, and K. H. Kuo
A High-Resolution Electron Microscopy Study of Domain Structure in Ni_3Nb. I. Translation Domain Boundaries
(61.70; 61.16; S1.2)
(a) Vol. **93**, No. 1, p. 35—44 (1986).

J. P. Zhang, H. Q. Ye, K. H. Kuo, and S. Amelinckx
A High-Resolution Electron Microscopy Study of the Domain Structure in Ni_3Nb. II. Orientation and General Domains
(61.70; 61.16; S.1.2)
(a) Vol. **93**, No. 2, p. 457—462 (1986).

K. Z. Botros and S. S. Sheinin
On Identifying Crystal Surfaces in Electron Microscope Images of Stacking Faults
(61.70 and 68.20; 61.16; S1.2; S1.3)
(a) Vol. **93**, No. 2, p. 515—521 (1986).

Yu-Jin Chang and Xiang-Rong-Qian
A High Resolution Microscopic Study on Nd–Fe–B Permanent Magnetic Alloys
(64.70 and 75.60; 61.16; 75.50; S1.1)
(a) Vol. **93**, No. 2, p. 573—583 (1986).

A. P. Filippov, G. N. Gaidukov, and S. K. Maksimov
*Electron Microscopy Images of Rectilinear Dislocations Crossing the Surface at an Arbitrary Angle
(61.16 and 61.70)
(a) Vol. **94**, No. 1, p. 53—60 (1986).

M. Berti, A. V. Drigo, G. Lulli, P. G. Merli, and M. Vittori Antisari
On the Dynamic Annealing Mechanism in P^+-Implanted Silicon
(61.55 and 61.80; 61.16; S5.11)
(a) Vol. **94**, No. 1, p. 95—106 (1986).

C. Boulesteix, Zenchuan Kang, and M. Lattiaux
HREM Study of Crystalline and Amorphous Regions of a TiO_2 Thin Layer. Modification of the Lattice Parameter inside Small Clusters
(68.55; 61.16; S10.1)
(a) Vol. **94**, No. 2, p. 499—506 (1986).

M. J. Whelan
*On the Theory of Transmission Electron Microscope Diffraction Contrast of Crystalline Specimens with Defects
(61.16 and 61.70)
(a) Vol. **94**, No. 2, p. 609—616 (1986).

D. Baither, R. Koegler, D. Panknin, and E. Wieser
Residual Defects in Implanted Silicon after Annealing with Incoherent Light
(61.70; 61.16; S5.11)
(a) Vol. **94**, No. 2, p. 767—772 (1986).

H. Bartsch, G. Andrä, and E. Glaser
TEM Cross-Sectional Imaging of Laser-Induced Crystallization after Ion Implantation
(68.55; 61.16; 61.80; S5.11)
(a) Vol. **94**, No. 2, p. 773—779 (1986).

B. Rauschenbach and C. Blochwitz
Electron Microscopical Investigations of Ion-Implanted and Fatigued Nickel
(61.70; 61.16; S1.2)
(a) Vol. **94**, No. 2, p. 827—831 (1986).

J. Mäki
Coincidence Patterns of Two Perpendicular Long-Period Domains in Ag_3Mg
(61.55; 61.16; S1.3; S3)
(a) Vol. **95**, No. 1, p. 51—55 (1986).

I. E. Bolotov, V. Yu. Kolosov, and V. B. Malkov
Electron Microscope Investigation of Crystals Based on Bend-Contour Arrangement. III. Formation of Subgrain Boundaries in Dislocation-Free Crystals of Selenium
(61.70; 61.16; S5)
(a) Vol. **95**, No. 2, p. 377—383 (1986).

L. I. Fedina and A. L. Aseev
Study of Interaction of Point Defects with Dislocations in Silicon by Means of Irradiation in an Electron Microscope
(61.70; 61.16; 61.80; S5.11)
(a) Vol. **95**, No. 2, p. 517—529 (1986).

B. Akamatsu, A. M. Jean-Louis, A. C. Papadopoulo and P. Henoc
Electrical Activity of Dislocations Introduced in Optically Degraded InP
(61.70; 61.16; 78.60; S7.11)
(a) Vol. **95**, No. 2, p. 557—564 (1986).

M. Kuwabara, M. Tomita, H. Hashimoto, and H. Endoh
Direct Observation of the Superstructure of the Nearly Commensurate Phase in 1T–TaS_2 by High Resolution Electron Microscopy
(61.50; 61.16; S1.63)
(a) Vol. **96**, No. 1, p. 39—51 (1986).

D. R. McKenzie, P. S. Turner, and D. J. H. Cockayne
Electron Imaging and Energy Loss Studies of the Crystallization of Hydrogenated Amorphous Silicon
(68.55; 61.16; 61.80; S5.11)
(a) Vol. **96**, No. 1, p. 67—73 (1986).

D. I. Piskunov, Yu. A. Danilov, S. K. Maksimov, and E. A. Pitirimova
Nonstoichiometry in Ion Implantation of Gallium Arsenide. Two Competing Mechanisms
(61.70; 61.16; 72.80; S7.12)
(a) Vol. **96**, No. 1, p. 141—153 (1986).

C. Manolikas, J. van Tendeloo, and S. Amelinckx
The "Devil's Staircase" in CdP_2 and ZnP_2
(61.70; 61.14; 61.16; S7)
(a) Vol. **97**, No. 1, p. 87—102 (1986).

K. Z. Botros, S. S. Sheinin, and S. M. Saleh
*On Distinguishing between Electron Microscope Images of Overlapping and Intrinsic Faults in F.C.C. Materials
(61.16 and 61.70; S1.2)
(a) Vol. **97**, No. 1, p. 143—151 (1986).

Y. Bouazra
*Sur certains effets multifaisceaus en faisceau convergent
(On Some Multi-Beam Effects in a Convergent Beam)
(61.16; S1.2; S1.3; S4)
(a) Vol. **98**, No. 1, p. K5—K9 (1986).

A. I. Chumakov and G. V. Smirnov
*On the Possibility of Multiple Increase of the Resolution of Layer-by-Layer Analysis of Materials
(61.16; 76.80; 79.20)
(a) Vol. **98**, No. 2, p. 339—343 (1986).

61.20. Classical, semiclassical, and quantum theories of liquid structure (for electronic states, see 71; for liquid helium, see 67)

Original Paper

R. Tognato
Interfacial Free Energies per Unit Area of Liquid Flat Aggregates Embedded in a Crystal Matrix
(68.45; 61.20; S4)
(a) Vol. **96**, No. 2, p. 451—461 (1986).

61.25. Studies of specific liquid structures

Original Papers and Short Notes

R. V. Gopala Rao and Ratna Das
*Structure Factor Studies of Like Charges in the One-Component Plasma Approximation
(61.25; S9.11)
(a) Vol. **93**, No. 2, p. 443—448 (1986).

P. V. Geld, V. V. Singer, I. Z. Radovskii, and S. V. Konovalov
Variable Valence of Ce and Pr in Liquid Alloys with Fe, Co, and Ni
(71.25; 61.25; 72.15; 75.30; S1.1; S1.2; S1.4)
(b) Vol. **136**, No. 1, p. 153—158 (1986).

K. N. Khanna and G. Shanker
*One-Component Plasma Structure Factor of Liquid Alkali Metals
(61.25; S2)
(b) Vol. **137**, No. 2, p. K107—K110 (1986).

N. Mattern, A. G. Ilinskii, H. Hermann, and A. V. Romanova
*X-Ray Diffraction Study on Amorphous and Liquid Fe–B Alloys
(61.25 and 61.40; 61.10; S1.1)
(a) Vol. **97**, No. 2, p. 397—401 (1986).

61.30. Liquid crystals

Original Papers and Short Notes

S. M. M. Rahman and D. K. Sarker
Temperature Dependence of Structural and Transport Properties of Less-Simple Liquid Metals
(62.10 and 66.10; 61.30; S1.3; S4)
(b) Vol. **133**, No. 2, p. 517—524 (1986).

M. Petrov and P. Simova
*Influence of the Boundary Conditions on the Characteristic Temperature in the Nematic Preceding Smectic C Phase of 4,n-Heptyloxybenzoic Acid
(61.30 and 72.80; 68.45; S12)
(a) Vol. **95**, No. 2, p. K165—K167 (1986).

A. M. Biradar, S. S. Bawa, S. B. Samanta, and S. Chandra
*Measurements of a Free Value of Helical Pitch in a Thin Sample of Ferroelectric Liquid Crystal
(61.30 and 77.80; S12)
(a) Vol. **97**, No. 2, p. 427—430 (1986).

L. Komitov, G. Hauck, and H. D. Koswig
*Initially Splayed Nematic Liquid Crystal Layer under an Electric Field. Induction of π-Twist Metastable State
(61.30; 77.90; 78.65; S12)
(a) Vol. **97**, No. 2, p. 645—655 (1986).

O. Uemura, A. Shishido, and T. Satow
*The X-Ray Diffraction of Molten AgTlTe
(61.30; S8.16)
(a) Vol. **97**, No. 2, p. K113—K117 (1986).

H. P. Hinov
*On the Theory of Partial and Total Internal Reflection of a Plane TM Wave from Symmetrically Realigned Nematic Liquid Crystal Layers
(61.30; 78.65)
(b) Vol. **138**, No. 1, p. 31—46 (1986).

M. P. Petrov and P. D. Simova
*Textures in Twisted Nematics with Smectic C Ordering and in Smectic C Liquid Crystals
(61.30; S12)
(a) Vol. **98**, No. 2, p. 399—407 (1986).

61.40. Amorphous and polymeric materials

Review Article

J. KLIAVA
EPR of Impurity Ions in Disordered Solids. Distributions of the Spin Hamiltonian Parameters
(76.30; 61.40)
(b) Vol. **134**, No. 2, p. 411—455 (1986).
Erratum
(b) Vol. **137**, No. 2, p. 709 (1986).

Original Papers and Short Notes

K. SUMIYAMA, H. EZAWA, and Y. NAKAMURA
Metastable $Fe_{1-x}Ti_x$ Alloys Produced by Vapor Quenching
(64.75; 61.40; S1.1)
(a) Vol. **93**, No. 1, p. 81—86 (1986).

K. H. HEINIG and H.-D. GEILER
*Phenomenological Theory of Explosive Solid-Phase Crystallization of Amorphous Silicon. II. Dynamical Processes
(61.40 and 61.50; 64.70; S5.11)
(a) Vol. **93**, No. 1, p. 99—104 (1986).

M. DERNO, G. HERMS, and H. STEIL
*X-Ray Study on the Boron Coordination in Sodium Borate Glasses at Higher Temperatures
(61.40; 61.10; S11.1)
(a) Vol. **93**, No. 1, p. K5—K8 (1986).

N. ZEKRI and A. BREZINI
*Anderson Localization on a Disordered Series-Parallel Lattice
(61.40)
(b) Vol. **133**, No. 2, p. 463—468 (1986).

A. A. BOKOV, I. P. RAYEVSKII, V. G. SMOTRAKOV, and O. I. PROKOPALO
Kinetics of Compositional Ordering in $Pb_2B'B''O_6$ Crystals
(77.80; 61.40; S11.1)
(a) Vol. **93**, No. 2, p. 411—417 (1986).

A. ZENTKO, P. DUHAJ, M. TIMKO, and V. KAVEČANSKÝ
The Effect of Neutron Irradiation on the Electrical Resistivity of Amorphous $Co_{47}Ni_{25}B_{18}Si_{10}$ Alloys
(72.15; 61.40; 61.80; S1.2)
(a) Vol. **93**, No. 2, p. 685—693 (1986).

V. YU. BALANDIN, A. V. DVURECHENSKII, and L. N. ALEKSANDROV
Intermediate and Self-Sustaining Crystallization of a-Si Layers during Pulsed Electron Beam Annealing
(68.55; 61.40; S5.11)
(a) Vol. **93**, No. 2, p. K105—K109 (1986).

D. M. VANDERWALKER
*The Glass Transition Phases in the Formation of PtSi
(61.40; 68.48; S1.61)
(a) Vol. **94**, No. 1, p. 77—81 (1986).

S. WILKE, J. MAŠEK, and B. VELICKÝ
*Electronic Structure of a-SiH_x with Arbitrary Hydrogen Concentration
(61.40 and 71.20; S5.11)
(b) Vol. **135**, No. 1, p. 309—319 (1986).

A. JEZIERSKI
*Chemical Short-Range Order in the Computer Simulated Transition Metal–Metalloid Alloys
(61.40; S1)
(b) Vol. **135**, No. 2, p. 467—473 (1986).

N. LENGE and H. KRONMÜLLER
Low Temperature Magnetization of Sputtered Amorphous Fe–Ni–B Films
(75.70; 61.40; S1.1)
(a) Vol. **95**, No. 2, p. 621—633 (1986).

N. K. GOBRAN, S. SALEH, B. F. HANNA, and R. KAMEL
Kinetics of Crystallization for $Fe_{40}Ni_{40}B_{20}$ Metallic Glass
(64.70; 61.40; 62.20; 75.60; S1.1)
(a) Vol. **96**, No. 1, p. 33—37 (1986).

S. P. CHENAKIN, V. T. CHEREPIN, A. L. PIVOVAROV, and M. A. VASILEV
*Secondary Ion Emission from Amorphous Metallic Alloys
(61.40 and 61.55; 79.20; S1.1)
(a) Vol. **96**, No. 1, p. K21—K26 (1986).

S. S. KASYMOVA, ZH. K. KOMILOV, and D. M. YUDIN
*Diagrams of Glass Radiation-Induced Center Potential Curves
(61.40 and 61.80)
(a) Vol. **96**, No. 1, p. K49—K52 (1986).

K. WINER and F. WOOTEN
Vibrational Properties of Metastable Phases of Si and Ge
(63.50; 61.40; S5.11; S5.12)
(b) Vol. **136**, No. 2, p. 519—527 (1986).

P. BLAUDECK
Cluster Embedding in an Effective Shell and Calculation of the Density of States for Amorphous Copper
(71.20; 61.40; S1.2)
(b) Vol. **136**, No. 2, p. 673—678 (1986).

M. A. MOSTAFA, J. BALOGH, and E. KUZMANN
*Mössbauer Study of $(Fe_{1-x}Co_x)_{77}B_{13}Si_{10}$ Amorphous Alloys
(61.40; 76.80; S1.1)
(a) Vol. **96**, No. 2, p. 445—449 (1986).

W. GRÄFE
*A Method for the Determination of a Dynamic Fatigue Limit of Glasses
(61.40 and 62.20; 68.25)
(a) Vol. **96**, No. 2, p. 501—507 (1986).

K. SUMIYAMA, Y. KAWAWAKE, and Y. NAKAMURA
*Amorphous Fe–Cu–Ag–Alloys Produced by Vapor Quenching
(61.40 and 75.50; 76.80; S1.1; S1.2; S1.3)
(a) Vol. **96**, No. 2, p. K107—K110 (1986).

B. A. VAID and K. C. SHARMA
*Structural Disorder Model for Amorphous Semiconductors: a-Se
(61.40; 65.50; 78.20; S5)
(b) Vol. **137**, No. 2, p. 433—440 (1986).

N. MATTERN, A. G. ILINSKII, H. HERMANN, and A. V. ROMANOVA
*X-Ray Diffraction Study on Amorphous and Liquid Fe–B Alloys
(61.25 and 61.40; 61.10; S1.1)
(a) Vol. **97**, No. 2, p. 397—401 (1986).

M. KRAUSE, C. MICHALK, B. LIPPOLD, M. BENEDIX, and J. SUWALSKI
Slow Paramagnetic Relaxation in Oxalate (2,2′-Bipyridine) Iron
(75.20; 61.40; 76.80; S1.1)
(a) Vol. **97**, No. 2, p. 521—526 (1986).

J. MENDOLIA and D. LEMOINE
Electrical and Optical Properties of Evaporated Thin Films of a-CdTe
(73.60; 61.40; 72.40; S8.13)
(a) Vol. **97**, No. 2, p. 601—607 (1986).

G. KAMPFRATH, D. DUSCHL, and C. HAMANN
*Doped Polymer Films Prepared by Glow Discharge Polymerization
(61.40 and 73.60; 61.65; S12)
(a) Vol. **97**, No. 2, p. 619—626 (1986).

LI ZONGQUAN and HE YIZHEN
*Crystallization of Metallic Glass $Pd_{80}Si_{20}$
(61.40; S1.61)
(a) Vol. **98**, No. 1, p. 91—105 (1986).

P. DANESH, I. SAVATINOVA, E. ANACHKOVA, and ST. GEORGIEV
Raman Scattering in Glow Discharge Si:H:Cl Films
(78.30; 61.40; 68.55; S5.11)
(a) Vol. **98**, No. 1, 127—132 (1986).

J. SHIRAFUJI, S. NAGATA, and K. SHIRAKAWA
*γ-Irradiation Effect on Electronic Properties in Hydrogenated Amorphous Silicon
(61.40 and 73.60; 61.80; S5.11)
(a) Vol. **98**, No. 1, p. 311—318 (1986).

D. Bahadur
Synthesis and Characterization of NASICON Based Glass
(66.30; 61.40; S10.15)
(a) Vol. **98**, No. 1, p. K23—K26 (1986).

A. A. Bahgat, E. E. Shaisha, S. F. El-Desouki, and A. I. Sabry
*On the Determination on the Partial Pair Distribution Functions in Amorphous Alloys
(61.40; S1.1)
(a) Vol. **98**, No. 2, p. 351—357 (1986).

R. Manaila and M. Zaharescu
*Local Order in High Surface Area Amorphous Silica
(61.40 and 68.20; S10.1)
(a) Vol. **98**, No. 2, p. 377—382 (1986).

P. Simonnin, R. Tourbot, B. Boucher, M. Perrin, and J. Vanhaute
Thermal, Electrical, and Magnetic Properties of the Amorphous Alloys RE_xSi_{1-x} (RE = Gd or Tb; $x = 0.87, 0.59, 0.18$)
(75.50; 61.40; 64.75; 72.60; S1.4; S1.61)
(a) Vol. **98**, No. 2, p. 551—566 (1986).

61.50. Crystalline state (including molecular motions in solids) (for magnetic structure and spin systems, see 75.25)

Original Papers and Short Notes

D. FARKAS and A. RAN
Space Group Theoretical Analysis of Grain Boundaries in Ordered Alloys
(61.70; 61.50)
(a) Vol. **93**, No. 1, p. 45—55 (1986).

V. K. YANOVSKII and V. I. VORONKOVA
*Polymorphism and Properties of Bi_2WO_6 and Bi_2MoO_6
(61.50; 64.70; 77.80; S11.1)
(a) Vol. **93**, No. 1, p. 57—66 (1986).

K. H. HEINIG and H.-D. GEILER
*Phenomenological Theory of Explosive Solid-Phase Crystallization of Amorphous Silicon. II. Dynamical Processes
(61.40 and 61.50; 64.70; S5.11)
(a) Vol. **93**, No. 1, p. 99—104 (1986).

G. HÜBNER and H. LÖFFLER
*The Growth of Guinier-Preston Zones in an Al–12 at% Zn Alloy during the Ageing at RT in Dependence on the Homogenization Temperature
(61.50; S4)
(a) Vol. **93**, No. 1, p. 113—118 (1986).

S. K. SHADANGI, M. SINGH, S. C. PANDA, and S. BHAN
*On the Structure of Equilibrium and Metastable Phases in $Ni_{75}Sn_{25-x}Ge_x$ System
(61.50; 64.70; S1.2)
(a) Vol. **93**, No. 1, p. K1—K4 (1986).

R. JAKUBAS, Z. CZAPLA, Z. GALEWSKI, L. SOBCZYK, O. J. ZOGAŁ, and T. LIS
*Structure and Phase Transition in $(CH_3NH_3)_3Sb_2Cl_9$
(61.50 and 64.70; 76.60; 77.20; S11)
(a) Vol. **93**, No. 2, p. 449—455 (1986).

U. PIETSCH
*Bond Charges and Electronic Charge Transfer in Ternary Semiconductors
(61.50 and 71.50; S7.1; S7.15; S7.16)
(b) Vol. **134**, No. 1, p. 21—27 (1986).

B. V. KUVALDIN, E. V. ZOLOTOYABKO, E. M. IOLIN, and E. A. RAITMAN
*Neutron Topography Study of Large Single Crystals by the Local Rocking Curve Method
(61.12 and 61.50; S1.2; S5.11)
(a) Vol. **94**, No. 1, p. 45—52 (1986).

P. V. Pavlov, D. I. Tetelbaum, V. D. Skupov, Yu. A. Semin, and G. V. Zorina
*Abnormally Deep Structural Changes in Ion-Implanted Silicon
(61.50 and 61.70; S5.11)
(a) Vol. **94**, No. 1, p. 395—402 (1986).
Erratum
(a) Vol. **98**, No. 2 p. 649 (1986).

A. M. Samuel, Meera Rao, R. S. Tiwari, and O. N. Srivastava
*On the Energetics of MX_2 Type Short Period Polytypes with Special Reference to CdI_2 Structures
(61.50; S9)
(a) Vol. **94**, No. 1, p. K5—K10 (1986).

K. Pawlik
*Determination of the Orientation Distribution Function from Pole Figures in Arbitrarily Defined Cells
(61.50)
(b) Vol. **134**, No. 2, p. 477—483 (1986).

H. Wiesner, W. Cordts, and J. Monecke
*Self-Consistent Calculation of the Equilibrium Lattice Constant and of the Bulk Modulus of Silicon with Local and Non-Local Exchange-Correlation Functionals
(61.50 and 71.45; S5.11)
(b) Vol. **134**, No. 2, p. K93—K95 (1986).

D. K. Neb, R. K. Singh, and S. P. Sanyal
*On Stability of Crystal Structures of Rare Gas Solids
(61.50; S5.2)
(b) Vol. **134**, No. 2, p. K97—K100 (1986).

M. Lenglet, P. Foulatier, J. Dürr et J. Arsène
*Caracterisation de la liaison Cu–O dans les oxydes mixtes $CuMM'O_4$ (M = Fe, Cr; M' = Al, Ga, Mn). Corrélation avec l'effect Jahn-Teller
(Characterization of Cu–O Bonds in $CuMM'O_4$ (M = Fe, Cr; M' = Al, Ga, Mn) Mixed Oxydes. Correlation with the Jahn-Teller Effect)
(61.50; 78.40; 78.70; S11.2)
(a) Vol. **94**, No. 2, p. 461—466 (1986).

J. Härtwig and V. Holý
*The Absolute "Lattice Parameter" of Crystals with Defects Measured with X-Ray Methods
(61.50; 61.10)
(b) Vol. **135**, No. 1, p. 37—47 (1986).

M. Shahmiri, D. J. Vaughan, and S. Murphy
*Observation of Twin-Related Order Domains in $CuFePt_2$
(61.50; S1.1; S1.3)
(a) Vol. **95**, No. 1, p. 63—72 (1986).

Ch. Geisser and A. Shluger
Application of Molecular Models to Electronic Structure Calculations of Defects in Oxide Crystals. I. Structural Model for Silicon Dioxide
(71.25; 61.50; S10.1)
(b) Vol. **135**, No. 2, p. 669—680 (1986).

K. Sigrist and H. Fichtner
*Investigation of the Influence of Lattice Distortions on the Energy Inherent in Microcrystalline Materials
(61.50 and 62.20; 65.50; S10.1)
(a) Vol. **95**, No. 2, p. 473—481 (1986).

M. Kuwabara, M. Tomita, H. Hashimoto, and H. Endoh
*Direct Observation of the Superstructure of the Nearly Commensurate Phase in 1 T–TaS_2 by High Resolution Electron Microscopy
(61.50; 61.16; S1.63)
(a) Vol. **96**, No. 1, p. 39—51 (1986).

R. A. Evarestov and V. P. Smirnov
*Application of Band Representations of Space Groups in the Theory of Phase Transitions and Point Defects in Crystals
(61.50 and 71.25; 61.70; 64.60)
(b) Vol. **136**, No. 2, p. 409—415 (1986).

A. K. M. A. Islam, K. Islam, M. N. Islam, and G. Mortuza
*Polarizabilities, Energies, and Pressure Derivatives of Bulk Moduli of Some Rutile-Type Oxide Crystals
(61.50 and 62.20; S10.1)
(b) Vol. **136**, No. 2, p. 443—451 (1986).

O. P. Singh and V. P. Gupta
Comments on Levin's Modifications in the Phillips and Van Vechten Theory. d-Electron Effect on Ionicity
(71.25; 61.50; S8.16; S9; S10.1)
(b) Vol. **137**, No. 1, p. 97—101 (1986).

N. Dutt, A. J. Kaur, and J. Shanker
*Effective Potentials and Crystalline State Properties of Some Fluorite-Structure Crystals
(61.50 and 64.30; 62.20; 63.90; 65.50; 77.20; S9.12)
(b) Vol. **137**, No. 2, p. 459—468 (1986).

L. Misiak, P. Mikołajczak, and M. Subotowicz
*Lithium Rare-Earth Flouride Crystal Growth and Thermal Variation of Lattice Constants
(61.50; 65.70; S9.16)
(a) Vol. **97**, No. 2, p. 353—359 (1986).

C. Jourdan and J. Gastaldi
*Synchroton X-Ray Topography Study of Stresses Appearing to Crystal Boundaries during the Aluminium Recrystallization Process
(61.50 and 61.70; 61.10; S4)
(a) Vol. **97**, No. 2, p. 361—369 (1986).

S. García and E. Altshuler
*Reactivity Study of Ferric Oxides from Different Origins
(61.50; S10.1)
(a) Vol. **97**, No. 2, p. K119—K121 (1986).

M. Ohta, T. Kanadani, and A. Sakakibara
*Effect of Quenching Conditions on the Growth of GP Zones in Al–Zn Alloys
(61.50 and 72.15; S4)
(a) Vol. **97**, No. 2, p. K193—K196 (1986).

R. Kaschner and P. Ziesche
*Force Sum Rules at Surfaces
(61.50 and 68.90; 71.10)
(b) Vol. **138**, No. 1, p. 65—71 (1986).

J. Olives
*The Electrostatic Lattice Energy
(61.50; 68.40)
(b) Vol. **138**, No. 2, p. 457—464 (1986).

A. I. Potekaev, V. E. Egorushkin, and N. S. Golosov
*Formation of Superstructures with Long Period. Microscopic Description
(61.50 and 64.60)
(a) Vol. **98**, No. 2, p. 345—349 (1986).

P. R. Sarode
*EXAFS in Niobium Dichalcogenides Intercalated with First-Row Transition Metals
(61.50; 78.70; S1.63)
(a) Vol. **98**, No. 2, p. 391—397 (1986).

61.55. Specific structures of elements and alloys

Original Papers and Short Notes

J. Dutkiewicz, J. Salawa, H. Löffler, and E. Korngiebel
*The Effect of Small Additions of Si on the Structure of an Al–2 at% Zn–1 at% Mg Mother Alloy
(61.55; S4)
(a) Vol. **93**, No. 2, p. 429—441 (1986).

G. Wendrock and H. Löffler
*Information about the Position of the Coherent Spinodal Curve in the Al–Zn System by Means of the Course of the SAXS Intensity at Elevated Temperatures
(61.55; 64.70; S1.2; S4)
(a) Vol. **93**, No. 2, p. 487—493 (1986).

A. P. Dudka, A. A. Kaminskii, and V. I. Simonov
*Refinement of $NaGdGeO_4$, $NaYGeO_4$, and $NaLuGeO_4$ Single-Crystal Structures
(61.55; 71.70; 78.45; 78.55; S11.1)
(a) Vol. **93**, No. 2, p. 495—502 (1986).

M. Berti, A. V. Drigo, G. Lulli, P. G. Merli, and M. Vittori Antisari
*On the Dynamic Annealing Mechanism in P^+-Implanted Silicon
(61.55 and 61.80; 61.16; S5.11)
(a) Vol. **94**, No. 1, p. 95—106 (1986).

W. Bażela, A. Szytuła, and J. Leciejewicz
Magnetic Properties of ZrMnSi, ZrMnGe, and HfMnSi Intermetallics
(75.20; 61.55; S1.61)
(a) Vol. **94**, No. 1, p. 207—211 (1986)

S. Manhas, A. Manoogian, G. Lamarche, and J. C. Wolley
Lattice Parameters and Magnetic Properties of the $Cd_xZn_yMn_zSe$ Alloy System
(75.50; 61.55; 76.30; S1.63)
(a) Vol. **94**, No. 1, p. 213—221 (1986).

Wang Rong-Yao
*A Further Study of Superconducting Critical Temperature of $L1_2$ Type Superconductors
(61.55 and 74.70; S1.2; S1.3; S1.4; S4)
(a) Vol. **94**, No. 2, p. 445—452 (1986).

M. Hervieu, D. Groult, and B. Raveau
*HREM Study of Hexaferrite $BaFe_{12}O_{19}$ Samples Irradiated by 1.8 GeV Argon Ions
(61.55; 61.80; 75.50; S11.2)
(a) Vol. **94**, No. 2, p. 467—481 (1986).

Sz. B. Török
*Chemical Characterization of Contaminating Airborne Dust Particles by X-Ray Microprobes
(61.55; 79.20)
(a) Vol. **94**, No. 2, p. 849—854 (1986).

J. Mäki
*Coincidence Patterns of Two Perpendicular Long-Period Domains in Ag_3Mg
(61.55; 61.16; S1.3; S3)
(a) Vol. **95**, No. 1, p. 51—55 (1986).

U. Schmidt, A. Zahra, R. Schülbe, and H. Löffler
*On the Stability of Guinier-Preston Zones in an Al–1.87 at% Cu Alloy at Elevated Temperature
(61.55; 62.20; S4)
(a) Vol. **95**, No. 1, p. 73—80 (1986).

T. Ito
*A Pseudopotential Approach to the Discorder Effects in III–V Ternary Semiconductor Alloys
(61.55 and 65.50; S7.15)
(b) Vol. **135**, No. 2, p. 493—502 (1986).

N. Frangis, G. van Tendeloo, C. Manolikas, J. van Landuyt, and S. Amelinckx
*On the Structure of Layered β-$CuIn_5Se_8$. Electron Diffraction and High Resolution Electron Microscopy Study
(61.55; 64.70; S8.16)
(a) Vol. **96**, No. 1, p. 53—65 (1986).

B. S. Sidzhimov, N. V. Stanev, P. T. Tscholakov, and S. D. Neov
*Structure and Magnetic Properties of the Pseudobinary Intermetallic Compounds $Tb(Fe_xCo_{1-x})_3$
(61.55; 75.50; S1)
(a) Vol. **96**, No. 1, p. K15—K19 (1986).

S. P. Chenakin, V. T. Cherepin, A. L. Pivovarov, and M. A. Vasilev
*Secondary Ion Emission from Amorphous Metallic Alloys
(61.40 and 61.55; 79.20; S1.1)
(a) Vol. **96**, No. 1, p. K21—K26 (1986).

Wang Rong-yao and Luo Qi-guang
*Structure of A15 Compounds
(61.55; 74.70; S1; S4)
(a) Vol. **96**, No. 2, p. 397—406 (1986).

M. Berti, A. V. Drigo, G. Lulli, P. G. Merli, and M. Vittori Antisari
*Dose Rate Effects on the Dynamic Annealing Mechanism in P^+-Implanted Silicon
(61.55 and 61.80; S5.11)
(a) Vol. **97**, No. 1, p. 77—85 (1986).

F. Corni and G. M. Antonini
EXAFS Investigation of Interdiffusive Kinetics in Cu–Au Thin Films
(66.30; 61.55; 78.70; S1.2; S1.3)
(a) Vol. **97**, No. 1, p. K21—K24 (1986).

E. V. Chulkov and I. Yu. Sklyadneva
*Phenomenological Description of the Lattice Location of Alloying Additions in Intermetallic Compounds
(61.55; S1.2)
(b) Vol. **137**, No. 2, p. K101—K105 (1986).

W. Pfeiler and R. Reihsner
*Concentration Dependence of Short-Range Order in Some Concentrated Alloys
(61.55; 64.75; S1.1; S1.2; S1.3)
(a) Vol. **97**, No. 2, p. 377—385 (1986).

R. Schülbe, U. Schmidt, and P. Siebert
*On the Stability of Metastable Precipitates in an Al–1.87 at% Cu Alloy at Elevated Temperature
(61.55; 62.20; S4)
(a) Vol. **98**, No. 1, p. 81—89 (1986).

L. Kertész, A. Szász, M. I. Zakharova, and A. G. Khundzhua
*Study of Crystal and Electronic Structure of Ti–Cr Alloys during Decomposition of β-Solid Solution
(61.55; 64.75; 71.25; S1.2)
(a) Vol. **98**, No. 1, p. 107—115 (1986).

61.60. Specific structures of inorganic compounds

Original Papers and Short Notes

M. Ben Salem, B. Yangui, G. Schiffmacher, and C. Boulesteix
*Erratum to: Twinning of the Hexagonal (A) Structure of Rare Earth Sesquioxides
(61.60; S10)
(a) Vol. **93**, No. 1, p. 397 (1986).

G. L. Sharma, S. D. Sharma, and V. K. Agrawal
*Polytypism in Mixed Crystals of $CdBr_{2-x}I_x$
(61.60; S9.15)
(a) Vol. **95**, No. 1, p. 57—61 (1986).

D. Degenhardt, P. Rabe, and R. Haensel
*EXAFS Investigations of TiS_2 and Cu_xTiS_2 Single Crystals ($x = 0.35, 0.4$)
(61.60; 78.70; S8; S8.16)
(a) Vol. **95**, No. 2, p. 439—445 (1986).

A. Klimakow, M. Schenk, and B. Christ
Incorporation of Tl in Monocrystalline n-PbS
(61.70; 61.60; 66.30; S8)
(a) Vol. **95**, No. 2, p. 573—578 (1986).

S. E. Sarkisov and A. A. Kaminskii
Luminescence of Nd^{3+} Ions in Semiconductor $Bi_{12}GeO_{20}$ Crystals
(71.55 and 78.55; 61.60; 78.20; 78.50; S11.1)
(a) Vol. **95**, No. 2, p. 641—649 (1986).

A. Collomb, B. Lambert-Andron, J. X. Boucherle, and D. Samaras
*Crystal Structure and Cobalt Location in the W-Type Hexagonal Ferrite $[Ba]Co_2$–W
(61.60; S11.2)
(a) Vol. **96**, No. 2, p. 385—395 (1986).

O. S. Rajora and A. E. Curzon
*Crystal Structure and Staging of Cu_xNbSe_2 and Ag_xNbSe_2
(61.60; S8.16)
(a) Vol. **97**, No. 1, p. 65—76 (1986).

M. Shiojiri, T. Maeda, T. Isshiki, Y. Hirota, S. Sekimoto, Y. Ikeda, and K. Iwauchi
*Crystal Structures and Magnetic Properties of $BaTi_{2-x}Sn_xFe_4O_{11}$
(61.60 and 75.50; S11)
(a) Vol. **97**, No. 1, p. 231—241 (1986).

A. A. Kaminskii, E. L. Belokoneva, B. V. Mill, S. E. Sarkisov, and K. Kurbanov
Crystal Structure, Absorption, Luminescence Properties, and Stimulated Emission of Ga Gehlenite ($Ca_{2-x}Nd_xGa_{2+x}Si_{1-x}O_7$)
(78.30 and 78.55; 61.60; 78.45; S10)
(a) Vol. **97**, No. 1, p. 279—290 (1986).

D. P. SINGH, K. D. KUNDRA and K. C. NAGPAL
*X-Ray Study of Semiconducting Alloy Ga_2Te_3–In_2Se_3
(61.60; S8.16)
(a) Vol. **97**, No. 1, p. K5—K7 (1986).

M. K. BOROMBAEV, R. Z. LEVITIN, A. S. MARKOSYAN, Z. SMETANA, V. V. SNEGIREV, and P. SVOBODA
*Magnetic and Crystallographic Properties of $Gd(Cu_{1-x}Ni_x)_2$ and $Gd(Cu_{1-x}Al_x)_2$ Intermetallic Compounds
(61.60 and 75.50; S1.2; S1.4)
(a) Vol. **97**, No. 2, p. 501—509 (1986).

M. K. BOROMBAEV, R. Z. LEVITIN, A. S. MARKOSYAN, Z. SMETANA, P. SVOBODA, and V. ŠÍMA
*Magnetic and Crystallographic Properties of $Gd(Cu_{1-x}Co_x)_2$
(61.60 and 75.50; S1.2; S1.4)
(a) Vol. **98**, No. 1, p. 221—227 (1986).

61.65. Specific structures of organic compounds

Original Papers

G. Kampfrath, D. Duschl, and C. Hamann
Doped Polymer Films Prepared by Glow Discharge Polymerization
(61.40 and 73.60; 61.65; S12)
(a) Vol. **97**, No. 2, p. 619—626 (1986).

Chang Youming, Lü Shengqing, and Qian Renyuan
*Polymorphism and Electronic Properties of 3-Ethyl-5-[2-(3-Ethyl-2-Benzothiazolinylidene)-Ethylidene]-Rhodanine
(61.65; 72.40; 78.30; S12)
(a) Vol. **98**, No. 1, p. 37—42 (1986).

M. S. D. Elyoubi, M. H. Ben Ghozlen, Y. Mlik, and A. Daoud
*Crystal Structure and Theoretical Symmetry Analysis of the Phase Transition Sequence in $[(C_2H_5)_4N]_2PtCl_6$
(61.65; 64.70; S12)
(a) Vol. **98**, No. 2, p. 435—440 (1986).

61.70. Defects in crystals

(see also 61.80. Radiation damage, 62. Mechanical and acoustical properties, 71.55. Impurities and defect levels, 76.30. EPR of color centers and other defects, 78.50. Impurity and defect absorption in solids)

Review Articles

B. X. Liu
Ion Mixing and Metallic Alloy Phase Formation
(61.80 and 64.75; 61.70; S1.1; S1.2; S1.3; S4)
(a) Vol. **94**, No. 1, p. 11—34 (1986).

R. Z. Valiev, V. Yu. Gertsman, and O. A. Kaibyshev
*Grain Boundary Structure and Properties under External Influence
(61.70)
(a) Vol. **97**, No. 1, p. 11—56 (1986).

Original Papers and Short Notes

J. P. Zhang, H. Q. Ye, and K. H. Kuo
*A High-Resolution Electron Microscopy Study of Domain Structure in Ni_3Nb. I. Translation Domain Boundaries
(61.70; 61.16; S1.2)
(a) Vol. **93**, No. 1, p. 35—44 (1986).

D. Farkas and A. Ran
*Space Group Theoretical Analysis of Grain Boundaries in Ordered Alloys
(61.70; 61.50)
(a) Vol. **93**, No. 1, p. 45—55 (1986).

K. Shahi, W. Weppner, and A. Rabenau
*Defects and the First-Order Phase Transition in AgI
(61.70 and 64.70; 66.30; S9)
(a) Vol. **93**, No. 1, p. 171 176 (1986).

T. Kawakubo
*Color Centers in Irradiated Para-Dichlorobenzene Crystals
(61.70; 76.30; 78.50; S12)
(a) Vol. **93**, No. 1, p. 177—184 (1986).

H. Reuther
*Conversion Electron Mössbauer Spectroscopy of Nitrogen and Phosphorus Implanted Iron
(61.70; 76.80; S1.1)
(a) Vol. **93**, No. 1, p. 201—205 (1986).

E. Wendler, W. Wesch, and G. Götz
*Near-Edge Optical Absorption Behaviour in Weakly Damaged Ion-Implanted GaAs
(61.70 and 78.50; S7.12)
(a) Vol. **93**, No. 1, p. 207—212 (1986).

J. FILIPECKI, E. JAKUBCZYK, B. CALUSIŃSKI, and D. GAUDYN
*Electron Phenomena Connected with Hydriding of the Metallic Glass $Fe_{81}B_{13.5}Si_{3.5}C_2$
(61.70; 72.15; 78.70; S1.1)
(a) Vol. **93**, No. 1, p. K25—K28 (1986).

G. BOTTONI, D. CANDOLFO, A. CECCHETTI, F. MASOLI, E. MELLO CERESA, and A. PAPOIAN
*Magnetic Parameters and Crystal Lattice Modifications of Doped CrO_2 Particles Subjected to Thermal Treatment
(61.70 and 75.50; 61.10; S10.1)
(a) Vol. **93**, No. 1, p. K41—K45 (1986).

A. KUCZKOWSKI and J. LIWO
Change in Electrical Conductivity of Polymers with Conjugated Bonds by Means of H_2^+ Ion Implantation
(72.80; 61.70; S12)
(a) Vol. **93**, No. 1, p. K69—K73 (1986).

S. G. GAGARIN, A. V. MUDRYI, A. L. PUSHKARCHUK, A. G. ULYASHIN, and YU. A. TETERIN
*Theoretical Analysis of the Chemical Activity of Noble Gas Atoms in Silicon
(61.70 and 71.55; S5.11)
(b) Vol. **133**, No. 2, p. 499—503 (1986).

YU. M. ANNENKOV, S. B. NURMAGAMBETOV, and A. B. CHORMONOV
*Nature of the Centres of Positron Annihilation in Intensively Irradiated Alkali Halides
(61.70 and 78.70; S9.11)
(b) Vol. **133**, No. 2, p. 505—509 (1986).

V. A. RADTSIG and A. A. BOBYSHEV
Twofold Coordinated Si and Ge Atoms and "Hydrogen" Paramagnetic Centers in Amorphous SiO_2 and GeO_2
(71.55; 61.70; 76.30; S10.1)
(b) Vol. **133**, No. 2, p. 621—627 (1986).

K. D. NIERZEWSKI, B. MACALIK, and H. OPYRCHAŁ
*Effect of Low Temperature γ-Irradiation on I–V Dipoles in KCl: Eu^{2+} Crystals
(61.70 and 61.80; 78.50; S9.11)
(b) Vol. **133**, No. 2, p. K91—K93 (1986).

B. T. DESHMUKH, B. H. GHIYA, and R. B. PODE
Study of the Particle Size Effect on the Thermoluminescence of KCl
(78.60; 61.70; S9.11)
(b) Vol. **133**, No. 2, p. K125—K128 (1986).

J. P. ZHANG, H. Q. YE, K. H. KUO, and S. AMELINCKX
*A High-Resolution Electron Microscopy Study of the Domain Structure in Ni_3Nb. II. Orientation and General Domains
(61.70; 61.16; S1.2)
(a) Vol. **93**, No. 2, p. 457—462 (1986).

K. Z. BOTROS and S. S. SHEININ
*On Identifying Crystal Surfaces in Electron Microscope Images of Stacking Faults
(61.70 and 68.20; 61.16; S1.2; S1.3)
(a) Vol. **93**, No. 2, p. 515—521 (1986).

C. ZALDO and F. AGULLÓ-LÓPEZ
*Optical Monitoring of the Precipitation-Induced Hardening in NaCl:Sr and NaCl:Ca
(61.70; 78.50; 78.55; S9.11)
(a) Vol. **93**, No. 2, p. 535—542 (1986).

Cs. SZELES, A. VÉRTES, and G. BRAUER
*The Effect of Deposition Temperature on the Defect Structure of Ni Electrodeposits Studied by Positron Annihilation
(61.70; 78.70; S1.2)
(a) Vol. **93**, No. 2, p. 559—564 (1986).

K. D. GLINCHUK, N. M. LITOVCHENKO, and V. YU. PTITSIN
*A Variety of Oxygen-Induced Recombination Centres in 450 to 600 °C Heat-Treated Silicon
(61.70; 71.55; S5.11)
(a) Vol. **93**, No. 2, p. 565—571 (1986).

F.-G. KIRSCHT, V. I. NIKITENKO, H. RICHTER, E.-A. STEINMANN, and E. B. YAKIMOV
*Photoluminescence of Preannealed Plastically Deformed Silicon Crystals
(61.70 and 78.55; S5.11)
(a) Vol. **93**, No. 2, p. K143—K146 (1986).

L. I. MURIN
*On the Nature of Interstitial Defect with E_c — 0.16 eV Level in Irradiated Silicon
(61.70 and 71.55; S5.11)
(a) Vol. **93**, No. 2, p. K147—K149 (1986).

N. I. AKULOVICH, V. M. MAKSIMENKO, and V. I. UTENKO
*On Localization of Phosphorus in Neutron-Doped Silicon
(61.70; 72.80; S5.11)
(a) Vol. **93**, No. 2, p. K157—K159 (1986).

J. Q. SHEN, C. W. LUNG, and K. L. WANG
*Dislocation Core Models and Their Positron Annihilation Effects
(61.70; 78.70)
(b) Vol. **134**, No. 1, p. 97—102 (1986).

A. KLAMT and H. TEICHLER
*Quantum-Mechanical Calculations for Hydrogen in Niobium and Tantalum. I. Model Potential and Properties of Self-Trapped States
(61.70; 66.30; S1)
(b) Vol. **134**, No. 1, p. 103—114 (1986).

L. Bosi and M. Nimis
*Perturbative Effects on the F Centre Lifetime in NaF Crystals Doped with Magnesium
(61.70; 78.50; S9.11)
(b) Vol. **134**, No. 1, p. K17—K21 (1986).

R. B. Pode and B. H. Ghiya
*On the Deformation Effect on Z_1 Centres in KCl:Sr
(61.70 and 78.60; S9.11)
(b) Vol. **134**, No. 1, p. K23—K26 (1986).

Chen Guanghua, Zhang Fangqing, and Xu Xixiang
*Doping Effect on the Hydrogen Content of a-$Si_{1-x}C_x$:H Films
(61.70; S5.11)
(b) Vol. **134**, No. 1, p. K27—K29 (1986).

A. P. Filippov, G. N. Gaidukov, and S. K. Maksimov
*Electron Microscopy Images of Rectilinear Dislocations Crossing the Surface at an Arbitrary Angle
(61.16 and 61.70)
(a) Vol. **94**, No. 1, p. 53—60 (1986).

O. A. Aleksandrova, M. I. Kamchatka, M. S. Miropolskii, and V. V. Passynkov
Diffusion of Native Defects in $(Pb_{1-x}Sn_x)_{1-y}Te_y$ during Liquid Phase Epitaxy
(66.30; 61.70; S8.15)
(a) Vol. **94**, No. 1, p. 139—145 (1986).

N. V. Moseev, V. G. Chudinov, B. N. Goshchitskii, and V. I. Protasov
*Molecular Dynamics Study of Atomic Configuration of Point Defects in Intermetallic Compound Mo_3Si
(61.70; S1.61)
(a) Vol. **94**, No. 1, p. 147—152 (1986).

T. V. Saunina, D. B. Chesnokova, and D. A. Yaskov
*The Equilibrium of Native Defects in the Crystals of $(Pb_{1-x}Sn_x)_{1-y}Te_y$
(61.70; 72.20; S8.15)
(a) Vol. **94**, No. 1, p. 161—170 (1986).

A. M. Mazzone, M. Servidori, and G. Cembali
*Monte Carlo Simulation of the Annealing Kinetics of Oxygen Recoils Formed in Through-Oxide Implants on Silicon
(61.70; S5.11)
(a) Vol. **94**, No. 1, p. 171—176 (1986).

D. Enakshi and K. V. Rao
Influence of High AC Electric Field on Dielectric Properties of Quenched and X-Ray Irradiated NaCl and KCl Single Crystals
(77.40; 61.70; S9.11)
(a) Vol. **94**, No. 1, p. 177—182 (1986).

C. M. Su and T. S. Kê
Further Experiments on the High-Temperature Internal Friction Peak in High-Purity Single Crystal Aluminium
(62.40; 61.70; S4)
(a) Vol. **94**, No. 1, p. 191—202 (1986).

K. Mochizuki and M. Takakusaki
Growth of In-Doped ZnS_xSe_{1-x} Single Crystals and Their Photoluminescence
(78.55; 61.70; 72.80; S8.15)
(a) Vol. **94**, No. 1, p. 243—250 (1986).

O. Goede, W. Heimbrodt, T. Lau, G. Matzkeit, and B. Selle
Optical and RBS Studies of ZnS: Te Thin Films
(78.65; 61.70; S8.11)
(a) Vol. **94**, No. 1, p. 259—265 (1986).

F. Cembali, M. Servidori, E. Landi, and S. Solmi
Influence of Damage Depth Profile on the Characteristics of Shallow p^+/n Implanted Junctions
(73.40; 61.70; S5.11)
(a) Vol. **94**, No. 1, p. 315—319 (1986).

V. S. Vavilov, I. P. Akimchenko, and V. V. Krasnopevtsev
*Ion Implantation as a Tool to Control Properties of Amorphous Hydrogenated Silicon
(61.70 and 73.60; 78.65; S5.11)
(a) Vol. **94**, No. 1, p. 343—350 (1986).

Cs. Hajdu, F. Pászti, G. Mezey, and I. Lovas
Stress Model for the Formation of Wave-Like Structures on High-Dose Ion Implanted Materials
(61.80 and 68.20; 61.70)
(a) Vol. **94**, No. 1, p. 351—352 (1986).

J. Hlávka
*Excess Carrier Recombination Centre Distribution in Implanted and Annealed Silicon
(61.70; 72.40; S5.11)
(a) Vol. **94**, No. 1, p. 353—355 (1986).

D. S. Karpuzov
*Profile Evolution in High-Dose Ion Implantation. A Computer Simulation Study
(61.70; 79.20; S5.11)
(a) Vol. **94**, No. 1, p. 365—370 (1986).

I. B. Khaibullin, G. G. Zakirov, M. M. Zaripov, T. Lohner, L. Pogány, G. Mezey, M. Fried, E. Kótai, F. Pászti, A. Manuaba, and J. Gyulai
*Effect of Heavy Ion Implantation and Laser Annealing on the Structural Properties of Germanium
(61.70; 61.80; S5.12)
(a) Vol. **94**, No. 1, p. 371—377 (1986).

A. F. Khokhlov, A. I. Mashin, A. V. Ershov, and Yu. A. Mordvinova
The Properties of Amorphous Silicon Doped with the Isovalent Germanium Impurity
(73.60 and 78.65; 61.70; S5.11)
(a) Vol. **94**, No. 1, p. 379—384 (1986).

J. Kotliński, K. Mojejko-Kotlińska, M. Subotowicz, and I. Bryłowska
*Deep Levels in Si(p) Implanted with C^+, N^+, O^+, and Ne^+ Ions Measured by DLTS Method
(61.70; S5.11)
(a) Vol. **94**, No. 1, p. 385—389 (1986).

K. Paprocki and I. Bryłowska
Recoil Implantation of Al into Disordered Silicon
(61.80; 61.70; S5.11)
(a) Vol. **94**, No. 1, p. 391—394 (1986).

P. V. Pavlov, D. I. Tetelbaum, V. D. Skupov, Yu. A. Semin, and G. V. Zorina
*Abnormally Deep Structural Changes in Ion-Implanted Silicon
(61.50 and 61.70; S5.11)
(a) Vol. **94**, No. 1, p. 395—402 (1986).
Erratum
(a) Vol. **98**, No. 2, p. 649 (1986).

G. Vízkelethy, M. Fried, G. Mezey, F. Pászti, and J. Gyulai
*Simulations of Energetic Ions in Solids by Monte-Carlo Method (SEISM) and Its Comparison with Experiments
(61.70; S5.11)
(a) Vol. **94**, No. 1, p. 413—418 (1986).

T. Barbaszewski, M. Drwiega, E. Lipinska, and F. Starzyk
*Iodine Implantation in Poly-N-Vinylcarbazole — an Organic Semiconductor
(61.70; 61.80; 78.65; S12)
(a) Vol. **94**, No. 1, p. 419—422 (1986).

P. P. Pal-Val and S. Kadečková
Amplitude Dependent Internal Friction in High-Purity Iron Single Crystals at Low Temperatures
(62.20; 61.70; S1.1)
(a) Vol. **94**, No. 1, p. K29—K32 (1986).

P. Gaworzewski and E. Hild
*Maximum Concentration of Thermal Donors Formed at 450 °C in Silicon in Dependence on the Oxygen Concentration
(61.70; S5.11)
(a) Vol. **94**, No. 1, p. K33—K38 (1986).

M. J. Whelan
*Comment on the Contrast of Stacking Fault Fringes
(61.70)
(a) Vol. **94**, No. 1, p. K39—K40 (1986).

H. Lemke
Doping Properties of Silver in Silicon
(71.55; 61.70; S5.11)
(a) Vol. **94**, No. 1, p. K55—K59 (1986).

I. Kosacki, K. Hibner, A. P. Litvinchuk, and M. Ya. Valakh
Lattice Dynamics and Ionic Conductivity of PbF_2: 6% LiF Superionic Crystals
(66.30 and 78.30; 61.70; S9)
(b) Vol. **134**, No. 2, p. 495—501 (1986).

A. Klamt and H. Teichler
Quantum-Mechanical Calculations for Hydrogen in Niobium and Tantalum. II. Diffusion Coefficients
(66.30; 61.70; S1)
(b) Vol. **134**, No. 2, p. 533—544 (1986).

J. Golczewski, K. Maier, and W. Frank
*Defect Reactions in Electron-Irradiated α-$Fe^{111}In$ Studied by Means of PAC Measurements
(61.70; 78.90; S1.1)
(b) Vol. **134**, No. 2, p. 551—559 (1986).

R. C. Bhuniya and S. E. Rao
*Colour Centres and Luminescence in KCl:Ni Crystals
(61.70 and 78.60; 78.50; S9.11)
(b) Vol. **134**, No. 2, p. 561—568 (1986).

K. Schiele
φ^3, φ^4 Model with Quenched Impurities
(64.60; 61.70)
(b) Vol. **134**, No. 2, p. K89—K91 (1986).

M. J. Marcinkowski
*Relationship between Rotation, Deformation, and Cell Structure
(61.70)
(a) Vol. **94**, No. 2, p. 489—498 (1986).

M. S. Sakr, A. E. E. Abdel-Rahiem, and A. A. El-Daly
Steady State Creep during Transformation in Zn–40 wt% Al and Zn–0.5 wt% Al Alloys
(62.20; 61.70; S1.2)
(a) Vol. **94**, No. 2, p. 561—568 (1986).

V. P. Popov, A. V. Dvurechenskii, B. P. Kashinikov, and A. I. Popov
*Defects in Supersaturated Solid Solution of Arsenic in Silicon at Rapid Thermal Annealing
(61.70; 72.80; S5.11)
(a) Vol. **94**, No. 2, p. 569—572 (1986).

Qian Dingrong
Below-Gap Absorption and Precipitation in n-Type $Hg_{1-x}Cd_xTe$
(78.30; 61.70; 63.20; S8.15)
(a) Vol. **94**, No. 2, p. 573—579 (1986).

A. F. Vladimirov and E. N. Moos
*Energy of Dislocation Loops in Crystals
(61.70)
(a) Vol. **94**, No. 2, p. 581—586 (1986).

I. Mártil, N. de Diego, and C. Hidalgo
*Role of Defects in the Annealing Behaviour of RF Sputtered CdS Films
(61.70; 73.60; 78.70; S8.11)
(a) Vol. **94**, No. 2, p. 587—593 (1986).

S. Takamura and M. Kobiyama
*Effect of Low-Temperature Deformation on Internal Friction Peaks of Irradiated Al Alloys
(61.70; 62.40; S4)
(a) Vol. **94**, No. 2, p. 595—600 (1986).

K. Iwasaki
*Strain Amplitude Dependence of the Grain Boundary Internal Friction Peak of Pure Aluminium
(61.70; 62.40; S4)
(a) Vol. **94**, No. 2, p. 601—607 (1986).

M. J. Whelan
*On the Theory of Transmission Electron Microscope Diffraction Contrast of Crystalline Specimens with Defects
(61.16 and 61.70)
(a) Vol. **94**, No. 2, p. 609—616 (1986).

W. Wesch and G. Götz
*Rapid Annealing of Ion-Implanted GaAs
(61.70; 61.80; 73.60; 78.20; S7.12)
(a) Vol. **94**, No. 2, p. 745—766 (1986).

D. Baither, R. Koegler, D. Panknin, and E. Wieser
*Residual Defects in Implanted Silicon after Annealing with Incoherent Light
(61.70; 61.16; S5.11)
(a) Vol. **94**, No. 2, p. 767—772 (1986).

E. Glaser, G. Andrä, H. Bartsch, K. Drenda, and G. Götz
Formation of Monocrystalline Layers by Explosive Crystallization of Ion Implanted Amorphous Silicon
(68.55; 61.70; S5.11)
(a) Vol. **94**, No. 2, p. 781—786 (1986).

A. Turos
*Defect–Impurity Interactions in Ion-Implanted Metals
(61.70; S1.1; S1.2; S1.3; S4)
(a) Vol. **94**, No. 2, p. 809—822 (1986).

K. K. Kadyrzhanov, A. K. Zhetbaev, M. Sh. Aimanov, and A. N. Ozernoi
*Mössbauer Investigations of Phase Formation in Nitrogen-Implanted Iron
(61.70; 76.80; S1.1)
(a) Vol. **94**, No. 2, p. 823—825 (1986).

B. Rauschenbach and C. Blochwitz
*Electron Microscopical Investigations of Ion-Implanted and Fatigued Nickel
(61.70; 61.16; S1.2)
(a) Vol. **94**, No. 2, p. 827—831 (1986).

B. Rauschenbach and K. Hohmuth
*Synthesis of Compounds by High-Fluence Nitrogen Ion Implantation in Titanium
(61.70 and 68.55; S1.2)
(a) Vol. **94**, No. 2, p. 833—837 (1986).

T. Ungár and G. Szenes
*Investigation of N-Implanted Fe Single Crystals by X-Ray Line Broadening
(61.70; 61.10; S1.1)
(a) Vol. **94**, No. 2, p. 839—844 (1986).

J. Zemek and J. Král
*Chemical State of Ion-Implanted Nitrogen in Iron
(61.70; 79.20; 79.60; S1.1)
(a) Vol. **94**, No. 2, p. 845—848 (1986).

H. Opyrchał, K. D. Nierzewski, B. Macalik, and M. Mladenova
*γ-Ray Induced Coloration of KCl: Eu^{2+} Crystals
(61.70; 78.50; 78.60; S9.11)
(b) Vol. **135**, No. 1, p. 141—148 (1986).

E. Rzepka, S. Lefrant, and L. Taurel
*Optical Study of the Defect Stability in Tl-Doped Potassium Iodide
(61.70; 78.30; 78.50; 78.60; S9.11)
(b) Vol. **135**, No. 1, p. 149—154 (1986).

J. Mašek and J. J. Mareš
*Model of Inactive Nitrogen Incorporation in Amorphous Silicon: Nitridated Vacancy
(61.70 and 71.55; S5.11)
(b) Vol. **135**, No. 1, p. K33—K38 (1986).

K. Takaishi, T. Kikuchi, K. Furuya, I. Hashimoto, H. Yamaguchi, E. Yagi, and M. Iwaki
Thermal Extraction of Krypton in Aluminium Using Mass Spectrometer
(66.30; 61.70; S4)
(a) Vol. **95**, No. 1, p. 135—139 (1986).

F. H. Cocks and J. T. A. Pollock
*The Ion Implantation of Helium into Gold. A Combined X-Ray Reflection Spectroscopy and Rutherford Backscattering Study
(61.70; 68.55; S1.3)
(a) Vol. **95**, No. 1, p. 141—148 (1986).

A. M. Mazzone
Defect Distribution in Ion-Implanted Silicon. A Monte Carlo Simulation
(61.70; 61.80; S5.11)
(a) Vol. **95**, No. 1, p. 149—154 (1986).

S. Kishida, K. Matsuura, H. Nagase, H. Mori, F. Takeda, and I. Tsurumi
The Photosensitive Optical Absorption Bands in Zn-Treated and Neutron-Irradiated ZnSe Single Crystals
(78.50; 61.70; S8.12)
(a) Vol. **95**, No. 1, p. 155—164 (1986).

S. Takamura and M. Kobiyama
*Dislocation Pinning in Al and Ag Alloys after Low-Temperature Deformation
(61.70; 62.20; 72.15; S1.3; S4)
(a) Vol. **95**, No. 1, p. 165—172 (1986).

I. E. Bondarenko, H. Blumtritt, J. Heydenreich, V. V. Kazmiruk, and E. B. Yakimov
*Recombination Properties of Dislocation Slip Planes
(61.70; 71.55; S5.11)
(a) Vol. **95**, No. 1, p. 173—177 (1986).

J. L. Lindström, G. S. Oehrlein, and J. W. Corbett
*A Study of the Annealing of the 830 cm^{-1} IR Band Observed in Electron-Irradiated Silicon
(61.70 and 78.30; S5.11)
(a) Vol. **95**, No. 1, p. 179—184 (1986).

H. Indusekhar and V. Kumar
Properties of Iron Related Quenched-In Levels in p-Silicon
(71.55; 61.70; 73.40; S5.11)
(a) Vol. **95**, No. 1, p. 269—278 (1986).

W. R. Fahrner, C. P. Schneider, and E. F. Gorey
Electrical Properties of 1 MeV Helium Implanted Metal–Oxide–Semiconductor Structures
(73.40; 61.70; S5.11)
(a) Vol. **95**, No. 1, p. 343—351 (1986).

Z. A. Matysina, M. I. Milyan, and S. Yu. Zaginaichenko
The Solubility of Interstitial Impurities in Ordering Phases
(64.75; 61.70; S1.1; S1.3)
(a) Vol. **95**, No. 1, p. K9—K14 (1986).

K. Schmalz, V. V. Bolotov, H. Richter, and K. Tittelbach-Helmrich
*DLTS Investigation on the Interaction of Radiation Defects with Thermal Defects in Czochralski p-Si
(61.70 and 61.80; 71.55; S5.11)
(a) Vol. **95**, No. 1, p. K15—K20 (1986).

V. S. Varichenko, A. M. Zaitsev, and V. F. Stelmakh
Luminescence of Natural IIa Diamond Implanted with Nitrogen Ions
(78.55 and 78.60; 61.70; S5)
(a) Vol. **95**, No. 1, p. K25—K28 (1986).

A. M. Zaitsev, V. B. Shipilo, E. M. Shishonok, and A. A. Melnikov
Cathodoluminescence of Annealed Cubic Boron Nitride
(78.60; 61.70; S7)
(a) Vol. **95**, No. 1, p. K29—K32 (1986).

P. H. Bisio and A. M. Monti
*Small Vacancy Clusters in H.C.P. Metals
(61.70; S3)
(b) Vol. **135**, No. 2, p. 545—553 (1986).

I. E. Bolotov, V. Yu. Kolosov, and V. B. Malkov
*Electron Microscope Investigation of Crystals Based on Bend-Contour Arrangement. III. Formation of Subgrain Boundaries in Dislocation-Free Crystals of Selenium
(61.70; 61.16; S5)
(a) Vol. **95**, No. 2, p. 377—383 (1986).

G. Gawlik, J. Jagielski, and A. Podgórski
Fatigue-Life Enhancement of Nitrogen-Implanted Chromium Steel
(62.20; 61.70; S1.1)
(a) Vol. **95**, No. 2, p. 453—458 (1986).

L. I. Fedina and A. L. Aseev
*Study of Interaction of Point Defects with Dislocations in Silicon by Means of Irradiation in an Electron Microscope
(61.70; 61.16; 61.80; S5.11)
(a) Vol. **95**, No. 2, p. 517—529 (1986).

N. M. Nancheva, K. Saarinen, and G. S. Popov
*Positron Annihilation in Shock Loaded Titanium and Titanium Alloy BT14
(61.70; 78.70; S1.2)
(a) Vol. **95**, No. 2, p. 531—536 (1986).

B. G. Svensson and J. L. Lindström
*Annealing Studies of the 862 cm^{-1} Infrared Band in Silicon
(61.70; 78.30; S5.11)
(a) Vol. **95**, No. 2, p. 537—542 (1986).

J. N. Daou, P. Vajda, A. Lucasson, and J. P. Burger
*Defect Kinetics in Quenched and Electron-Irradiated α-$ScH(D)_x$ Solid Solutions
(61.70 and 72.15; 61.80; S1.2)
(a) Vol. **95**, No. 2, p. 543—555 (1986).

B. Akamatsu, A. M. Jean-Louis, A. C. Papadopoulo et P. Henoc
*Activité électrique de dislocations induites dans InP dégradé optiquement
(Electrical Activity of Dislocations Introduced in Optically Degraded InP)
(61.70; 61.16; 78.60; S7.11)
(a) Vol. **95**, No. 2, p. 557—564 (1986).

A. KLIMAKOW, M. SCHENK, and B. CHRIST
*Incorporation of Tl in Monocrystalline n-PbS
(61.70; 61.60; 66.30; S8)
(a) Vol. **95**, No. 2, p. 573—578 (1986).

K. T. GABRIELYAN, J. KUBĚNA, and V. HOLÝ
*A Study of the Nature of Dislocation Loops in Growth Striations in Silicon by X-Ray Double Crystal Topography
(61.10 and 61.70; S5.11)
(a) Vol. **95**, No. 2, p. 579—588 (1986).

S. GUIMARÃES, E. LANDI, and S. SOLMI
*Enhanced Diffusion Phenomena during Rapid Thermal Annealing of Preamorphized Boron-Implanted Silicon
(61.70 and 66.30; S5.11)
(a) Vol. **95**, No. 2, p. 589—598 (1986).

H. LEMKE
*Störstellenreaktionen bei Cu-dotierten Siliziumkristallen
(Defect Interactions in Cu-Doped Silicon Crystals)
(61.70 and 71.55; S5.11)
(a) Vol. **95**, No. 2, p. 665—677 (1986).

A. R. ALI and G. SAID
*On the Migration of Point Defects in Irradiated Pure Ni and Ni–Cu Alloys
(61.70; 61.80; S1.2)
(a) Vol. **95**, No. 2, p. K109—K112 (1986).

V. S. VARICHENKO, A. M. ZAITSEV, and V. F. STELMAKH
*Spatial Distribution of Impurities and Defects Introduced in Diamond by High Energy Ion Implantation
(61.70; S5)
(a) Vol. **95**, No. 2, K123—K126 (1986).

M. KOHYAMA, R. YAMAMOTO, and M. DOYAMA
*Energies and Structures of (111) Coincidence Twist Boundaries in 3C-SiC, Diamond, and Silicon
(61.70; S5; S5.11; S6)
(b) Vol. **136**, No. 1, p. 31—36 (1986).

I. HEYNDERICKX, E. GOOVAERTS, S. V. NISTOR, and D. SCHOEMAKER
*Electron-Spin-Resonance Study of $Pb^+(1)$ Centers of the Laser-Active Structure in KCl and RbCl
(61.70 and 76.30; S9.11)
(b) Vol. **136**, No. 1, p. 69—83 (1986).

T. SOMA, T. MATSUOKA, and H.-MATSUO KAGAYA
*Formation and Migration Energies of an Interstitial Atom in Si and Ge
(61.70; S5.11; S5.12)
(b) Vol. **136**, No. 1, p. 85—90 (1986).

M. B. Dutt, R. Kumar, R. Nath, M. N. Sen, and K. K. Laroia
*Depth Profiling of Ion Implanted Silicon by Electrical Methods
(61.70; 66.30; 73.40; S5.11)
(a) Vol. **96**, No. 1, p. 121—127 (1986).

V. P. Goltsev, S. I. Zhukova, and V. M. Anishchik
*Bismuth Crystal Hardening at High-Temperature Creep after Heavy-Ion Implantation
(61.70; 62.20; S4)
(a) Vol. **96**, No. 1, p. 135—139 (1986).

D. I. Piskunov, Yu. A. Danilov, S. K. Maksimov, and E. A. Pitirimova
*Nonstoichiometry in Ion Implantation of Gallium Arsenide. Two Competing Mechanisms
(61.70; 61.16; 72.80; S7.12)
(a) Vol. **96**, No. 1, p. 141—153 (1986).

A. V. Dvurechenskii, V. V. Suprunchik, and L. S. Smirnov
*Divacancy-Type Defect with Low Symmetry in Silicon
(61.70; 71.55; 76.30; S5.11)
(a) Vol. **96**, No. 1, p. 155—159 (1986).

I. Baker and D. J. Gaydosh
*Displacement Fringes in FeAl
(61.70; S1.1)
(a) Vol. **96**, No. 1, p. 185—190 (1986).

B. C. Grabmaier and R. Oberschmid
*Properties of Pure and Doped $Bi_{12}GeO_{20}$ and $Bi_{12}SiO_{20}$ Crystals
(61.70 and 71.55; 72.40; 78.50; 78.55; S10)
(a) Vol. **96**, No. 1, p. 199—210 (1986).

M. T. Sprackling and M. E. Bahrololoom
*Strain-Ageing in Silver Iodobromide
(61.70; 62.20; S9)
(a) Vol. **96**, No. 1, p. 217—224 (1986).

P. Astié, J. J. Couderc, P. Chomel, D. Quelard, and M. Duseaux
*Thermal Activation of Plastic Deformation of Undoped GaAs between 528 and 813 K
(61.70 and 62.20; S7.12)
(a) Vol. **96**, No. 1, p. 225—242 (1986).

L. A. Kazakewich, V. I. Kuznetsov, and P. F. Lugakov
*The Effect of Dislocations on the Activation Energy of Radiation Defect Thermal Ionization
(61.70 and 61.80; 71.55; 72.80; S5.11)
(a) Vol. **96**, No. 1, p. 243—247 (1986).

J. Kub
*A Study of the Lattice Relaxation around Cracks in LPE Garnet Films
(61.70 and 68.60; 75.70; S11.2)
(a) Vol. **96**, No. 1, p. 249—253 (1986).

T. WOSINSKI and O. BREITENSTEIN
Spatial Distribution of Dominant Electron and Hole Traps in Plastically Deformed GaAs
(71.55; 61.70; S7.12)
(a) Vol. **96**, No. 1, p. 311—315 (1986).

H. MENNIGER, M. BEER, and H. LANGE
*Surface Imperfection Induced Electrical Leakage Paths in Silicide Schottky Diodes Investigated by SEM/EBIC and AES
(61.70 and 73.30; 61.14; 79.20; S1.61)
(a) Vol. **96**, No. 1, p. 353—362 (1986).

A. PAWEŁEK
*The Expressions for the Cottrell-Stokes Law Derived from Seeger's Theory of Thermally-Activated Dislocation Motion
(61.70)
(a) Vol. **96**, No. 1, p. K39—K42 (1986).

R. A. EVARESTOV and V. P. SMIRNOV
Application of Band Representations of Space Groups in the Theory of Phase Transitions and Point Defects in Crystals
(61.50 and 71.25; 61.70; 64.60)
(b) Vol. **136**, No. 2, p. 409—415 (1986).

V. M. BELENKII, R. B. BEGZANOV, V. N. BOLOTOV, and A. V. KUZNICHENKO
*Influence of a Surface on Lifetime Spectra of Positrons
(61.70 and 68.20; 78.70)
(b) Vol. **136**, No. 2, p. 551—556 (1986).

R. B. BOSSOLI and L. E. HALLIBURTON
^{27}Al Hyperfine and Quadrupole Interactions for the $[AlO_4]^0$ Center in Quartz
(71.70 and 76.70; 61.70; S10.1)
(b) Vol. **136**, No. 2, p. 709—714 (1986).

K. INABE and N. TAKEUCHI
*On the X-Ray Production of F-Centers in BaFCl Doped with Mono- or Trivalent Impurities
(61.70; 78.50; S9.15)
(b) Vol. **136**, No. 2, p. K97—K99 (1986).

M. J. MARCINKOWSKI
*Grain Boundaries Revisited
(61.70)
(a) Vol. **96**, No. 2. p. 425—434 (1986).

T. S. KÊ and B. S. ZHANG
*Contribution of Bamboo Boundaries to the Internal Friction Peak in Macrocrystalline High-Purity Aluminium
(61.70; 62.40; S4)
(a) Vol. **96**, No. 2, p. 515—525 (1986).

I. M. Filippov, L. A. Kazakevich, P. F. Lugakov, and V. V. Shusha
*The Charge Carrier Recombination Processes in Dislocated Silicon
(61.70 and 71.55; 72.80; S5.11)
(a) Vol. **96**, No. 2, p. 527—531 (1986).

M. R. Dalmau, A. Paz, and J. A. Gorri
*The Electric Charge of Dislocations in Alkali Halides
(61.70; S9.11)
(a) Vol. **96**, No. 2, p. 533—540 (1986).

R. Angelucci, S. Solmi, and Q. Zini
Influence of Precipitation on Phosphorus Diffusivity in Silicon
(66.30; 61.70; S5.11)
(a) Vol. **96**, No. 2, p. 541—546 (1986).

J. D. Stephenson and B. Strocka
White Synchotron Radiation Topography of $Ba_{1.03}Fe_{11.97}O_{19}$ and $Ba_{1.06}Fe_{10.92}Mn_{1.02}O_{19}$ Single Crystals
(61.10 and 75.60; 61.70; S11.2)
(a) Vol. **96**, No. 2, p. 597—602 (1986).

M. Kittler, W. Seifert, K. Schmalz, and K. Tittelbach-Helmrich
*Comparison of EBIC and DLTS Measurements on Boron-Doped CZ Silicon Contaminated with Iron
(61.70 and 71.55; S5.11)
(a) Vol. **96**, No. 2, p. K133—K137 (1986).

R. B. Pode
*Formation of the Z_1 Colour Centre by Mechanical Bleaching in γ-Irradiated KCl: Ca
(61.70; 78.55; S9.11)
(a) Vol. **96**, No. 2, p. K147—K149 (1986).

V. P. Markevich and L. I. Murin
*The Nature of Intrinsic Defects in n-Type Silicon γ-Irradiated at 77 K
(61.70 and 71.55, 61.80; S5.11)
(a) Vol. **96**, No. 2, p. K151—K155 (1986).

A. F. Rakov
*Investigation of Colour Centres in γ-Irradiated $Y_3Al_5O_{12}$ Crystals at Room Temperature
(61.70; 78.50; S11.2)
(a) Vol. **96**, No. 2, p. K169—K173 (1986).

M. Dressel and H. W. Helberg
AC Conductivity of Deformed Germanium Single Crystals at $T = 4.2$ K
(72.20; 61.70; S5.12)
(a) Vol. **96**, No. 2, p. K199—K202 (1986).

M. Kohyama, R. Yamamoto, and M. Doyama
*Structures and Energies of Symmetrical $\langle 011 \rangle$ Tilt Grain Boundaries in Silicon
(61.70; S5.11)
(b) Vol. **137**, No. 1, p. 11—20 (1986).

A. Hamaïdia, A. Hachimi, J. Margerie, and J. F. Hémidy
*EPR of Colour Centres in Alkaline Metal Doped Calcium and Strontium Fluorides
(61.70; 76.30; 78.50; S9.12)
(b) Vol. **137**, No. 1, p. 47—56 (1986).

H. A. Kassim
The F^+-Centre in Alkali Oxides of Antifluorite Structure. A Model Potential Numerical Calculation
(71.55; 61.70; S10.1)
(b) Vol. **137**, No. 1, p. 199—206 (1986).

A. Sakai, H. Saka, and T. Imura
*Reduction of Dislocation Density in Si by Thermal Cyclic Annealing
(61.70; 61.10; S5.11)
(a) Vol. **97**, No. 1, p. 57—63 (1986).

C. Manolikas, J. van Tendeloo, and S. Amelinckx
*The "Devil's Staircase" in CdP_2 and ZnP_2
(61.70; 61.14; 61.16; S7)
(a) Vol. **97**, No. 1, p. 87—102 (1986).

K. Z. Botros, S. S. Sheinin, and S. M. Saleh
*On Distinguishing between Electron Microscope Images of Overlapping and Intrinsic Faults in F.C.C. Materials
(61.16 and 61.70; S1.2)
(a) Vol. **97**, No. 1, p. 143—151 (1986).

H. Stoyan and D. Stoyan
*Simple Stochastric Models for the Analysis of Dislocation Distributions
(61.70)
(a) Vol. **97**, No. 1, p. 163—172 (1986).

J. Kreissl
A Quantitative EPR Analysis of ZnS: Mn Powders of Different Crystal Structure
(76.30; 61.70; S8.11)
(a) Vol. **97**, No. 1, p. 191—198 (1986).

M. Ben Salem, B. Yangui, G. Schiffmacher, and C. Boulesteix
*Erratum to: Twinning of the Hexagonal (A) Structure of Rare Earth Sesquioxides
(61.70; S10.1)
(a) Vol. **97**, No. 1, p. 331 (1986).

P. Gall, A. Kallel, N. Lauret, J. P. Peyrade, B. Brousseau, and J. Mazzaschi
*An Experimental Evidence of the Destruction of Complex Defects by the Introduction of Dislocations in InP
(61.70; 78.55; S7.11)
(a) Vol. **97**, No. 1, p. K25—K27 (1986).

L. N. Kamysheva, S. N. Drozhdin, and O. M. Serdyuk
*The Influence of Defects on the Processes of Polarization Relaxation in TGS Crystals
(61.70 and 77.80; S12)
(a) Vol. **97**, No. 1, p. K29—K34 (1986).

M. Dębowska, A. Baranowski, K. Jerie, and W. Świątkowski
*The Effect of Impregnation with Gold on Positron Lifetime Spectra in NaY Zeolite
(61.70; 68.20; 78.70; S2)
(a) Vol. **97**, No. 1, p. K35—K38 (1986).

S. Shigetomi and T. Matsumori
Electrical Properties of p-Type Layer in Si-Implanted GaAs
(72.80; 61.70; S7.12)
(a) Vol. **97**, No. 1, p. K83—K86 (1986).

J. Bollmann, H. Klose, and A. Mertens
Formation of Schottky Junctions in Silicon by Ion Implantation
(73.30; 61.70; S5.11)
(a) Vol. **97**, No. 1, p. K95—K99 (1986).

L. A. Zilberman
*Fluctuation Unpinning of Dislocations
(61.70)
(b) Vol. **137**, No. 2, p. 469—478 (1986).

W. Pfeiler
*Vacancy Activation Enthalpies in Concentrated AgAl
(61.70; 66.30; S1.3)
(b) Vol. **137**, No. 2, p. K127—K130 (1986).

V. O. Sokolov and V. B. Sulimov
*Semiempirical Calculation of Fluorine Substitutional Impurity in Vitreous Silicon Dioxide
(61.70 and 71.55; S10.1)
(b) Vol. **137**, No. 2, p. K167—K170 (1986).

C. Jourdan and J. Gastaldi
*Synchroton X-Ray Topography Study of Stresses Appearing to Crystal Boundaries during the Aluminium Recrystallization Process
(61.50 and 61.70; 61.10; S4)
(a) Vol. **97**, No. 2, p. 361—369 (1986).

J. Ilczuk and H. Morawiec
*Change of Dislocation Density during Reversible Martensitic Transformation in Cu–Zn–Al
(61.70 and 64.70; S1.2; S4)
(a) Vol. **97**, No. 2, p. 371—376 (1986).

P. Yang, S. S. Jiang, H. N. Zhou, and D. Feng
*The X-Ray Dynamical Diffraction Contrast Images of Brazil Twin Boundaries in Natural Quartz
(61.10 and 61.70; S10.1)
(a) Vol. **97**, No. 2, p. 411—420 (1986).

B. Ya. Farber, Yu. L. Iunin, and V. I. Nikitenko
*Experimental Study of the Double Kink Formation Kinetics and Kink Mobility on the Dislocation Line in Si Single Crystals
(61.70; S5.11)
(a) Vol. **97**, No. 2, p. 469—478 (1986).

D. M. Vanderwalker
*Hydrides on Dislocations in Silicon
(61.70; S5.11)
(a) Vol. **97**, No. 2, p. K127—K130 (1986).

J. Bollmann, H. Klose, and A. Mertens
*Big Channeling Effects at Low-Energy Ion Implantation in Silicon
(61.70; 61.80; S5.11)
(a) Vol. **97**, No. 2, p. K135—K139 (1986).

V. P. Markevich, L. F. Makarenko, and L. I. Murin
*Some New Features of Thermal Donor Formation in Silicon at $T < 800$ K
(61.70 and 71.55; S5.11)
(a) Vol. **97**, No. 2, p. K173—K176 (1986).

H. Weinert, B. Diegner, and J. Kugler
*Photoluminescence Study on the Influence of Substrate Doping on the Properties of Nominally Undoped VPE-GaAs
(61.70 and 78.55; S7.12)
(a) Vol. **97**, No. 2, p. K177—K182 (1986).

M. Wautelet and P. B. Legrand
*Crystal Instability Due to Defect–Phonon Interactions
(61.70 and 63.20; 64.60)
(b) Vol. **138**, No. 1, p. 109—116 (1986).

J. Golczewski and W. Frank
*Rate-Equation Treatment of PAC-Monitored Defect Reactions in Irradiated Metals
(61.70; 76.80; S1)
(b) Vol. **138**, No. 1, p. 123—133 (1986).

A. A. Kusov and V. I. Vladimirov
*The Theory of Dynamic Annihilation of Dislocations
(61.70)
(b) Vol. **138**, No. 1, p. 135—142 (1986).

S. Benakki, A. Goltzene, C. Schwab, Wang Guangyu, and Zou Yuanxi
*Electron Paramagnetic Resonance Analysis of Strain-Induced Defects in Semi-Insulating GaAs
(61.70 and 76.30; S7.12)
(b) Vol. **138**, No. 1, p. 143—149 (1986).

K. Helmig and B. Rauschenbach
Determination of Implantation-Induced Textures by Electron Diffraction
(61.14 and 68.55; 61.70; S1.2; S1.62)
(b) Vol. **138**, No. 1, p. K1—K6 (1986).

M. J. MARCINKOWSKI
*Reanalysis of the Burgers Circuit
(61.70)
(a) Vol. **98**, No. 1, p. 25—35 (1986).

L. D. ZHANG, J. SHI, and T. S. KÊ
*High Temperature Internal Friction Peak in High Purity Single Crystal Aluminium Sheets and Electron Microscopic Observation on the Related Dislocation Configurations
(61.70; 62.40; S4)
(a) Vol. **98**, No. 1, p. 151—162 (1986).

K. ISHIDA and H. KAWANO
*Different Etch Pit Shapes Revealed by Molten KOH Etching on the (001) GaAs Surface and Their Dependence on the Burgers Vectors
(61.70; S7.12)
(a) Vol. **98**, No. 1, p. 175—181 (1986).

J. N. DAOU, P. VAJDA, J. P. BURGER, and A. LUCASSON
*Defect Study of β-TbH(D)$_{1.9+x}$
(61.70; 72.15; S1.4)
(a) Vol. **98**, No. 1, p. 183—194 (1986).

K. BENHADDANE and P. BEAUCHAMP
*Core Structure of the $a\langle 100\rangle\langle 001\rangle$ Edge Dislocation in a B2 Ordered Alloy
(61.70; S1.1)
(a) Vol. **98**, No. 1, p. 195—202 (1986).

V. I. NIKOLAICHIK and I. I. KHODOS
*Determination of the Sign of a Dissociated Dislocation in Weak Beam
(61.70; S5.11)
(a) Vol. **98**, No. 1, p. K19—K21 (1986).

K. P. AREFIEV, V. V. LOPATIN, and YU. P. SUROV
*Macrodefects of Pyrolitic Boron Nitride Structure
(61.70; 78.70; S7)
(a) Vol. **98**, No. 1, p. K27—K32 (1986).

K. H. GEORGY, N. L. TAWFIK, and T. H. YOUSSEF
Role of Grain Boundary Movements on the Creep Rate Sensitivities in Aluminium
(62.20; 61.70; S4)
(a) Vol. **98**, No. 1, p. K33—K38 (1986).

M. KOHYAMA, R. YAMAMOTO, and M. DOYAMA
*Reconstructed Structures of Symmetrical $\langle 011\rangle$ Tilt Grain Boundaries in Silicon
(61.70; S5.11)
(b) Vol. **138**, No. 2, p. 387—397 (1986).

AMAL K. DAS
*Damping of a Kinked Dislocation by Electrons in Metals at Low Temperatures
(61.70)
(b) Vol. **138**, No. 2, p. 483—492 (1986).

K. INABE, Y. UTSUI, and M. TAKEUCHI
Thermoluminescence of BaFCl:Dy and BaFCl:Cu X-Irradiated at Room Temperature
(78.60; 61.70; S9.16)
(b) Vol. **138**, No. 2, p. 493—500 (1986).

J. F. BOAS
*Electron Spin Resonance Studies of Light Sensitive Centers with Tetragonal Symmetry in CaO
(61.70 and 76.30; 71.55; S10.1)
(b) Vol. **138**, No. 2, p. 501—507 (1986).

O. N. SENKOV and V. A. LIKHACHEV
The Effect of Grain Growth on the Plastic Instabilities and Uniaxial Tensile Ductilities in Superplastic Alloys
(62.20; 61.70)
(a) Vol. **98**, No. 2, p. 441—452 (1986).

N. CH. MITRA and A. L. BHATTACHARYYA
*Specification of Dipole Complexes in Dy_2O_3–CaO Systems from Dielectric Relaxation Studies
(61.70; 77.40; S10.15)
(a) Vol. **98**, No. 2, p. 477—486 (1986).

J. NI and T. S. KÊ
*A Theory of Anomalously Amplitude-Dependent Internal Friction Peaks Due to the Interaction between Dislocation Kinks and Substitutional Solute Atoms in F.C.C. Metals
(61.70; 62.40; 66.30; S1.2; S3; S4)
(a) Vol. **98**, No. 2, p. 487—501 (1986).

H.-R. HÖCHE, H. S. LEIPNER, and G. STADERMANN
*Line Defects in $A^{III}B^{V}$ Semiconductors
(61.70; S7.1)
(a) Vol. **98**, No. 2, p. 503—510 (1986).

F. CEMBALI, M. SERVIDORI, S. SOLMI, Z. ŠOUREK, U. WINTER and P. ZAUMSEIL
*Structural and Electrical Characterization of Boron Implanted in Preamorphized Silicon Layers
(61.70; 61.10; 72.80; S5.11)
(a) Vol. **98**, No. 2, p. 511—516 (1986).

I. S. BRAUDE, U. HOFMANN, H.-J. KAUFMANN, and V. I. STARTSEV
*Effect of Point Defects on the Character of Distortions in High-Purity Molybdenum Single Crystals
(61.70; S1)
(a) Vol. **98**, No. 2, p. 521—526 (1986).

A. PAZ, M. R. DALMAU, and J. A. GORRI
*Influence of the Electric Charge of Dislocations on the Hardening of Alkali Halides
(61.70; 62.20; S9.11)
(a) Vol. **98**, No. 2, p. 535—540 (1986).

A. KESSLER and U. KLEIN
Conductivity and Polarization of NH_4Cl between 150 and 500 K
(66.90 and 77.30; 61.70; S9)
(a) Vol. **98**, No. 2, p. 617—627 (1986).

E. VETTER, T. GESSNER, E. WIESER, and C. WEISE
*Redistribution of Implanted Arsenic in a $MoSi_2$/Polysilicon Structure by Silicidation Tempering
(61.70; 68.55; S1.61)
(a) Vol. **98**, No. 2, p. K129—K132 (1986).

R. TOGNATO
*On the Formation of Swirl Defects in Silicon and Germanium
(61.70; S5.11; S5.12)
(a) Vol. **98**, No. 2, p. K133—K136 (1986).

61.80. Radiation damage and other irradiation effects (for techniques of structure determination, see 61.10 to 61.16; for electron and ion impact phenomena, see 79.20)

Review Article

B. X. Liu
*Ion Mixing and Metallic Alloy Phase Formation
(61.80 and 64.75; 61.70; S1.1; S1.2; S1.3; S4)
(a) Vol. **94**, No. 1, p. 11—34 (1986).

Original Papers and Short Notes

A. I. Melker, S. N. Romanov, and N. L. Tarasenko
*Defect Clusters and Subcascades. II. The Splitting and Subcascade Dimensions
(61.80; S1.3)
(b) Vol. **133**, No. 1, p. 111—118 (1986).

V. N. Baier, V. M. Katkov, and V. M. Strakhovenko
*Nondipole Radiation of Electrons at Planar Channeling
(61.80 and 78.70; S5; S5.11)
(b) Vol. **133**, No. 1, p. 211—222 (1986).

M. Kato and Y. H. Ohtsuki
*Stopping Power of Atoms near Metal Surfaces at Grazing Incidence Condition
(61.80; 71.35; S2)
(b) Vol. **133**, No. 1, p. 267—275 (1986).

M. Scrocco
*Electron-Energy-Loss Spectra of Ag Metal
(61.80; S1.3)
(b) Vol. **133**, No. 1, p. K35—K38 (1986).

V. N. Brudnyi, V. A. Charchenko, N. G. Kolin, V. A. Novikov, A. D. Pogrebnyak, and Sh. M. Ruzimov
*Electrical Properties and Positron Annihilation in Neutron-Irradiated n-InP
(61.80; 72.80; 78.70; S7.11)
(a) Vol. **93**, No. 1, p. 195—200 (1986).

E. A. Koptelov and A. A. Semenov
*Two Ways of Void Ordering under Irradiation
(61.80)
(a) Vol. **93**, No. 1, p. K33—K36 (1986).

A. M. Taratin and S. A. Vorobiev
*Computer Simulation of Proton Channeling Catch-Up in Bent Crystals
(61.80; S5.11)
(b) Vol. **133**, No. 2, p. 511—516 (1986).

V. N. BAIER, V. M. KATKOV, and V. M. STRAKHOVENKO
*Radiation Yield of High-Energy Electrons in Thick Crystals
(61.80 and 78.70; S1.1; S1.2; S5; S5.11; S5.12)
(b) Vol. **133**, No. 2, p. 583—592 (1986).

K. D. NIERZEWSKI, B. MACALIK, and H. OPYRCHAŁ
*Effect of Low Temperature γ-Irradiation on I–V Dipoles in KCl: Eu^{2+} Crystals
(61.70 and 61.80; 78.50; S9.11)
(b) Vol. **133**, No. 2, p. K91—K93 (1986).

J. D. STEPHENSON
The Tetragonal–Cubic Phase Transition and Coalescence of Double Topographs in Top-Seeded (001) $BaTiO_3$ Single Crystals Using White Synchrotron X-Radiation Topography (WSRT)
(61.10 and 77.80; 61.80; S11.1)
(a) Vol. **93**, No. 2, p. 463—470 (1986).

H. INDUSEKHAR, V. KUMAR, and D. SENGUPTA
Investigation of Deep Defects Due to α-Particle Irradiation in n-Silicon
(71.55; 61.80; S5.11)
(a) Vol. **93**, No. 2, p. 645—653 (1986).

A. ZENTKO, P. DUHAJ, M. TIMKO, and V. KAVEČANSKÝ
The Effect of Neutron Irradiation on the Electrical Resistivity of Amorphous $Co_{47}Ni_{25}B_{18}Si_{10}$ Alloys
(72.15; 61.40; 61.80; S1.2)
(a) Vol. **93**, No. 2, p. 685—693 (1986).

R. FLAGMEYER and B. SCHUMANN
Rutherford Backscattering/Channeling Analyses of Lattice-Matched $Ca_{0.43}Sr_{0.57}F_2$ Films on GaAs
(68.55; 61.80; 68.48; S9.15)
(a) Vol. **93**, No. 2, p. K117—K121 (1986).

A. ZENTKO, I. ŠKOROVÁNEK, V. V. JEREMENKO, and V. D. CHECHERSKII
Mössbauer Study of Neutron Irradiated $Fe_{60}Ni_{20}B_{20}$ Amorphous Alloys
(76.80; 61.80; S1.1)
(a) Vol. **93**, No. 2, p. K161—K164 (1986).

V. I. SHULGA
*Surface Semichanneling of Molecular Ions
(61.80; S1.2)
(b) Vol. **134**, No. 1, p. 87—95 (1986).

ANAND P. PATHAK and PRASANNA K. J. BALAGARI
*The Effects of Defects on Channeling Radiation
(61.80 and 78.70)
(b) Vol. **134**, No. 1, p. 115—118 (1986).

R. BUTZ and H. WAGNER
Reactions at the Ti/SiO_2 Interface Studied by EELS
(68.48; 68.30; 61.80; S1.2; S10.1)
(a) Vol. **94**, No. 1, p. 71—76 (1986).

M. Berti, A. V. Drigo, G. Lulli, P. G. Merli, and M. Vittori Antisari
*On the Dynamic Annealing Mechanism in P^+-Implanted Silicon
(61.55 and 61.80; 61.16; S5.11)
(a) Vol. **94**, No. 1, p. 95—106 (1986).

K. D. Glinchuk, V. I. Guroshev, and A. V. Prokhorovich
Photoluminescence in Chromium-Doped Semi-Insulating GaAs after Irradiation by 2.2 MeV Electrons and Annealing
(78.55; 61.80; S7.12)
(a) Vol. **94**, No. 1, p. 203—206 (1986).

M. Posselt
*A Short Overview on Monte Carlo Simulations of Ion Beam Penetration into Amorphous Solids
(61.80)
(a) Vol. **94**, No. 1, p. 337—341 (1986).

Cs. Hajdu, F. Pászti, G. Mezey, and I. Lovas
*Stress Model for the Formation of Wave-Like Structures on High-Dose Ion Implanted Materials
(61.80 and 68.20; 61.70)
(a) Vol. **94**, No. 1, p. 351—352 (1986).

K. Holldack, H. Kerkow, and W. Frentrup
*Anomalous Diffusion of Boron in Silicon Due to Heavy Ion Bombardement
(61.80 and 66.30; S5.11)
(a) Vol. **94**, No. 1, p. 357—363 (1986).

I. B. Khaibullin, G. G. Zakirov, M. M. Zaripov, T. Lohner, L. Pogány, G. Mezey, M. Fried, E. Kótai, F. Pászti, A. Manuaba, and J. Gyulai
Effect of Heavy Ion Implantation and Laser Annealing on the Structural Properties of Germanium
(61.70; 61.80; S5.12)
(a) Vol. **94**, No. 1, p. 371—377 (1986).

K. Paprocki and I. Bryłowska
*Recoil Implantation of Al into Disordered Silicon
(61.80; 61.70; S5.11)
(a) Vol. **94**, No. 1, p. 391—394 (1986).

M. Posselt
*Effect of Subcascades on Energy-Loss and Damage Profiles
(61.80; S5.11; S5.12)
(a) Vol. **94**, No. 1, p. 403—408 (1986).

B. Rauschenbach
*Ion Beam Mixing Effects in Cu/Au Multilayered Thin Films
(61.80 and 68.55; 65.50; S1.2; S1.3)
(a) Vol. **94**, No. 1, p. 409—412 (1986).

T. Barbaszewski, M. Drwiega, E. Lipinska, and F. Starzyk
Iodine Implantation in Poly-N-Vinylcarbazole — an Organic Semiconductor
(61.70; 61.80; 78.65; S12)
(a) Vol. **94**, No. 1, p. 419—422 (1986).

I. Bryłowska, K. Paprocki, and M. Subotowicz
*Obtaining Hafnium Silicides by the Mixing Method
(61.80 and 68.55; 73.30; S1.61)
(a) Vol. **94**, No. 1, p. 423—426 (1986).

M. Kulik and J. Żuk
*Ellipsometric Studies of Radiation Damage in Ion Bombarded GaP
(61.80 and 78.20; S7.11)
(a) Vol. **94**, No. 1, p. 427—431 (1986).

F. Berg, H. Pagnia, and N. Sotnik
Influence of Ion Bombardment on the Electrical Behaviour of Discontinuous Metal Film Diodes
(73.40; 61.80; S10.1)
(a) Vol. **94**, No. 1, p. K89—K92 (1986).

M. Hervieu, D. Groult, and B. Raveau
HREM Study of Hexaferrite $BaFe_{12}O_{19}$ Samples Irradiated by 1.8 GeV Argon Ions
(61.55; 61.80; 75.50; S11.2)
(a) Vol. **94**, No. 2, p. 467—481 (1986).

W. Wesch and G. Götz
Rapid Annealing of Ion-Implanted GaAs
(61.70; 61.80; 73.60; 78.20; S7.12)
(a) Vol. **94**, No. 2, p. 745—766 (1986).

H. Bartsch, G. Andrä, and E. Glaser
TEM Cross-Sectional Imaging of Laser-Induced Crystallization after Ion Implantation
(68.55; 61.16; 61.80; S5.11)
(a) Vol. **94**, No. 2, p. 773—779 (1986).

C. Kaschner, A. Witzmann, K. Gärtner, and G. Götz
Laser Annealing of Ion-Implanted NiSi Layers
(68.55; 61.80; S1.61)
(a) Vol. **94**, No. 2, p. 787—791 (1986).

H. Kerkow, G. Kreysch, G. Wolf, and K. Holldack
Ion-Beam-Assisted Annealing of Amorphized Silicon
(68.55; 61.80; S5.11)
(a) Vol. **94**, No. 2, p. 793—796 (1986).

G. Mezey
*Physics of Ion Implantation in Plasma–Surface Interactions
(61.80 and 68.20)
(a) Vol. **94**, No. 2, p. 797—807 (1986).

I. Krafcsik, L. Királyhidi, P. Riedl, J. Gyulai, and M. Fried
*Implantation with Ion Pulses
(61.80)
(a) Vol. **94**, No. 2, p. 855—858 (1986).

F. K. Naehring and G. Otto
*Design of an Electrostatic Octupole for Micro-Beam Deflection
(61.80)
(a) Vol. **94**, No. 2, p. 859—863 (1986).

A. M. Zaitsev, A. A. Melnikov, V. B. Shipilo, and E. M. Shishonok
*Cathodoluminescence of Electron Irradiated Cubic Boron Nitride
(61.80 and 78.60; S7.1)
(a) Vol. **94**, No. 2, p. K125—K127 (1986).

M. Bianconi and A. M. Mazzone
*On the Concept of Cascade Volume in Spike Effects
(61.80; S5.11)
(b) Vol. **135**, No. 1, p. K5—K7 (1986).

A. A. Alybakov, V. A. Gubanova, and M. M. Kidibaev
Anomalous Change of Absorption Spectra of X-Irradiated LiF:U and LiF:U, Na Crystals when Heating
(78.50; 61.80; S9.11)
(b) Vol. **135**, No. 1, p. K49—K52 (1986).

B. Panigrahi, A. K. Tyagi, and R. V. Nandedkar
*Surface Damage of Ferritic Steels by Helium Ion Irradiation
(61.80 and 64.80; S1.1)
(a) Vol. **95**, No. 1, p. 99—103 (1986).

A. M. Mazzone
Defect Distribution in Ion-Implanted Silicon. A Monte Carlo Simulation
(61.70; 61.80; S5.11)
(a) Vol. **95**, No. 1, p. 149—154 (1986).

S. D. Kouimtzi
*AC Conductivity Measurements in Electron-Irradiated Al-Doped Si
(61.80 and 72.20; S5.11)
(a) Vol. **95**, No. 1, p. 317—321 (1986).

K. Schmalz, V. V. Bolotov, H. Richter, and K. Tittelbach-Helmrich
*DLTS Investigation on the Interaction of Radiation Defects with Thermal Defects in Czochralski p-Si
(61.70 and 61.80; 71.55; S5.11)
(a) Vol. **95**, No. 1, p. K15—K20 (1986).

Yu. D. Kornyushkin
*Passing of Electron Beams through a Laminated Electronresist
(61.80)
(a) Vol. **95**, No. 1, p. K81—K86 (1986).

V. V. Kirsanov and E. M. Kislitsina
*Displacement Threshold Energy Change in Uniform Stress Fields
(61.80)
(b) Vol. **135**, No. 2, p. K113—K117 (1986).

S. V. Nistor, I. Ursu, and M. Velter-Stefănescu
Non-Cubic Fe^{3+}-Centers in Irradiated NaCl: Fe Crystals
(76.30; 61.80; S9.11)
(b) Vol. **135**, No. 2, p. K149—K152 (1986).

M. I. Guseva and G. V. Gordeeva
*Phase Transitions in the Surface Layers of Materials under Ion Implantation
(61.80 and 68.20; 64.70; S1.1)
(a) Vol. **95**, No. 2, p. 385—390 (1986).

V. Rybka, V. Odzhaev, J. Červená, V. Hnatowicz, J. Kvítek, and H. Jelínková
*Laser Annealing of Bismuth-Implanted (111) Silicon
(61.80; S5.11)
(a) Vol. **95**, No. 2, p. 511—515 (1986).

L. I. Fedina and A. L. Aseev
Study of Interaction of Point Defects with Dislocations in Silicon by Means of Irradiation in an Electron Microscope
(61.70; 61.16; 61.80; S5.11)
(a) Vol. **95**, No. 2, p. 517—529 (1986).

J. N. Daou, P. Vajda, A. Lucasson, and J. P. Burger
Defects Kinetics in Quenched and Electron-Irradiated α-$ScH(D)_x$ Solid Solutions
(61.70 and 72.15; 61.80; S1.2)
(a) Vol. **95**, No. 2, p. 543—555 (1986).

A. R. Ali and G. Said
On the Migration of Point Defects in Irradiated Pure Ni and Ni–Cu Alloys
(61.70; 61.80; S1.2)
(a) Vol. **95**, No. 2, p. K109—K112 (1986).

A. Balasiński, A. Jakubowski, T. Brożek, and K. Iniewski
*On γ-Radiation and Electron Beam Dose and Gate Oxide Thickness Dependence of Radiation Defects in MOS Capacitors
(61.80 and 73.40; S5.11)
(a) Vol. **95**, No. 2, p. K203—K206 (1986).

A. R. Urmanov, A. M. Linetskaya, and A. A. Puzanov
*A Kinetic Theory for Motion of Accelerated Particles in Solids. I. Derivation of Basic Equations
(61.80)
(b) Vol. **136**, No. 1, p. 37—49 (1986).

Tran Thoai Duy Bao
*Surface Excitations on Small Metal Particles by Fast Electrons
(61.80 and 73.20; S4)
(b) Vol. **136**, No. 1, p. 291—298 (1986).

Y. Kazumata, Y. Nakano, S. Yugo, and T. Kimura
Exchange Coupling Model in Neutron-Irradiated Graphite
(76.30; 61.80; 71.70; S5)
(b) Vol. **136**, No. 1, p. 325—329 (1986).

D. R. McKenzie, P. S. Turner, and D. J. H. Cockayne
Electron Imaging and Energy Loss Studies of the Crystallization of Hydrogenated Amorphous Silicon
(68.55; 61.16; 61.80; S5.11)
(a) Vol. **96**, No. 1, p. 67—73 (1986).

A. Armigliato, G. Queirolo, and G. Valdrè
Nitrogen Profiles of Ion-Implanted TiN Films by Electron Energy Loss Spectroscopy
(68.55; 61.80; S1.62; S5.11)
(a) Vol. **96**, No. 1, p. 89—99 (1986).
Erratum
(a) Vol. **98**, No. 2, p. 651 (1986).

V. V. Bolotov, A. V. Karpov, V. A. Stuchinskii, and K. Schmalz
*Accumulation of Radiation Defects in Oxygen-Rich n-Type Silicon Heat-Treated at Temperatures from 600 to 1000 °C
(61.80; 71.55; S5.11)
(a) Vol. **96**, No. 1, p. 129—134 (1986).

I. Abril, R. Garcia-Molinà, and A. Gras-Marti
*Simple Estimates of Recoil Implantation Quantities
(61.80)
(a) Vol. **96**, No. 1, p. 161—166 (1986).

L. A. Kazakevich, V. I. Kuznetsov, and P. F. Lugakov
*The Effect of Dislocations on the Activation Energy of Radiation Defect Thermal Ionization
(61.70 and 61.80; 71.55; 72.80; S5.11)
(a) Vol. **96**, No. 1, p. 243—247 (1986).

F. Cembali, E. Gabilli, R. Lotti, M. Servidori, and A. Zani
*Ambiguity in the Strain Profile Determination by Simulation of Double Crystal X-Ray Rocking Curves
(61.10 and 61.80; S5.11)
(a) Vol. **96**, No. 1, p. K31—K34 (1986).
Erratum
(a) Vol. **97**, No. 2, p. K215 (1986).

M. K. Georgieva, P. L. Mishev, and G. Ch. Toumbev
*Temperature Dependence of Positron Annihilation in LiF
(61.80 and 78.70; S9.11)
(a) Vol. **96**, No. 1, p. K35—K37 (1986).

S. S. Kasymova, Zh. K. Komilov, and D. M. Yudin
*Diagrams of Glass Radiation-Induced Center Potential Curves
(61.40 and 61.80)
(a) Vol. **96**, No. 1, p. K49—K52 (1986).

A. M. Mazzone
*Channeling of P^+ Ions at 1 MeV in Silicon. A Monte Carlo Simulation
(61.80; S5.11)
(b) Vol. **136**, No. 2, p. K81—K83 (1986).

V. P. Zhukov, A. V. Demidov, and M. N. Uspenskii
*On the Mössbauer Effect Probability Change in Irradiated Materials
(61.80 and 76.80)
(b) Vol. **136**, No. 2, p. K91—K96 (1986).

D. E. Popov, S. E. Savushkin, and S. A. Vorobiev
*Orientation and Energetic Dependence of Electron–Atom Close Collisions at Molecular Channeling
(61.80; S5.11)
(b) Vol. **136**, No. 2, p. K109—K112 (1986).

V. Yu. Petukhov, I. B. Khaibullin, M. M. Zaripov, R. Groetzschel, M. Voelskow, and R. Klabes
Investigation of Silicides Obtained by Fe^+ Implantation into Silicon
(68.20; 61.80; S1.61)
(a) Vol. **96**, No. 2, p. 463—468 (1986).

C. Ascheron, A. Schindler, R. Flagmeyer, and G. Otto
*Swelling, Strain, and Radiation Damage of He^+ Implanted GaP
(61.80; S7.11)
(a) Vol. **96**, No. 2, p. 555—562 (1986).

N. S. Bogdanyuk and S. S. Ostapenko
Structure Peculiarities and Annealing Mechanism of the Red Flash-Like Luminescent Centre in CdS Single Crystals Irradiated with 1.2 MeV Electron Beam
(78.60; 61.80; 71.55; S8.11)
(a) Vol. **96**, No. 2, p. 621—628 (1986).

C. A. Londos
Charge-Dependent Defect-Traces in the DLTS and MCTS Spectra of Silicon
(71.55; 61.80; S5.11)
(a) Vol. **96**, No. 2, p. 637—642 (1986).

A. M. Gue and A. Mazel
*a ⟨100⟩ Dislocation Loops in Electron-Irradiated Cadmium Telluride
(61.80; S8.13)
(a) Vol. **96**, No. 2, p. K145—K146 (1986).

V. P. Markewich and L. I. Murin
The Nature of Intrinsic Defects in n-Type Silicon γ-Irradiated at 77 K
(61.70 and 71.55; 61.80; S5.11)
(a) Vol. **96**, No. 2, p. K151—K155 (1986).

L. Pranevičius, S. Tamulevičius, L. Puodžiukynas, and A. Matiukas
*Application of Laser Interferometer Systems for the Study of Ion Interaction with Solids
(61.80; 68.25; 78.20; S10.1)
(a) Vol. **96**, No. 2, p. K157—K161 (1986).

CHEN GUANGHUA, XU XIXIANG, ZHANG FANGQING, and JIANG ZHICHEN
Valence-Electron Properties of Glow-Discharge a-C:H Films Studied by Electron Energy Loss Spectroscopy
(71.45; 61.80; S5; S6)
(b) Vol. **137**, No. 1, p. K55—K57 (1986).

A. M. TARATIN, YU. M. FILIMONOV, and S. A. VOROBIEV
*Computer Simulation of the Volume Capture Effect for Relativistic Electrons in Crystals
(61.80 and 71.90; S5.11; S5.12)
(b) Vol. **137**, No. 1, p. K63—K67 (1986).

M. M. KIDIBAEV, A. A. ALYBAKOV, and V. A. GUBANOVA
Absorption Bands of U^{4+} and U^{3+} Ions in X-Irradiated LiF:U, Me Crystals
(78.50; 61.80; S9.11)
(b) Vol. **137**, No. 1, p. K73—K75 (1986).

M. BERTI, A. V. DRIGO, G. LULLI, P. G. MERLI, and M. VITTORI ANTISARI
*Dose Rate Effects on the Dynamic Annealing Mechanism in P^+-Implanted Silicon
(61.55 and 61.80; S5.11)
(a) Vol. **97**, No. 1, p. 77—85 (1986).

C. ASCHERON, G. OTTO, R. FLAGMEYER, H.-E. ZSCHAU, V. N. BUGROV, and S. A. KARAMYAN
*Damage Profile on High Energy Ion Implanted GaP
(61.80; S7.11)
(a) Vol. **97**, No. 1, p. K15—K19 (1986).

E. N. VIDALKO, G. P. GAIDAR, V. A. GIRII, V. V. SAVYAK, and V. I. SHAKHOVTSOV
Mobility of Majority Carriers in γ-Irradiated Oxygen-Containing Crystals of n-Si and n-Ge
(72.80; 61.80; S5.11; S5.12)
(a) Vol. **97**, No. 2, p. 565—570 (1986).

J. BOLLMANN, H. KLOSE, and A. MERTENS
Big Channeling Effects at Low-Energy Ion Implantation in Silicon
(61.70; 61.80; S5.11)
(a) Vol. **97**, No. 2, p. K135—K139 (1986).

G. PRASAD and K. V. RAO
Spectral Characteristics of Luminescence Exhibited by X-Ray Irradiated KCl and NaCl Single Crystals Subsequently Excited with Laser Light
(78.55; 61.80; S9.11)
(a) Vol. **97**, No. 2, p. K183—K186 (1986).

J. SHIRAFUJI, S. NAGATA, and K. SHIRAKAWA
γ-Irradiation Effect on Electronic Properties in Hydrogenated Amorphous Silicon
(61.40 and 73.60; 61.80; S5.11)
(a) Vol. **98**, No. 1, p. 311—318 (1986).

F. P. Korshunov, I. G. Marchenko, and V. T. Troshchinskii
*Peculiarities of Minority Carrier Lifetime Variations in p–n–p–n Structures Fabricated upon Neutron-Transmutation Doped Silicon and Irradiated by Fast Electrons
(61.80 and 73.40; S5.11)
(a) Vol. **98**, No. 1, p. K39—K42 (1986).

S. Ashok, K. Giewont, and H. P. Vyas
*On the Ubiquity of Ion Bombardment Modification of Silicon Schottky Barriers
(61.80 and 73.40; S5.11)
(a) Vol. **98**, No. 1, p. K99—K104 (1986).

F. Garofalo and A. M. Mazzone
*Channeling in GaAs. A Monte Carlo Simulation of Low Energy Implants
(61.80; S7.12)
(a) Vol. **98**, No. 2, p. 517—520 (1986).

K. Holldack and H. Kerkow
*Flux Dependence of Damage Accumulation in Silicon during Ion Bombardment
(61.80; S5.11)
(a) Vol. **98**, No. 2, p. 527—534 (1986).

G. J. Papaioannou, P. C. Euthymiou, C. Carabatos, B. Lepley, and A. Bath
*On the Electron Irradiation Induced Defects in GaP:Zn
(61.80 and 71.55; S7.11)
(a) Vol. **98**, No. 2, p. K125—K127 (1986).

H. J. Blythe, E. Klugmann, and F. Dworschak
*Magnetic Relaxation in Cobalt Following Low-Temperature Electron Irradiation
(61.80 and 75.60; S1.2)
(a) Vol. **98**, No. 2, p. K147—K151 (1986).

61.90. Other topics in structure of liquids and solids

Short Note

Y. Yuge
*Renormierungsgruppen-Methode für die Platz-Perkolation im Bienenwaben-Gitter
(Renormalization Group Methods for Site Percolation in the Honeycomb Lattice)
(61.90)
(b) Vol. **137**, No. 1, p. K1—K4 (1986).

62. Mechanical and acoustical properties of condensed matter (see also 61.70. Defects in crystals, 68.30. Surfaces and interfaces)

62.10. Mechanical properties of liquids (for viscosity of liquids, see also 66.20)

Original Paper

62.20. Mechanical properties of solids (related to microscopic structure)

Review Article

J. SHANKER and W. N. BHENDE
*Higher-Order Elastic Constants and Thermoelastic Properties of Ionic Solids
(62.20 and 63.90; 64.30; 65.70; S7.1; S9.11; S9.12; S10.1)
(b) Vol. **136**, No. 1, p. 11—30 (1986).

Original Papers and Short Notes

V. K. JINDAL and J. KALUS
Calculation of Thermal Expansion and Phonon Frequency Shift in Deuterated Naphthalene
(63.20 and 65.70; 62.20; S12)
(b) Vol. **133**, No. 1, p. 89—99 (1986).

E. BIAGI, S. FROSINI, L. MASOTTI, and E. BORCHI
The Propagation of Acoustic Waves in Polycrystals with Axial Texture
(62.65; 62.20; S1.2; S4; S5.11)
(a) Vol. **93**, No. 1, p. 151—162 (1986).

M. R. NAGY, M. S. SAKR, M. M. MOSTAFA, and R. KAMEL
*Recovery Behaviour after Stress Reductions during the Steady State Creep of Al–14 wt% Zn
(62.20; S1.2; S4)
(a) Vol. **93**, No. 1, p. 185—193 (1986).

R. LACROIX
The Ligand Field on d^n Ions in a Strained Corundum Crystal
(71.70; 62.20; S10.1)
(b) Vol. **133**, No. 2, p. 669—682 (1986).

T. SOMA, K. WADA, and H.-MATSUO KAGAYA
*Temperature Effect on the Elastic Stiffness Constants of Al
(62.20; S4)
(b) Vol. **133**, No. 2, p. K87—K90 (1986).

H.-J. WEISS, A. KRELL, and B. MÜLLER
*Comparison of Vickers Hardness Numbers VH and LVH in the Case of Alumina and Doped Sapphire
(62.20; S10.1)
(a) Vol. **93**, No. 2, p. 509—514 (1986).

A. K. GIRI and G. B. MITRA
*Theoretical Prediction of Elastic Constants of Mixed Systems
(62.20; S7.15; S9.15)
(b) Vol. **134**, No. 1, p. K11—K16 (1986).

M. S. SAKR, A. E. E. ABDEL-RAHIEM, and A. A. EL-DALY
*Steady State Creep during Transformation in Zn–40 wt% Al and Zn–0.5 wt% Al Alloys
(62.20; 61.70; S1.2)
(a) Vol. **94**, No. 2, p. 561—568 (1986).

U. PURI and V. KUMAR
Pressure-Induced Polymorphic Phase Transitions of Potassium and Rubidium Halides
(64.60; 62.20; S9.11)
(b) Vol. **135**, No. 1, p. 127—131 (1986).

N. M. PLAKIDA
Superionic Phase Transitions in Hydrogen Bonded Crystals
(64.60; 62.20; S11.1)
(b) Vol. **135**, No. 1, p. 133—139 (1986).

U. SCHMIDT, A. ZAHRA, R. SCHÜLBE, and H. LÖFFLER
On the Stability of Guinier-Preston Zones in an Al–1.87 at% Cu Alloy at Elevated Temperature
(61.55; 62.20; S4)
(a) Vol. **95**, No. 1, p. 73—80 (1986).

E. V. ZAROCHENTSEV, S. M. OREL, and I. V. KOCHERGIN
*Elastic Constants of a Stressed Crystal. V. Harmonic Generation
(62.20; 62.30)
(a) Vol. **95**, No. 1, p. 105—111 (1986).

P. ANURADHA and I. V. K. BHAGAVAN RAJU
*Study of Plastic Flow in $CaCO_3$ and $NaNO_3$ Crystals under Concentrated Loads
(62.20; S11.1)
(a) Vol. **95**, No. 1, p. 113—119 (1986).

S. TAKAMURA and M. KOBIYAMA
Dislocation Pinning in Al and Ag Alloys after Low-Temperature Deformation
(61.70; 62.20; 72.15; S1.3; S4)
(a) Vol. **95**, No. 1, p. 165—172 (1986).

G. SINGH and R. P. S. RATHORE
*Generalised Morse Potential for B. C. C. Complex Metals
(62.20; S1; S4)
(b) Vol. **135**, No. 2, p. 513—518 (1986).

T. SOMA and H.-MATSUO KAGAYA
*Pressure-Derivatives of Elastic Constants and Mode Grüneisen Parameters under Pressure for Tetrahedral Compounds
(62.20 and 65.50; S7.11; S8.11; S8.12; S8.13)
(b) Vol. **135**, No. 2, p. K103—K106 (1986).

G. SORGE, J. JANICH, and L. A. SHUVALOV
Influence of Deuteration on Dielectric, Elastic, and Electromechanical Properties of NaH_3SeO_3
(64.70; 62.20; 77.80; S11)
(a) Vol. **95**, No. 2, p. 433—438 (1986).

R. K. Singh, N. Kumar, and G. B. Varshney
*Multipole Interaction Effects on Cohesive and Anharmonic Properties of Alkali Cyanide Crystals
(62.20; S11)
(b) Vol. **136**, No. 2, p. 457—465 (1986).

W. Gräfe
*A Method for the Determination of a Dynamic Fatigue Limit of Glasses
(61.40 and 62.20; 68.25)
(a) Vol. **96**, No. 2, p. 501—507 (1986).

H. Neumann
*Bulk Modulus–Volume Relationship in Ternary Chalcopyrite Compounds
(62.20; S8.16)
(a) Vol. **96**, No. 2, p. K121—K125 (1986).

M. R. Nagy, M. I. El-Agrab, M. H. Mahmood, and A. A. Mohamed
*Hardness and Creep Behaviour Associated with Isothermal Aging of Aluminium–Silver Base Alloys
(62.20; S1.3; S4)
(a) Vol. **96**, No. 2, p. K139—K143 (1986).

K. Srinivasan and S. Lakshmi Raghavan
Strain Derivatives of Dielectric Constants of MgO and Thallous Halides
(77.20; 62.20; S9; S10.1)
(b) Vol. **137**, No. 1, p. 33—38 (1986).

H. Siethoff, H. G. Brion, K. Ahlborn, and W. Schröter
*Dynamical Recovery of ⟨111⟩ Germanium
(62.20; S5.11; S5.12)
(a) Vol. **97**, No. 1, p. 153—162 (1986).

N. V. K. Prabhakar and S. P. Sanyal
*Anharmonic Properties of the Ternary Semiconductor $CuGe_2P_3$
(62.20; 64.30; S7.16)
(b) Vol. **137**, No. 2, p. 453—458 (1986).

N. Dutt, A. J. Kaur, and J. Shanker
Effective Potentials and Crystalline State Properties of Some Fluorite-Structure Crystals
(61.50 and 64.30; 62.20; 63.90; 65.50; 77.20; S9.12)
(b) Vol. **137**, No. 2, p. 459—468 (1986).

D. O. Northwood and I. O. Smith
*The Creep Behaviour of High Purity Polycrystalline Magnesium at 0.32 to 0.51 T_m
(62.20; S3)
(a) Vol. **97**, No. 2, p. 441—450 (1986).

R. Schülbe, U. Schmidt, and P. Siebert
On the Stability of Metastable Precipitates in an Al–1.87 at% Cu Alloy at Elevated Temperature
(61.55; 62.20; S4)
(a) Vol. **98**, No. 1, p. 81—89 (1986).

62.30. Mechanical and elastic waves

Original Papers

E. V. Zarochentsev, S. M. Orel, and I. V. Kochergin
Elastic Constants of a Stressed Crystal. V. Harmonic Generation
(62.20; 62.30)
(a) Vol. **95**, No. 1, p. 105—111 (1986).

K. Franke and W. Weihnacht
Surface Acoustic Wave Damping in Piezoelectric Ceramics by Scattering and Dissipation
(68.25; 62.30; 62.80)
(a) Vol. **95**, No. 2, p. 483—494 (1986).

G. Martin and K. Franke
An Examination of the Validity of a Common Model Used for the Analysis of Surface Acoustic Wave Interdigital Transducers
(68.25; 62.30; 77.60)
(a) Vol. **96**, No. 2, p. 489—499 (1986).

62.40. Anelasticity, internal friction, and mechanical resonances
For thermomechanical effects, see 65.70
For magnetomechanical effects, see 75.80
For piezoelectric effects, see 77.60
For elasto-optical effects, see 78.20

Original Papers

C. M. SU and T. S. KÊ
*Further Experiments on the High-Temperature Internal Friction Peak in High-Purity Single Crystal Aluminium
(62.40; 61.70; S4)
(a) Vol. **94**, No. 1, p. 191—202 (1986).

S. TAKAMURA and M. KOBIYAMA
Effect of Low-Temperature Deformation on Internal Friction Peaks of Irradiated Al Alloys
(61.70; 62.40; S4)
(a) Vol. **94**, No. 2, p. 595—600 (1986).

K. IWASAKI
Strain Amplitude Dependence of the Grain Boundary Internal Friction Peak of Pure Aluminium
(61.70; 62.40; S4)
(a) Vol. **94**, No. 2, p. 601—607 (1986).

LI XIAO-GUANG and HE YIZHEN
*Internal Friction of Metallic Glass $Ni_{74}P_{16}B_6Al_4$ near T_x
(62.40; S1.2)
(a) Vol. **95**, No. 2, p. 467—472 (1986).

T. S. KÊ and B. S. ZHANG
Contribution of Bamboo Boundaries to the Internal Friction Peak in Macrocrystalline High-Purity Aluminium
(61.70; 62.40; S4)
(a) Vol. **96**, No. 2, p. 515—525 (1986).

L. D. ZHANG, J. SHI, and T. S. KÊ
High Temperature Internal Friction Peak in High Purity Single Crystal Aluminium Sheets and Electron Microscopic Observation on the Related Dislocation Configurations
(61.70; 62.40; S4)
(a) Vol. **98**, No. 1, p. 151—162 (1986).

J. NI and T. S. KÊ
A Theory of Anomalously Amplitude-Dependent Internal Friction Peaks Due to the Interaction between Dislocation Kinks and Substitutional Solute Atoms in F.C.C. Metals
(61.70; 62.40; 66.30; S1.2; S3; S4)
(a) Vol. **98**, No. 2, p. 487—501 (1986).

62.65. Acoustical properties of solids
For magnetoacoustic effects in metals, see 75.80
For acoustoelectric effects, see 72.50
For acousto-optical effects, see 78.20

Original Papers and Short Notes

E. Biagi, S. Frosini, L. Masotti, and E. Borchi
*The Propagation of Acoustic Waves in Polycrystals with Axial Texture
(62.65; 62.20; S1.2; S4; S5.11)
(a) Vol. **93**, No. 1, p. 151—162 (1986).

R. Nava
*The Effect of Energy Conservation Surface Overlap Breakdown on the Low-Temperature Ultrasonic Attenuation in Dielectrics
(62.65 and 77.50)
(b) Vol. **133**, No. 2, p. K81—K85 (1986).

E. V. Zarochentsev, S. M. Orel, and A. Yu. Yakovets
Elastic Constants of a Stressed Crystal. IV. Sound Propagation Matrix
(62.20; 62.65)
(a) Vol. **94**, No. 2, p. 515—521 (1986).

J. Bracinik and J. Štelina
*Sound Generation in Photoresistive Semiconductors
(62.65 and 77.60; S7.12)
(a) Vol. **95**, No. 2, p. K179—K182 (1986).

G. M. Shmelev, Nguyen Hong Shon, and G. I. Tsurkan
*The Amplification of Hypersound by Photoelectrons in Quantizing Magnetic Field
(62.65; 72.40)
(b) Vol. **136**, No. 1, p. K17—K22 (1986).

A. Brilingas, R. L. Davidovich, J. Grigas, S. Lapinskas, M. A. Medkov, V. Samulionis, and V. Skritskii
Relaxational Dynamics of the Improper Ferroelectric $K_3BiCl_6 \cdot 2\,KCl \cdot KH_3F_4$
(77.80; 62.65; S9.16)
(a) Vol. **96**, No. 1, p. 101—109 (1986).

S. Ghosh and R. B. Saxena
*Parametric Instabilities of Laser Beams in Material with Strain-Dependent Dielectric Constant
(62.65 and 77.60; S11.1)
(a) Vol. **96**, No. 1, p. 111—119 (1986).

M. Krzesińska
Three-Phonon-Interactions in Bismuth Germanium Oxide
(63.20; 62.65; S10)
(b) Vol. **136**, No. 2, p. 475—479 (1986).

V. V. Krasnogolovets and P. M. Tomchuk
The Influence of External Effects on the Proton Conductivity of an Hydrogen-Bonded Chain
(66.30; 62.65; 71.38; S12.1)
(b) Vol. **138**, No. 2, p. 727—733 (1986).

62.80. Ultrasonic relaxation
(see also 74.30. Ultrasonic attenuation in superconductors)

Original Papers

V. G. BARYSHEVSKII and V. V. SKADOROV
*Mössbauer Radiation Dynamical Diffraction in Crystals Being Subjected to the Action of External Variable Fields
(62.80 and 76.80)
(b) Vol. **133**, No. 1, p. 75—87 (1986).

L. T. TSANKOV, N. I. OGNJANOV, and R. P. CHAKUROVA
Time Dependence of Mössbauer Radiation Transmitted through a RF Vibrating Resonant Medium
(78.70; 62.80)
(a) Vol. **94**, No. 1, p. 131—138 (1986).

K. FRANKE and W. WEIHNACHT
Surface Acoustic Wave Damping in Piezoelectric Ceramics by Scattering and Dissipation
(68.25; 62.30; 62.80)
(a) Vol. **95**, No. 2, p. 483—494 (1986).

63. Lattice dynamics and crystal statistics (see also 65. Thermal properties, 66.70. Thermal conduction, 68.30. Dynamics of surface and interface vibrations, 78.30. Infrared and Raman spectra)

63.20. Phonons and vibrations in crystal lattices

Original Papers and Short Notes

V. K. Jindal and J. Kalus
*Calculation of Thermal Expansion and Phonon Frequency Shift in Deuterated Naphthalene
(63.20 and 65.70; 62.20; S12)
(b) Vol. **133**, No. 1, p. 89—99 (1986).

A. E. Mefed and V. A. Atsarkin
The "Cascade" Narrowing of the NMR Line in Solids by Radio-Frequency Fields and Slow Molecular Motions
(76.60; 63.20; S12)
(a) Vol. **93**, No. 1, p. K21—K24 (1986).

Yu. L. Shitikov, N. A. Tulina, and M. G. Zemlyanov
*Special Features of Lattice Dynamics and Electron–Phonon Interaction in Re
(63.20 and 71.38; S1)
(b) Vol. **133**, No. 2, p. 469—474 (1986).

J. F. Higgs, W. Y. Zeng, and A. Anderson
*Dynamical Model for the Vibrations of the Hydrogen Halide Crystals
(63.20; S9)
(b) Vol. **133**, No. 2, p. 475—482 (1986).

K. T. Stoychev
Non-Linear Optical Susceptibilities of Molecular Crystals in the Region of Vibronic Excitations. III. Non-Totally Symmetric Vibrations
(78.20; 63.20; 71.35)
(b) Vol. **133**, No. 2, p. 557—562 (1986).

Luo Zun-Du
*One Possible Mechanism of Deexitation for Localized Electronic States — Lattice Relaxation in Virtual Photon Process
(63.20 and 78.20)
(b) Vol. **133**, No. 2, p. K77—K80 (1986).

W. Mutscheller and M. Wagner
Thermal Conductivity of Cooperative Jahn-Teller E-Systems
(66.70; 63.20; 64.90; S11.1)
(b) Vol. **134**, No. 1, p. 39—52 (1986).

A. Parisini, P. P. Lottici, C. Razzetti, and R. Bacewicz
Resonance Effects in the Raman Spectrum of $CdGa_2Se_4$
(78.30; 63.20; S8.16)
(b) Vol. **134**, No. 1, p. 81—86 (1986).

Yu. V. Naboikin, S. N. Andrianov, P. V. Zinoviev, Yu. V. Malyukin, E. N. Rudenko, V. V. Samartsev, N. B. Silaeva, and Yu. E. Sheibut
*Optical Superradiance in Pyrene-Doped Biphenyl Crystals and Effect of Phonons on Its Formation
(63.20 and 78.45; S12)
(b) Vol. **135**, No. 2, p. 503—512 (1986).

T. Juhász, J. S. Bakos, and Cs. Kuti
*Direct Lifetime Measurement of the A_{1g} Phonon Mode in Calcite
(63.20; 78.30; S11.1)
(b) Vol. **135**, No. 2, p. K99—K102 (1986).

N. M. Gasanly, N. N. Melnik, V. I. Tagirov, and A. A. Yushin
Special Features of First- and Second-Order Raman and Infrared Spectra of $GaS_{1-x}Se_x$ Mixed Crystals
(78.30; 63.20; S8)
(b) Vol. **135**, No. 2, p. K107—K112 (1986).

G. Singh and R. P. S. Rathore
*Two- and Three-Body Interactions and Phonons in Some F.C.C. Metals
(63.20; 62.20; S1; S3)
(b) Vol. **136**, No. 1, p. 57—61 (1986).

B. A. Danilchenko, V. N. Poroshin, M. I. Slutskii, and M. Asche
*Investigation of Nonequilibrium Phonons in GaAs
(63.20; S7.12)
(b) Vol. **136**, No. 1, p. 63—68 (1986).

P. V. Smith, J. E. Szymanski, and J. A. D. Matthew
A Reformulation of the LUC-CNDO Approach to the Properties of Solids. Application to Diamond and Silicon
(71.10; 63.20; S5; S5.11)
(b) Vol. **136**, No. 1, p. 261—271 (1986).

B. Pődör
*On Three-Phonon Processes in Gallium Phosphide
(63.20, S7.11)
(b) Vol. **136**, No. 1, p. K9—K12 (1986).

A. G. Abdullaev and T. D. Ibragimov
*One-Phonon Absorption in GaS Thin Films
(63.20 and 78.30; S8)
(b) Vol. **136**, No. 1, p. K13—K15 (1986).

L. E. Reznik
*Electron–Phonon Interaction and Structure of Molecular Luminescence Center in Crystals
(63.20 and 71.70; 71.38; 78.50; 78.55; S11.1)
(b) Vol. **136**, No. 2, p. 417—424 (1986).

P. Reineker, R. G. Winkler, and H. Port
*Temperature Dependence of Resonance Splittings in Molecular Pairs. Comparison of Optical Experiments and Theoretical Results for the Phenazine Triplet State
(63.20 and 71.35; S12)
(b) Vol. **137**, No. 1, p. 137—147 (1986).

V. Z. Polinger and S. I. Boldyrev
*The Multimode Jahn-Teller Effect in the Luminescence Spectrum of the Tetrahedral $^2T_2 \otimes (e + t_2 + t_2)$ Impurity System ZnS: Cu^{2+}
(63.20 and 71.70; 78.55; S8.11)
(b) Vol. **137**, No. 1, p. 241—253 (1986).

A. Memon and D. B. Tanner
*Long Wavelength Optical Phonons in Mixed Alkali Halide Powder Crystals
(63.20 and 78.30; S9.11)
(b) Vol. **137**, No. 1, p. K9—K13 (1986).

M. C. Al-Edani and K. S. Dubey
*Lattice Thermal Conductivity of Gadolinium Monosulphide above Room Temperature in the Frame of Two-Mode Conduction of Phonons
(63.20 and 66.70; S8)
(b) Vol. **137**, No. 2, p. 449—452 (1986).

A. P. Kovchavtsev, G. L. Kurishev, and K. O. Postnikov
*Inelastic Electron Tunneling Spectroscopy of InAs–Superthin Insulator–Au Structures
(63.20 and 73.40; 78.30; S7.12)
(a) Vol. **97**, No. 2, p. 421—425 (1986).

S. Ichinose
*Dynamical Structure of Alpha Helical Proteins
(63.20 and 71.45; S12)
(b) Vol. **138**, No. 1, p. 73—81 (1986).

M. T. Primatarova
*Solitary Excitations in Molecular Chains Due to the Anharmonicity of the Intramolecular Vibrations
(63.20; S12.1)
(b) Vol. **138**, No. 1, p. 101—108 (1986).

M. Wautelet and P. B. Legrand
*Crystal Instability Due to Defect–Phonon Interactions
(61.70 and 63.20; 64.60)
(b) Vol. **138**, No. 1, p. 109—116 (1986).

L. Wendler and R. Pechstedt
Dynamical Screening, Collective Excitations, and Electron–Phonon Interaction in Heterostructures and Semiconductor Quantum Wells. General Theory
(71.38 and 71.45; 63.20; 73.40; 77.30; 78.20)
(b) Vol. **138**, No. 1, p. 197—217 (1986).

63.50. Vibrational states in disordered systems

Original Papers

NGUYEN VAN DEN, A. HEINRICH, K. KLOSTERMANN, and G. SOBE
IR Absorption in CrSiO Thin Films
(78.30; 63.50; 78.65; S10.15)
(a) Vol. **93**, No. 1, p. 163—170 (1986).

H. C. GUPTA, G. S. N. REDDY, and B. B. TRIPATHI
*Phonon Density of States in Amorphous $Si_{0.5}Ge_{0.5}$
(63.50; S6)
(b) Vol. **134**, No. 1, p. 17—20 (1986).

J. KALUS, H. GERLACH, J. SCHLEIFER, F. WÖRLEN, G. VOSS, M. GODLEWSKA, I. NATKANIEC, N. KARL, and M. PRAGER
*Inelastic Incoherent Neutron Scattering and Raman Scattering of 2,3-Dimethylnaphthalene
(63.50; 78.30; S12)
(b) Vol. **134**, No. 1, p. 53—62 (1986).

B. K. GHOSH and BAL K. AGRAWAL
*Fluorinated Hydrogenated Amorphous Silicon Alloys. Vibrational Excitations of Chains
(63.50; S5.11)
(b) Vol. **136**, No. 1, p. 51—55 (1986).

S. GARG and H. C. GUPTA
*Phonon Dispersion of Disordered Fe–Al Alloy
(63.50; S1.1)
(b) Vol. **136**, No. 2, p. 453—456 (1986).

J. KALUS, J. SCHLEIFER, R. HEMPELMANN, and W. RICHTER
*Rotational Excitations of Methyl Groups in 2,3-Dimethylnaphthalene
(63.50; S12)
(b) Vol. **136**, No. 2, p. 513—518 (1986).

K. WINER and F. WOOTEN
*Vibrational Properties of Metastable Phases of Si and Ge
(63.50; 61.40; S5.11; S5.12)
(b) Vol. **136**, No. 2, p. 519—527 (1986).

D. WEGENER and H. BÖTTGER
*A Self-Consistent Theory of Phonon-Like Excitations in Structurally Disordered Solids
(63.50)
(b) Vol. **138**, No. 1, p. 83—91 (1986).

63.70. Statistical mechanics of lattice vibrations (see also 65. Thermal properties of condensed matter, and 66.70. Thermal conduction)

Original Paper

V. I. Zubov and J. N. Teixeira Rabelo
*On the Self-Consistent Statistical Theory of Anharmonic Effects in Imperfect Crystals
(63.70; 65.50; 68.30)
(b) Vol. **138**, No. 2, p. 433—445 (1986).

63.75. Statistical mechanics of displacive phase-transitions For order–dasorder and statistical mechanics of model systems, see 64.60. For crystallographic aspects of polymorphic and order–disorder transformations, see 61.50.

Original Papers and Short Notes

L. Dąbrowski
*Multiparticle Correlations in Alloys
(63.75 and 65.50; S1.2)
(b) Vol. **135**, No. 1, p. 61—65 (1986).

L. Dąbrowski
*Intermediate-Temperature Expansion in the Theory of Atomic Ordering in Alloys
(63.75 and 65.50)
(b) Vol. **136**, No. 2, p. K71—K73 (1986).

E. V. Kholopov
*Thermodynamic Features of the System of Packed Antiferroelectric Chains
(63.75 and 77.80; S9.16)
(b) Vol. **138**, No. 2, p. 415—422 (1986).

63.90. Other topics in lattice dynamics and crystal statistics

Rewiev Article

J. Shanker and W. N. Bhende
*Higher-Order Elastic Constants and Thermoelastic Properties of Ionic Solids
(62.20 and 63.90; 64.30; 65.70; S7.1; S9.11; S9.12; S10.1)
(b) Vol. **136**, No. 1, p. 11—30 (1986).

Original Papers and Short Notes

M. Kumari and N. Dass
*On the Pressure Dependence of Grüneisen Parameter in Solids
(63.90; 65.50; S2; S9.11; S9.12; S10.1)
(b) Vol. **133**, No. 1, p. 101—110 (1986).

A. Dhoble and M. P. Verma
Thermodynamic Analysis of the Anderson-Grüneisen Parameters
(65.50; 63.90; S9.11)
(b) Vol. **133**, No. 2, p. 491—497 (1986).

H.-Matsuo Kagaya and T. Soma
Volume-Dependence of Phonon Frequencies for III–V Covalent Crystals
(63.20; 63.90; S7.12; S7.13)
(b) Vol. **134**, No. 2, p. K105—K110 (1986).

N. Dutt, A. J. Kaur, and J. Shanker
Effective Potentials and Crystalline State Properties of Some Fluorite-Structure Crystals
(61.50 and 64.30; 62.20; 63.90; 65.50; 77.20; S9.12)
(b) Vol. **137**, No. 2, p. 459—468 (1986).

A. Kumar, B. K. Srivastava, and M. A. Ansari
Lattice Thermal Conductivity of Deformed Lithium Fluoride Crystals at Low Temperatures
(66.70; 63.90; S9.11)
(b) Vol. **138**, No. 2, p. 423—432 (1986).

64. Equations of state, phase equilibria, and phase transitions

64.10. General theory of equations of state and phase equilibria

Original Papers

V. M. Talanov and G. V. Bezrukov
*Thermodynamical Model of Interacting Multilevel Systems with a Single Order Parameter. I. State Equation
(64.10)
(a) Vol. **96**, No. 2, p. 475—482 (1986).

M. Delannoy-Coutris and G. Perrin
Fourth-Order Thermal Expansion for Cubic Solids. Application to Aluminium, Copper, Germanium, and Silicon
(65.70; 64.10; S1.2; S4; S5.11; S5.12)
(b) Vol. **138**, No. 1, p. 93—99 (1986).

64.30. Equations of state of specific substances (see also 65.70. Thermal expansion)

64.60. General studies of phase transitions (for magnetic, superconducting, and quantum fluid critical phenomena, 75, 74, and 67, respectively)

Original Papers and Short Notes

A. I. Gusev and A. A. Rempel
*Order–Disorder Phase Transition Channel in Niobium Carbide
(64.60; S1.61)
(a) Vol. **93**, No. 1, p. 71—80 (1986).

O. Simmich
*Diffusion-Controlled Phase Transformations in Finite Regions. An Exact Time-Dependent Analytic Solution
(64.60; 66.30)
(a) Vol. **93**, No. 1, p. 105—112 (1986).

K. Schiele
*Critical Behaviour of the φ^3, φ^4 Model
(64.60)
(b) Vol. **134**, No. 1, p. K1—K6 (1986).

F. El-Kabbany, Y. Badr, M. Tosson, S. Taha, and S. Mahrous
*A Detailed IR Study of the Order–Disorder Phase Transition of $AgNO_3$
(64.60; 78.30; S11.1)
(a) Vol. **94**, No. 1, p. 35—43 (1986).

K. Schiele
*φ^3, φ^4 Model with Quenched Impurities
(64.60; 61.70)
(b) Vol. **134**, No. 2, p. K89—K91 (1986).

U. Puri and V. Kumar
*Pressure-Induced Polymorphic Phase Transitions of Potassium and Rubidium Halides
(64.60; 62.20; S9.11)
(b) Vol. **135**, No. 1, p. 127—131 (1986).

N. M. Plakida
*Superionic Phase Transitions in Hydrogen Bonded Crystals
(64.60; 62.20; S11.1)
(b) Vol. **135**, No. 1, p. 133—139 (1986).

P. K. Dixit, B. A. Vaid, and K. C. Sharma
*Structural Disorder Model. Pressure-Induced Transitions in Sn and Semiconducting Compounds
(64.60; 65.50; S4; S7.1; S8.1)
(b) Vol. **135**, No. 2, p. 523—528 (1986).

M. Wautelet
*Cohesion of Solids under Very High Electronic Excitation Conditions
(64.60 and 79.20; 63.20; 71.35)
(b) Vol. **138**, No. 2, p. 447—456 (1986).

S. Trimper
*On a Noise Suppressed Phase Transition
(64.60)
(b) Vol. **138**, No. 2, p. K77—K81 (1986).

A. I. Potekaev, V. E. Egorushkin, and N. S. Golosov
*Formation of Superstructures with Long Period. Microscopic Description
(61.50 and 64.60)
(a) Vol. **98**, No. 2, p. 345—349 (1986).

K. V. S. Prasad and C. Bansal
*Resistivity and Thermoelectric Power Measurements on AgCd Shape Memory Alloys
(64.60; 72.15; S1.3)
(a) Vol. **98**, No. 2, p. 453—464 (1986).

64.70. Phase equilibria, phase transitions, and critical points of specific substances

Original Papers and Short Notes

G. Wendrock and H. Löffler
Information about the Position of the Coherent Spinodal Curve in the Al–Zn System by Means of the Course of the SAXS Intensity at Elevated Temperatures
(61.55; **64.70**; S1.2; S4)
(a) Vol. **93**, No. 2, p. 487—493 (1986).

Yu-Jin Chang and Xiang-Rong Qian
*A High Resolution Microscopic Study on Nd–Fe–B Permanent Magnetic Alloys
(**64.70** and 75.60; 61.16; 75.50; S1.1)
(a) Vol. **93**, No. 2, p. 573—583 (1986).

M. M. El-Zaidia, A. A. Ammar, and A. A. El-Rhmany
*Isothermal Crystallization and Kinetics of Amorphous $As_{34.2}Ge_xTe_{65.8-x}$
(**64.70**; 72.80; S8.15)
(a) Vol. **93**, No. 2, p. K111—K116 (1986).

H. J. Fischbeck
*The Incommensurate CDW Phase of a Quasi-One-Dimensional Peierls-Fröhlich System with a Nearly Quarter-Filled Band. II. Numerical Results
(**64.70** and 71.38)
(b) Vol. **134**, No. 1, p. 223—238 (1986).

K. N. Jog, S. P. Sanyal, and R. K. Singh
*Effect of Three-Body Interactions on the High-Pressure Behaviour of MgO and CaO Crystals
(62.20 and **64.70**; S10.1)
(a) Vol. **94**, No. 1, p. 107—114 (1986).

K. K. Mamedov, A. M. Abdullaev, and E. M. Kerimova
Heat Capacities of $TlInS_2$ and $TlInSe_2$ Crystals at Low Temperatures
(65.40; **64.70**, S8.16)
(a) Vol. **94**, No. 1, p. 115—119 (1986).

H. B. Gon
Growth of AgI Single Crystals in Silicate Gel and DC Conductivity of AgI
(66.30; **64.70**; S9)
(a) Vol. **94**, No. 1, p. K61—K64 (1986).

I. I. Sasovskaya and V. P. Korabel
Optical Properties of α- and β-CuZn Brasses in the Region of Quantum Absorption
(78.20; **64.70**; 71.25; 72.30; S1.2)
(b) Vol. **134**, No. 2, p. 621—630 (1986).

E. Wolska and J. Baszyński
*Prereactional Transformations in the Topotactic Conversion γ-FeOOH → $\gamma\text{-}Fe_2O_3$
(**64.70**; 61.10; 76.50; 78.30; S11.2)
(a) Vol. **95**, No. 1, p. 87—92 (1986).

P. Studzinski and J.-M. Spaeth
ENDOR Investigation of Cr^{3+} Centres in the Tetragonal Phase of $RbCdF_3$
(76.70; 64.70; 71.70; S9.16)
(b) Vol. **136**, No. 2, p. 735—742 (1986).

A. Tressaud, S. Khaïroun, L. Rabardel, T. Kobayashi, T. Matsuo, and H. Suga
*Phase Transitions in Ammonium Hexafluorometallates(III)
(64.70; 65.40; S11)
(a) Vol. **96**, No. 2, p. 407—414 (1986).

M. Hidaka, H. Fujii, B. J. Garrard, and B. M. Wanklyn
*Structural Phase Transitions in the Layer Compound $CsVF_4$
(64.70; S9.16)
(a) Vol. **96**, No. 2, p. 415—423 (1986).

U. Seipold and H. A. Ullner
*Martensitic Character of the Pressure Induced B1–B2 Phase Transition of Potassium Bromide
(64.70; 66.70; S9.11)
(a) Vol. **96**, No. 2, p. 469—474 (1986).

C. N. W. Darlington and D. A. O'Connor
On the Crystal Dynamics of $BaTiO_3$ near T_c
(63.20; 64.70; 76.80; S11.1)
(a) Vol. **96**, No. 2, p. 509—513 (1986).

T. Tylinszczak, A. Budkowski, and J. A. Sawicki
Study of Magnetism and Interface Processes between Thin Iron Film and Nickel Substrate by in-Situ Conversion Electron Mössbauer Spectroscopy
(75.70; 64.70; 76.80; S1.1; S1.2)
(a) Vol. **96**, No. 2, p. 573—579 (1986).

W. Kuczyński
*Refractometric Measurements in Cholesteric Blue Phases
(64.70 and 78.20; S12)
(a) Vol. **96**, No. 2, p. K127—K131 (1986).

T. Soma, K. Hosaka, and H.-Matsuo Kagaya
*Compression Effect on the Phase Diagrams at Low Temperatures of Alkali Alloys
(64.70; S2)
(b) Vol. **137**, No. 2, p. K111—K114 (1986).

E. V. Kholopov, A. M. Panich, N. K. Moroz, and Yu. G. Kriger
*One-Dimensional Mechanism of Antiferroelectric Ordering in a Non-Polar Crystal
(64.70 and 77.80; 65.40; 76.30; S9.16)
(b) Vol. **137**, No. 2, p. K121—K125 (1986).

64.75. Solubility, segregation, and mixing

Review Article

B. X. Liu
*Ion Mixing and Metallic Alloy Phase Formation
(61.80 and 64.75; 61.70; S1.1; S1.2; S1.3; S4)
(a) Vol. **94**, No. 1, p. 11—34 (1986).

Original Papers and Short Notes

K. Sumiyama, H. Ezawa, and Y. Nakamura
*Metastable $Fe_{1-x}Ti_x$ Alloys Produced by Vapour Quenching
(64.75; 61.40; S1.1)
(a) Vol. **93**, No. 1, p. 81—86 (1986).

A. S. Litovchenko, V. A. Chernenko, I. V. Matyash, and O. D. Ishutina
*Influence of Pressure on Dehydroxylation Effect in Muscovite
(64.75; 76.60; 77.20; S11)
(a) Vol. **93**, No. 1, p. K9—K13 (1986).

B. Labulle and C. Petipas
Thermodynamic Study of Rare-Earth–Transition Metal Systems by the Model of Miederma
(65.50; 64.75; S1.2; S1.4)
(a) Vol. **93**, No. 2, p. 419—428 (1986).

G. Graiss, A. Rizk, and A. H. Wafik
*Influence of Carbon Content and Quenching Rate on the Amount of Martensite Formed in Two Kinds of Steel
(64.75; S1.1)
(a) Vol. **93**, No. 2, p. K123—K124 (1986).

K. Masuda-Jindo
*On the Grain Boundary Segregation of sp-Valence Impurities in B.C.C. Transition Metal
(64.75; 71.55; S1.1)
(b) Vol. **134**, No. 2, p. 545—550 (1986).

Zhi-Hua Yan, Bai-Xin Liu, and Heng-De Li
*Ion Mixing to Form Amorphous Alloys in the Co–Gd System
(64.75 and 68.55; S1.2; S1.4)
(a) Vol. **94**, No. 2, p. 483—488 (1986).

Z. A. Matysina, M. I. Milyan, and S. Yu. Zaginaichenko
*The Solubility of Interstitial Impurities in Ordering Phases
(64.75; 61.70; S1.1; S1.3)
(a) Vol. **95**, No. 1, p. K9—K14 (1986).

64.80. Other phase properties of systems

Original Papers

V. N. NISHANOV and A. A. SOBYANIN
*Diffusive Decay of the Metastable State in Periodic Field
(64.80)
(a) Vol. **95**, No. 1, p. 41—50 (1986).

B. PANIGRAHI, A. K. TYAGI, and R. V. NANDEDKAR
*Surface Damage of Ferritic Steels by Helium Ion Irradiation
(61.80 and 64.80; S1.1)
(a) Vol. **95**, No. 1, p. 99—103 (1986).

V. GRÖGER
Influence of Decomposition and Coarsening on the Electrical Resistivity of Cu–2wt%Be–0.11 wt%Co Aged at 320 °C
(72.15; 64.80; S1.2)
(a) Vol. **96**, No. 1, p. 325—335 (1986).

64.90. Other topics in equations of state, phase equilibria, and phase transitions

Original Paper

W. MUTSCHELLER and M. WAGNER
Thermal Conductivity of Cooperative Jahn-Teller E-Systems
(66.70; 63.20; 64.90; S11.1)
(b) Vol. **134**, No. 1, p. 39—52 (1986).

65. Thermal properties of condensed matter
(see also 63. Lattice dynamics, for thermodynamic properties of quantum fluids, see 67.40; for thermal properties of solid helium, see 67.80)

65.40. Heat capacities of solids
(for specific heat of superconductors, see 74.30; for specific heat of magnetic systems, see 75.40)

Original Papers and Short Notes

A. G. GUSEINOV, A. YU. YANGIROV, K. K. MAMEDOV, K. M. MAMEDOVA, M. I. MEKHTIEV, and D. O. GUMBATOV
*Heat Capacity and Thermodynamic Properties of Tl_2Te at Low Temperatures
(65.40 and 65.50; S8)
(a) Vol. **93**, No. 1, p. K15—K19 (1986).

A. S. LITOVCHENKO, V. E. YACHMENEV, and V. A. VILKOVSKII
*Temperature Dependence of the Heat Capacity of Ferriferrous Micas
(65.40; S11)
(a) Vol. **93**, No. 2, p. K133—K135 (1986).

A. CZOPNIK, R. POTT, H. MÄDGE, N. ILIEV, and B. STALINSKI
*Specific Heat and Thermal Expansion of $LaIn_3$
(65.40 and 65.70; S1)
(a) Vol. **93**, No. 2, p. K137—K141 (1986).

K. K. MAMEDOV, A. M. ABDULLAEV, and E. M. KERIMOVA
*Heat Capacities of $TlInS_2$ and $TlInSe_2$ Crystals at Low Temperatures
(65.40; 64.70; S8.16)
(a) Vol. **94**, No. 1, p. 115—119 (1986).

H.-MATSUO KAGAYA and T. SOMA
*Temperature Dependence of Specific Heat and Debye Temperature of Tetrahedral Compounds
(65.40; S7.11; S8.11; S8.12; S8.13)
(b) Vol. **134**, No. 2, p. K101—K104 (1986).

V. K. THAKUR and T. N. SINGH
Lattice Dynamics of F.C.C. Metals
(63.20; 65.40; S1.2; S1.3)
(b) Vol. **135**, No. 1, p. 67—73 (1986).

Y. SHIBATA, K. ISHI, and S.-I. TAKAHASHI
Lattice Dynamics of Solid N_2O
(63.20; 65.40; S5.2)
(b) Vol. **135**, No. 1, p. 85—91 (1986).

M. MERTIG, G. POMPE, and E. HEGENBARTH
*Specific Heat of Nb-Doped RuO_2 Single Crystals
(65.40 and 71.20; S10.1)
(b) Vol. **135**, No. 1, p. 335—342 (1986).

A. V. Deryabin, A. V. Tkov, V. K. Kazantsev, Yu. A. Chirkov, and I. V. Zakharov
Concentration Magnetic Transitions in Alloys with Mixed Exchange Interaction
(75.30; 65.40; 72.15; S1.1)
(a) Vol. **95**, No. 1, p. 229—237 (1986).

B. A. Strukov, S. A. Taraskin, and Z. V. Savilova
Thermodynamical Properties of Diglycine Nitrate (DGN) Crystals
(77.80; 65.40; 65.50; S12)
(a) Vol. **95**, No. 2, p. 447—451 (1986).

P. Vaněk, B. Březina, M. Havránková, and J. Biroš
Room-Temperature Phase Transition in Rb_2CoCl_4. Heat Capacity Measurements
(64.70; 65.40; S9.16)
(a) Vol. **95**, No. 2, p. K101—K103 (1986).

T. Soma, K. Wada, and H.-Matsuo Kagaya
*Anharmonic Coefficient for the Specific Heat of Al
(65.40; S4)
(b) Vol. **136**, No. 1, p. K5—K8 (1986).

A. Tressaud, S. Khaïroun, L. Rabardel, T. Kobayashi, T. Matsuo, and H. Suga
Phase Transitions in Ammonium Hexafluorometallates(III)
(64.70; 65.40; S11)
(a) Vol. **96**, No. 2, p. 407—414 (1986).

T. Soma, N. Suzuki, and H.-Matsuo Kagaya
*Specific Heat of α-Phase Cu–Al Alloys
(65.40; S1.2; S4)
(b) Vol. **137**, No. 1, p. K15—K19 (1986).

E. Fischer
Dielectric and Thermal Properties of Relaxation Ferroelectrics at Low Temperatures
(77.80; 65.40; S11.1)
(a) Vol. **97**, No. 1, p. 121—128 (1986).

E. V. Kholopov, A. M. Panich, N. K. Moroz, and Yu. G. Kriger
One-Dimensional Mechanism of Antiferroelectric Ordering in a Non-Polar Crystal
(64.70 and 77.80; 65.40; 76.30; S9.16)
(b) Vol. **137**, No. 2, p. K121—K125 (1986).

V. K. Thakur and T. N. Singh
Lattice Dynamics of Transition Metals
(63.20; 62.20; 65.40; S1; S1.2; S1.5)
(b) Vol. **138**, No. 2, p. 407—414 (1986).

65.50. Thermodynamic properties and entropy

Original Papers and Short Notes

T. Ito
*A Pseudopotential Approach to the Disorder Effects in III–V Ternary Semiconductor Alloys
(61.55 and 65.50; S7.15)
(b) Vol. **135**, No. 2, p. 493—502 (1986).

P. K. Dixit, B. A. Vaid, and K. C. Sharma
Structural Disorder Model. Pressure-Induced Transitions in Sn and Semiconducting Compounds
(64.60; 65.50; S4; S7.1; S8.1)
(b) Vol. **135**, No. 2, p. 523—528 (1986).

T. Soma and H.-Matsuo Kagaya
*Pressure-Derivatives of Elastic Constants and Mode Grüneisen Parameters under Pressure for Tetrahedral Compounds
(62.20 and 65.50; S7.11; S8.11; S8.12; S8.13)
(b) Vol. **135**, No. 2, p. K103—K106 (1986).

B. A. Strukov, S. A. Taraskin, and Z. V. Savilova
Thermodynamical Properties of Diglycine Nitrate (DGN) Crystals
(77.80; 65.40; 65.50; S12)
(a) Vol. **95**, No. 2, p. 447—451 (1986).

R. Poprawski and W. Salejda
The Phenomenological Model of Phase Transitions in NH_4HSeO_4 Crystals
(77.80; 65.50; 77.70; S11.1)
(a) Vol. **95**, No. 2, p. 459—465 (1986).

K. Sigrist and H. Fichtner
Investigation of the Influence of Lattice Distortions on the Energy Inherent in Microcrystalline Materials
(61.50 and 62.20; 65.50; S10.1)
(a) Vol. **95**, No. 2, p. 473—481 (1986).

A. Dhoble and M. P. Verma
*On the Temperature Variation of the Anderson-Grüneisen Parameters of NaCl
(65.50 and 65.70; S9.11)
(b) Vol. **136**, No. 2, p. 497—504 (1986).

L. Dąbrowski
*Intermediate-Temperature Expansion in the Theory of Atomic Ordering in Alloys
(63.75 and 65.50)
(b) Vol. **136**, No. 2, p. K71—K73 (1986).

H. A. Brown
*The Three-Spin Box Product Interaction
(65.50 and 75.10)
(b) Vol. **137**, No. 1, p. 91—95 (1986).

65.70. Thermal expansion and thermomechanical effects

(see also 64.30. Equations of state)
For thermal conduction in nonmetallic liquids, see 66.60; for nonmetallic solids, see 66.70
For electronic thermal conduction, see 72.10 and 72.20
For thermal conductivity of superconductors, see 74.30
For pyroelectric and electrocaloric effects, see 77.70

Review Article

J. Shanker and W. N. Bhende
Higher-Order Elastic Constants and Thermoelastic Properties of Ionic Solids
(62.20 and 63.90; 64.30; 65.70; S7.1; S9.11; S9.12; S10.1)
(b) Vol. **136**, No. 1, p. 11—30 (1986).

Original Papers and Short Notes

V. K. Jindal and J. Kalus
*Calculation of Thermal Expansion and Phonon Frequency Shift in Deuterated Naphthalene
(63.20 and 65.70; 62.20; S12)
(b) Vol. **133**, No. 1, p. 89—99 (1986).

K. R. K. Rao, J. V. Rao, P. Venkatacharyulu, and V. Baliah
Phase Transition Studies of the Liquid Crystals Heptyloxybenzylidene Phenylazoaniline
(64.70; 65.70; S12)
(a) Vol. **93**, No. 1, p. 93—97 (1986).

U. Pietsch and D. Marlow
*The Thermal Expansion Coefficient of $Ga_xIn_{1-x}As_yP_{1-y}$ Epitaxial Layers Grown on InP Substrate
(65.70; 68.25; S7.15)
(a) Vol. **93**, No. 1, p. 143—149 (1986).

U. Pietsch
*Thermal Expansion Coefficients of Ternary $A^{III}B^{V}$ Solid Solutions
(65.70; S7.15)
(b) Vol. **133**, No. 2, p. 483—489 (1986).

A. Czopnik, R. Pott, H. Mädge, N. Iliev, and B. Stalinski
*Specific Heat and Thermal Expansion of $LaIn_3$
(65.40 and 65.70; S1)
(a) Vol. **93**, No. 2, p. K137—K141 (1986).

T. Soma and H.-Matsuo Kagaya
*Lattice Anharmonicity and Thermal Expansion of Alkali Metals
(65.70; S2)
(b) Vol. **134**, No. 1, p. K7—K9 (1986).

P. ENDERS
*Non-Fourierian Heat Conduction in Semiconductor Multilayer Systems with Thin Layers
(65.70; 66.70)
(b) Vol. **138**, No. 2, p. K87—K89 (1986).

H.-MATSUO KAGAYA and T. SOMA
*Thermal Expansion Coefficient of III–V Covalent Crystals
(65.70; 63.20; S7.12; S7.13)
(b) Vol. **138**, No. 2, p. K91—K93 (1986).

66. Transport properties of condensed matter (nonelectronic)

66.10. Diffusion and ionic conduction in liquids

Original Paper

66.30. Diffusion in solids

Original Papers and Short Notes

M. G. E. Brand and A. Lodder
*Electromigration in Niobium. A Finite-Cluster-Model Study
(66.30; S1)
(b) Vol. **133**, No. 1, p. 119—125 (1986).

O. Simmich
Diffusion-Controlled Phase Transformations in Finite Regions. An Exact Time-Dependent Analytic Solution
(64.60; 66.30)
(a) Vol. **93**, No. 1, p. 105—112 (1986).

K. Shahi, W. Weppner, and A. Rabenau
Defects and the First-Order Phase Transition in AgI
(61.70 and 64.70; 66.30; S9)
(a) Vol. **93**, No. 1, p. 171—176 (1986).

D. Shaw
*Erratum to: The Chemical Diffusion of In in $Hg_{0.8}Cd_{0.2}Te$
(66.30; S8.15)
(a) Vol. **93**, No. 1, p. 398 (1986).

E. Scheid and P. Chenevier
*1D Simulation of Oxidation-Enhanced and Oxidation Retarded Diffusion in Silicon and Validity of the Physical Model. Transition to the 2D Case
(66.30; S5.11)
(a) Vol. **93**, No. 2, p. 523—533 (1986).

P. Metsch, F. H. M. Spit, and H. Bakker
*Computer Simulation of Simultaneous Bulk, Grain-Boundary, and Surface Diffusion
(66.30)
(a) Vol. **93**, No. 2, p. 543—547 (1986).

S. Guimarães, R. Kögler, E. Landi, S. Solmi, and E. Wieser
*Anomalous Enhanced Diffusion and Electrical Activation of Boron in Silicon after Rapid Isothermal Annealing
(66.30; 72.80; S5.11)
(a) Vol. **93**, No. 2, p. 549—557 (1986).

V. F. Petrenko, T. Ebinuma, and N. Maeno
Protonic Photoconductivity of Ice
(72.40; 66.30; S10.1)
(a) Vol **93**, No. 2, p. 695—702 (1986).

A. Klamt and H. Teichler
Quantum-Mechanical Calculations for Hydrogen in Niobium and Tantalum. I. Model Potential and Properties of Self-Trapped States
(61.70; 66.30; S1)
(b) Vol. **134**, No. 1, p. 103—114 (1986).

O. A. Aleksandrova, M. I. Kamchatka, M. S. Miropolskii, and V. V. Passynkov
*Diffusion of Native Defects in $(Pb_{1-x}Sn_x)_{1-y}Te_y$ during Liquid Phase Epitaxy
(66.30; 61.70; S8.15)
(a) Vol. **94**, No. 1, p. 139—145 (1986).

V. V. Kolomytkin and U. R. Kevorkyan
*A Computer Simulation of Self-Interstitial Migration along the [100] (010) Edge Dislocation Core in α-Fe
(66.30; S1.1)
(a) Vol. **94**, No. 1, p. 153—160 (1986).

H. Bachert, W. Pittroff, I. Rechenberg, S. Stoeff, and K. Vogel
*The Influence of Mg Doping on the Optical and Electrical Properties of GaAs/(Al; Ga)As Injection Lasers
(66.30 and 68.55; 73.40; 78.45; S7.12; S7.15)
(a) Vol. **94**, No. 1, p. 321—327 (1986).

K. Holldack, H. Kerkow, and W. Frentrup
*Anomalous Diffusion of Boron in Silicon Due to Heavy Ion Bombardment
(61.80 and 66.30; S5.11)
(a) Vol. **94**, No. 1, p. 357—363 (1986).

J. Klöber, H. A. Schneider, and P. Schneider
*Measurement of Low-Temperature Cu Diffusion along Free-Surface Short Circuits in Thin Ni Films
(66.30; S1.2)
(a) Vol. **94**, No. 1, p. K25—K28 (1986).

H. B. Gon
*Growth of AgI Single Crystals in Silicate Gel and DC Conductivity of AgI
(66.30; 64.70; S9)
(a) Vol. **94**, No. 1, p. K61—K64 (1986).

I. Kosacki, K. Hibner, A. P. Litvinchuk, and M. Ya. Valakh
*Lattice Dynamics and Ionic Conductivity of PbF_2: 6% LiF Superionic Crystals
(66.30 and 78.30; 61.70; S9)
(b) Vol. **134**, No. 2, p. 495—501 (1986).

A. Klamt and H. Teichler
*Quantum-Mechanical Calculations for Hydrogen in Niobium and Tantalum. II. Diffusion Coefficients
(66.30; 61.70; S1)
(b) Vol. **134**, No. 2, p. 533—544 (1986).

Yu. S. Nechaev, S. A. Vladimirov, N. A. Olshevskii, and V. S. Khlomov
*On the Effect of High-Rate Strain on Diffusion Mass Transfer in Metals
(66.30; S1.1; S1.2; S1.3; S4)
(a) Vol. **94**, No. 2, p. 553—560 (1986).

K. Takaishi, T. Kikuchi, K. Furuya, I. Hashimoto, H. Yamaguchi, E. Yagi, and M. Iwaki
*Thermal Extraction of Krypton in Aluminium Using Mass Spectrometer
(66.30; 61.70; S4)
(a) Vol. **95**, No. 1, p. 135—139 (1986).

Yu. M. Shirshov, A. V. Nabok, O. N. Kopylov, and D. I. Kropman
*Investigation of Composition Profiles in Thin Phosphorus-Silicate Glass Layers
(66.30 and 68.55; S11)
(a) Vol. **95**, No. 1, p. 185—192 (1986).

R. Kutner
*Nonhomogeneous Generalized Master Equation and Nonhomogeneous Second-Order Integrodifferential Equation for the Time-Dependent Back-Jump Effect
(66.30)
(b) Vol. **135**, No. 2, p. 535—543 (1986).

B. Tasche und H. Richtering
*Kernspin–Gitter-Relaxation von ^{115}In und ^{75}As in InAs zwischen 300 und 1000 K und die Diffusion von Kupfer in InAs
(Spin–Lattice Relaxation of ^{115}In and ^{75}As in InAs between 300 and 1000 K and Diffusion of Copper in InAs)
(66.30 and 76.60; S7.12)
(a) Vol. **95**, No. 2, p. 565—571 (1986).

A. Klimakow, M. Schenk, and B. Christ
Incorporation of Tl in Monocrystalline n-PbS
(61.70; 61.80; 66.30; S8)
(a) Vol. **95**, No. 2, p. 573—578 (1986).

S. Guimarães, E. Landi, and S. Solmi
*Enhanced Diffusion Phenomena during Rapid Thermal Annealing of Preamorphized Boron-Implanted Silicon
(61.70 and 66.30; S5.11)
(a) Vol. **95**, No. 2, p. 589—598 (1986).

A. Simanovskis, S. V. Stolyarova, and J. Teteris
*Effect of Environment on Thermal Diffusion in Ag–As_2S_3
(66.30; S1.3; S8)
(a) Vol. **95**, No. 2, p. K105—K107 (1986).

M. B. Dutt, R. Kumar, R. Nath, M. N. Sen, and K. K. Laroia
Depth Profiling of Ion Implanted Silicon by Electrical Methods
(61.70; 66.30; 73.40; S5.11)
(a) Vol. **96**, No. 1, p. 121—127 (1986).

R. Hähnel, W. Miekeley, and H. Wever
*Diffusion Studies on the B8 Phase of the Ni/Sb System
(66.30; S1.2)
(a) Vol. **97**, No. 1, p. 181—190 (1986).

F. Corni and G. M. Antonini
*EXAFS Investigation of Interdiffusive Kinetics in Cu–Au Thin Films
(66.30; 61.55; 78.70; S1.2; S1.3)
(a) Vol. **97**, No. 1, p. K21—K24 (1986).

W. Pfeiler
Vacancy Activation Enthalpies in Concentrated AgAl
(61.70; 66.30; S1.3)
(b) Vol. **137**, No. 2, p. K127—K130 (1986).

A. Fischer, D. Knoll, T. Grabolla, and F. Fischer
*Redistribution of Implanted and Thermally Predeposited Dopants during High Pressure Oxidation
(66.30; S5.11)
(a) Vol. **97**, No. 2, p. K141—K144 (1986).

Hu An, Wang Yuan-hang, Li Xiang-jin, Yu. Zheng, and Feng Duan
*The Low-Temperature Interdiffusion in Compositionally Modulated Nb–Ti Films
(66.30; S1.2)
(a) Vol. **98**, No. 1, p. 215—220 (1986).

D. Bahadur
*Synthesis and Characterization of NASICON Based Glass
(66.30; 61.40; S10.15)
(a) Vol. **98**, No. 1, p. K23—K26 (1986).

V. V. Krasnogolovets and P. M. Tomchuk
*The Influence of External Effects on the Proton Conductivity of an Hydrogen-Bonded Chain
(66.30; 62.65; 71.38; S12.1)
(b) Vol. **138**, No. 2, p. 727—733 (1986).

J. Ni and T. S. Kê
A Theory of Anomalously Amplitude-Dependent Internal Friction Peaks Due to the Interaction between Dislocation Kinks and Substitutional Solute Atoms in F.C.C. Metals
(61.70; 62.40; 66.30; S1.2; S3; S4)
(a) Vol. **98**, No. 2, p. 487—501 (1986).

66.70. Nonelectronic thermal conduction and heat-pulse propagation in nonmetallic solids
(for thermal conduction in solid metals, see 72.15; for statistical mechanics of lattice vibrations, see 63.70)

Original Papers and Short Notes

W. MUTSCHELLER and M. WAGNER
*Thermal Conductivity of Cooperative Jahn-Teller E-Systems
(66.70; 63.20; 64.90; S11.1)
(b) Vol. **134**, No. 1, p. 39—52 (1986).

D. V. KAZAKOVTSEV and Y. B. LEVINSON
The Effect of Phonon–Phonon Scattering in the Substrate on Temperature Dynamics of a Phonon Film Injector
(63.20; 66.70; S5.11; S5.12; S7.12; S9.12)
(b) Vol. **136**, No. 2, p. 425—434 (1986).

U. SEIPOLD and H. A. ULLNER
Martensitic Character of the Pressure Induced B1–B2 Phase Transition of Potassium Bromide
(64.70; 66.70; S9.11)
(a) Vol. **96**, No. 2, p. 469—474 (1986).

M. C. AL-EDANI and K. S. DUBEY
*Lattice Thermal Conductivity of Gadolinium Monosulphide above Room Temperature in the Frame of Two-Mode Conduction of Phonons
(63.20 and 66.70; S8)
(b) Vol. **137**, No. 2, p. 449—452 (1986).

A. KUMAR, B. K. SRIVASTAVA, and M. A. ANSARI
*Lattice Thermal Conductivity of Deformed Lithium Fluoride Crystals at Low Temperatures
(66.70; 63.90; S9.11)
(b) Vol. **138**, No. 2, p. 423—432 (1986).

P. ENDERS
Non-Fourierian Heat Conduction in Semiconductor Multilayer Systems with Thin Layers
(65.70; 66.70)
(b) Vol. **138**, No. 2, p. K87—K89 (1986).

66.90. Other topics in nonelectronic transport properties

Original Papers

H. Vogt, R. Ulm, P. Würfel, and W. Ruppel
Anomalous Charge Storage in $NaNO_2$ Layers
(77.50; 66.90; S11.1)
(a) Vol. **93**, No. 1, p. 299—308 (1986).

A. Kessler and U. Klein
*Conductivity and Polarization of NH_4Cl between 150 and 500 K
(66.90 and 77.30; 61.70; S9)
(a) Vol. **98**, No. 2, p. 617—627 (1986).

67. Quantum fluids and solids, liquid and solid helium

67.40. Boson degeneracy and superfluidity of helium-4

Original Paper

R. Bahadur, N. Bhardwaj, and C. M. Kachhava
*Excitation Spectrum and Virial Coefficients of Superfluid He^4
(67.40; S5.2)
(b) Vol. **134**, No. 2, p. 671—674 (1986).

67.50. Fermi fluids, liquid helium-3

Original Paper

R. A. Singh and C. M. Kachhava
*Pressure Effects of Superfluidity in Liquid ^3He
(67.50)
(b) Vol. **138**, No. 2, p. 637—645 (1986).

67.80. Solid helium and related quantum crystals

Original Paper

D. I. Pushkarov
*On the Defecton Statistics in Quantum Crystals
(67.80; S5.2)
(b) Vol. **133**, No. 2, p. 525—531 (1986).

68. Surfaces and interfaces; thin films and whiskers (for impact phenomena, see 79; for crystal growth, see 61.50)

68.20. Solid surface structures

Review Article

T. Terzibaschian and R. Enderlein
*The Irreducible Representations of the Two-Dimensional Space Groups of Crystal Surfaces. Theory and Applications
(68.20 and 73.20)
(b) Vol. **133**, No. 2, p. 443—461 (1986).

Original Papers and Short Notes

G. Rudakoff and B. Müller
*RHEED Study on Surfaces of Amorphous Materials
(68.20; 61.14; S10.1)
(a) Vol. **93**, No. 2, p. 471—478 (1986).

K. Z. Botros and S. S. Sheinin
*On Identifying Crystal Surfaces in Electron Microscope Images of Stacking Faults
(61.70 and 68.20; 61.16; S1.2; S1.3)
(a) Vol. **93**, No. 2, p. 515—521 (1986).

A. E. Smith and D. F. Lynch
Calculated Low Energy Electron Total Current Spectra for the Al(111) Surface
(61.14; 68.20; S4)
(b) Vol. **134**, No. 1, p. 11—15 (1986).

Cs. Hajdu, F. Pászti, G. Mezey, and I. Lovas
*Stress Model for the Formation of Wave-Like Structures on High-Dose Ion Implanted Materials
(61.80 and 68.20; 61.70)
(a) Vol. **94**, No. 1, p. 351—352 (1986).

G. Mezey
*Physics of Ion Implantation in Plasma–Surface Interactions
(61.80 and 68.20)
(a) Vol. **94**, No. 2, p. 797—807 (1986).

V. Fritzsche and P. Rennert
A Reduced Angular Momentum Expansion in the Multiple-Scattering Theory
(79.20 and 79.60; 61.10; 61.14; 68.20)
(b) Vol. **135**, No. 1, p. 49—60 (1986).

V. V. Aristov, A. Yu. Nikulin, A. A. Snigirev, and P. Zaumseil
*Experimental Investigation of X-Ray Bragg Diffraction on the Periodic Surface Relief of a Perfect Crystal
(61.10 and 68.20; S5.11)
(a) Vol. **95**, No. 1, p. 81—86 (1986).

M. I. Guseva and G. V. Gordeeva
*Phase Transitions in the Surface Layers of Materials under Ion Implantation
(61.80 and 68.20; 64.70; S1.1)
(a) Vol. **95**, No. 1, p. 385—390 (1986).

S. Mróz and A. Mróz
*Quantitative AES Analysis of PbTe and SnTe Samples with and without Matrix Effect Correction
(68.20; 79.20; S8)
(a) Vol. **95**, No. 2, p. 407—412 (1986).

V. M. Belenkii, R. B. Begzanov, V. N. Bolotov, and A. V. Kuznichenko
*Influence of a Surface on Lifetime Spectra of Positrons
(61.70 and 68.20; 78.70)
(b) Vol. **136**, No. 2, p. 551—556 (1986).

V. Yu. Petukhov, I. B. Khaibullin, M. M. Zaripov, R. Groetzschel, M. Voelskow, and R. Klabes
*Investigation of Silicides Obtained by Fe^+ Implantation into Silicon
(68.20; 61.80; S1.61)
(a) Vol. **96**, No. 2, p. 463—468 (1986).

M. Dębowska, A. Baranowski, K. Jerie, and W. Świątkowski
The Effect of Impregnation with Gold on Positron Lifetime Spectra in NaY Zeolite
(61.70; 68.20; 78.70; S2)
(a) Vol. **97**, No. 1, p. K35—K38 (1986).

A. Chassé and P. Rennert
*Determination of Adsorbate Geometries from Angle-Resolved X-Ray Photoelectron Spectra. Nitrogen on Fe(111)
(68.20; 79.60; S1.1)
(b) Vol. **138**, No. 1, p. 53—64 (1986).

E. Chrzanowski, W. Doliński, S. Mróz, and E. Woźniak
*Investigation of $Pb_{1-x}Sn_xTe$ Crystal Surfaces by Quantitative AES, LEED, and Kikuchi Pattern Observation
(68.20; 61.14; 79.20; S8.15)
(a) Vol. **98**, No. 2, p. 359—366 (1986).

R. Manaila and M. Zaharescu
*Local Order in High Surface Area Amorphous Silica
(61.40 and 68.20; S10.1)
(a) Vol. **98**, No. 2, p. 377—382 (1986).

68.25. Mechanical and acoustical properties of solid surfaces and interfaces

Original Papers and Short Notes

U. Pietsch and D. Marlow
The Thermal Expansion Coefficient of $Ga_xIn_{1-x}As_yP_{1-y}$ Epitaxial Layers Grown on InP Substrate
(65.70; 68.25; S7.15)
(a) Vol. **93**, No. 1, p. 143—149 (1986).

H. Cerva and W. Graeff
*Contrast Investigations of Surface Acoustic Waves by Stroboscopic Topography. III. Contrast in Transmission Case
(61.10 and 68.25; S11.1)
(a) Vol. **93**, No. 2, p. K129—K132 (1986).

K. Franke and M. Weihnacht
*Surface Acoustic Wave Damping in Piezoelectric Ceramics by Scattering and Dissipation
(68.25; 62.30; 62.80)
(a) Vol. **95**, No. 2, p. 483—494 (1986).

G. Martin and K. Franke
*An Examination of the Validity of a Common Model Used for the Analysis of Surface Acoustic Wave Interdigital Transducers
(68.25; 62.30; 77.60)
(a) Vol. **96**, No. 2, p. 489—499 (1986).

W. Gräfe
A Method for the Determination of a Dynamic Fatigue Limit of Glasses
(61.40 and 62.20; 68.25)
(a) Vol. **96**, No. 2, p. 501—507 (1986).

L. Pranevičius, S. Tamulevičius, L. Puodžiukynas, and A. Matiukas
Application of Laser Interferometer Systems for the Study of Ion Interaction with Solids
(61.80; 68.25; 78.20; S10.1)
(a) Vol. **96**, No. 2, p. K157—K161 (1986).

68.30. Dynamics of solid surfaces and interface vibrations

Original Papers and Short Notes

68.40. Surface energy of solids; thermodynamic properties

Original Papers and Short Notes

T. YANAGIHARA and H. YAMAGUCHI
*Potential Energies of a Silver Atom Adsorbed on Flat and Stepped Surfaces of Sodium Chloride Crystals
(68.40; S9.11)
(a) Vol. **93**, No. 1, p. 67—70 (1986).

P. STREITENBERGER
*On the Force between High-Density Jellium-Metal Surfaces at Small Separation
(68.40 and 73.40)
(b) Vol. **135**, No. 1, p. 321—333 (1986).

S. K. LAI
*Density Functional Approach to the Surface Tension of Liquid Metals
(68.40; 71.10; S2)
(b) Vol. **136**, No. 2, p. 685—691 (1986).

J. OLIVES
The Electrostatic Lattice Energy
(61.50; 68.40)
(b) Vol. **138**, No. 2, p. 457—464 (1986).

68.45. Solid–fluid interface processes

Original Papers and Short Notes

V. Rakovics, T. Görög, and E. Lendvay
*Semitransparent LPE Reactor and Its Application for the Growth of the InP/InGaAsP Heterosystem
(68.45 and 68.55; S7.11; S7.15)
(a) Vol. **94**, No. 2, p. 727—730 (1986).

M. Somogyi
Dopant Concentration Determination at Frequencies above 10 kHz on GaP–Electrolyte System
(73.40; 68.45; S7.11; 7.16)
(a) Vol. **94**, No. 2, p. 731—734 (1986).

M. Petrov and P. Simova
Influence of the Boundary Conditions on the Characteristic Temperature in the Nematic Preceding Smectic C Phase of 4,n-Heptyloxybenzoic Acid
(61.30 and 72.80; 68.45; S12)
(a) Vol. **95**, No. 2, p. K165—K167 (1986).

A. A. Karpushin, A. N. Sorokin, I. Yu. Semenova, and M. Tomášek
Method of Generalized Bethe Lattices in Chemisorption Theory
(71.10 and 73.20; 68.45; S5.11)
(b) Vol. **136**, No. 1, p. 331—339 (1986).

R. Tognato
*Interfacial Free Energies per Unit Area of Liquid Flat Aggregates Embedded in a Crystal Matrix
(68.45; 61.20; S4)
(a) Vol. **96**, No. 2, p. 451—461 (1986).

C. Engler and E. Rabe
*Contribution of Medium Phonons to the Transition Probability of Reactive Processes on Solid/Solution Interfaces. Joint Treatment of Classical and Quantum Parts of the Medium
(68.30 and 68.45)
(b) Vol. **138**, No. 2, p. 477—482 (1986).

A. Wilson and R. A. Collins
*Gas–Surface Reactions on Phthalocyanine Thin Films
(68.45 and 73.60; 79.60; S12)
(a) Vol. **98**, No. 2, p. 633—644 (1986).

68.48. Solid–solid interfaces (including bicrystals) (for grain boundaries, see 61.70)

Original Papers and Short Notes

R. FLAGMEYER and B. SCHUMANN
Rutherford Backscattering/Channeling Analyses of Lattice-Matched $Ca_{0.43}Sr_{0.57}F_2$ Films on GaAs
(68.55; 61.80; 68.48; S9.15)
(a) Vol. **93**, No. 2, p. K117—K121 (1986).

E. I. KAFEDIISKA, I. MOJZES, R. VERESEGYHÁZY, and B. PÉCZ
*Mass-Spectrometric Study of the Dissociation of GaSb Covered with Contact Layers
(68.48; S7.13)
(a) Vol. **93**, No. 2, p. K125—K128 (1986).

R. BUTZ and H. WAGNER
*Reactions at the Ti/SiO_2 Interface Studied by EELS
(68.48; 68.30; 61.80; S1.2; S10.1)
(a) Vol. **94**, No. 1, p. 71—76 (1986).

D. M. VANDERWALKER
The Glass Transition Phases in the Formation of PtSi
(61.40; 68.48; S1.61)
(a) Vol. **94**, No. 1, p. 77—81 (1986).

R. VERESEGYHÁZY, B. PÉCZ, and I. MOJZES
*The Influence of a Gold Layer on the Thermal Decomposition of InAs
(68.48; S1.3; S7.12)
(a) Vol. **94**, No. 1, p. K11—K12 (1986).

B. PÉCZ, E. JÁROLI, G. RADNÓCZI, R. VERESEGYHÁZY, and I. MOJZES
*Pyramidal Pit Formation at the Au/GaAs Interface during Heat Treatment
(68.48; S1.3; S7.12)
(a) Vol. **94**, No. 2, p. 507—513 (1986).

K. HÜBNER, E.-D. KLINKENBERG, and A. STERN
*Theoretical Study of the Gradual Chemical Transition at the $Si–SiO_2$ Interface. I. Structural Models for the $Si(111)–SiO_2$ and $Si(100)–SiO_2$ Interregions
(68.48; S5.11; S10.1)
(b) Vol. **135**, No. 2, p. 475—485 (1986).

N. V. RUMAK and V. V. KHATKO
Stimulated Crystallization of Thermal Silica Films. I. Heat Treatment
(68.55; 68.48; 73.40; S10.1)
(a) Vol. **95**, No. 2, p. 391—395 (1986).

68.55. Thin film growth, structure, and epitaxy

Original Papers and Short Notes

V. M. Koleshko, V. F. Belitskii, and A. A. Khodin
*Electrical Conductance and Phase Transformations in Dy and Tb Thin Films on Si Substrates Subjected to Heat Treatment
(68.55; 73.60; S1.4; S1.61)
(a) Vol. **93**, No. 1, p. 29—33 (1986).

V. E. Borisenko and V. P. Parkhutik
*RBS Study of Transient Thermal and Anodic Oxidation of Tantalum Films
(68.55; S10)
(a) Vol. **93**, No. 1, p. 123—130 (1986).

U. Pittermann and S. Ripper
*Crystallization Behaviour of Thin NiP Alloy Films Prepared by Electroless or Galvanic Deposition
(68.55; S1.62)
(a) Vol. **93**, No. 1, p. 131—142 (1986).

N. Saito and N. Tanaka
Preparation and Properties of Magnetron Sputtered Fluorinated and Hydrogenated Amorphous Si–C Films
(73.60 and 78.65; 68.55; S6)
(a) Vol. **93**, No. 1, p. 341—345 (1986).

T. Jung
*Kinetics and Mechanism of Plasma Nitridation of Thin Aluminium Films
(68.55; S7.1)
(a) Vol. **93**, No. 2, p. 479—485 (1986).

V. Yu. Balandin, A. V. Dvurechenskii, and L. N. Aleksandrov
*Intermediate and Self-Sustaining Crystallization of a-Si Layers during Pulsed Electron Beam Annealing
(68.55; 61.40; S5.11)
(a) Vol. **93**, No. 2, p. K105—K109 (1986).

R. Flagmeyer and B. Schumann
*Rutherford Backscattering/Channeling Analyses of Lattice-Matched $Ca_{0.43}Sr_{0.57}F_2$ Films on GaAs
(68.55; 61.80; 68.48; S9.15)
(a) Vol. **93**, No. 2, p. K117—K121 (1986).

N. R. Stratieva and E. M. Konstantinova
*Structural and Photoelectrical Properties of $Zn_xCd_{1-x}S$ Layers Prepared by the $Zn_xCd_{1-x}S + Cr_2O_3$ System
(68.55 and 73.60; S8.15)
(a) Vol. **93**, No. 2, p. K191—K193 (1986).

E. GLASER, G. ANDRÄ, H. BARTSCH, K. DRENDA, and G. GÖTZ
*Formation of Monocrystalline Layers by Explosive Crystallization of Ion-Implanted Amorphous Silicon
(68.55; 61.70; S5.11)
(a) Vol. **94**, No. 2, p. 781—786 (1986).

C. KASCHNER, A. WITZMANN, K. GÄRTNER, and G. GÖTZ
*Laser Annealing of Ion-Implanted NiSi Layers
(68.55; 61.80; S1.61)
(a) Vol. **94**, No. 2, p. 787—791 (1986).

H. KERKOW, G. KREYSCH, G. WOLF, and K. HOLLDACK
*Ion-Beam-Assisted Annealing of Amorphized Silicon
(68.55; 61.80; S5.11)
(a) Vol. **94**, No. 2, p. 793—796 (1986).

B. RAUSCHENBACH and K. HOHMUTH
*Synthesis of Compounds by High-Fluence Nitrogen Ion Implantation in Titanium
(61.70 and 68.55; S1.2)
(a) Vol. **94**, No. 2, p. 833—837 (1986).

V. K. GUEORGUIEV, L. I. POPOVA, I. N. PETROV, and I. G. STOEV
*Anisotropic Etchants Inhibiting Properties of Silicon-Nitride Compound Layers Produced by Ion Implantation
(68.55; S7)
(a) Vol. **94**, No. 2, p. 865—869 (1986).

B. TILLACK, P. MÖCK, R. BANISCH, R. REINBOTH, H. H. RICHTER, M. VOELSKOW, J. MATTHÄI, and E. WIESER
*Thick Monocrystalline Silicon on Oxidized Silicon Wafers Produced by a Zone Melting Process Using a Scanning Halogen Lamp
(68.55; S5.11)
(a) Vol. **94**, No. 2, p. 871—877 (1986).

D. DOBREV, V. GADJOKOV, V. VELEV, and A. GITTIS
*Texture of Silver Films Condensed at High Rates onto Glass Substrates in Vacuum
(68.55; S1.3)
(a) Vol. **95**, No. 1, p. 93—97 (1986).

F. H. COCKS and J. T. A. POLLOCK
The Ion Implantation of Helium into Gold. A Combined X-Ray Reflection Spectroscopy and Rutherford Backscattering Study
(61.70; 68.55; S1.3)
(a) Vol. **95**, No. 1, p. 141—148 (1986).

YU. M. SHIRSHOV, A. V. NABOK, O. N. KOPYLOV, and D. I. KROPMAN
*Investigation of Composition Profiles in Thin Phosphorus-Silicate Glass Layers
(66.30 and 68.55; S11)
(a) Vol. **95**, No. 1, p. 185—192 (1986).

N. V. RUMAK and V. V. KHATKO
*Stimulated Crystallization of Thermal Silica Films. II. Treatment with Metal Particles and/or Inert Gas Ions
(68.55; 68.48; S10.1)
(a) Vol. **96**, No. 1, p. 19—24 (1986).

D. R. MCKENZIE, P. S. TURNER, and D. J. H. COCKAYNE
*Electron Imaging and Energy Loss Studies of the Crystallization of Hydrogenated Amorphous Silicon
(68.55; 61.16; 61.80; S5.11)
(a) Vol. **96**, No. 1, p. 67—73 (1986).

T. JUNG and M. SCHMIDT
*Properties of AlN_xO_y/GaAs Structures Made by Plasma Nitridation of Evaporated Aluminium Films
(68.55 and 73.60; 79.20; S7.1; S7.12)
(a) Vol. **96**, No. 1, p. 75—81 (1986).

A. ARMIGLIATO, G. QUEIROLO, and G. VALDRÈ
*Nitrogen Profiles of Ion-Implanted TiN Films by Electron Energy Loss Spectroscopy
(68.55; 61.80; S1.62; S5.11)
(a) Vol. **96**, No. 1, p. 89—99 (1986).
Erratum
(a) Vol. **98**, No. 2, p. 651 (1986).

S. A. AZIMOV, T. M. RAZYKOV, V. I. VYALY, and B. KH. KADYROV
*Physical Properties of $(ZnSe)_x(CdTe)_{1-x}$ Multicomponent System Films Fabricated by CVD in Hydrogen Flow
(68.55 and 71.25; 73.60; S8.15)
(a) Vol. **96**, No. 1, p. 281—284 (1986).

A. N. Y. SAMAAN, S. M. WASIM, A. E. HILL, D. G. ARMOUR, and R. D. TOMLINSON
*Preparation and Properties of $CuInS_2$ Thin Films Deposited by RF Sputtering
(68.55 and 73.60; S8.16)
(a) Vol. **96**, No. 1, p. 317—324 (1986).

B. DIEGNER, J. KUGLER, and H. WEINERT
On the Correlations between Electrical Properties and Growth Conditions for VPE GaAs Grown in the Hydride System
(73.60; 68.55; S7.12)
(a) Vol. **96**, No. 1, p. 345—351 (1986).

D. AUGIE, M. BELHADJI et P. VIGIER
*Contraintes internes de couches minces Ni–Ag amorphes
(Internal Strains in Amorphous Ni–Ag Thin Films)
(68.55; S1.2; S1.3)
(a) Vol. **96**, No. 1, p. K5—K8 (1986).

P. K. BHAT, J. J. DUBOWSKI, and D. F. WILLIAMS
*Emission of Neutral and Charged Species from InSb Induced by Pulsed Laser Evaporation
(68.55 and 79.20; S7.13)
(a) Vol. **96**, No. 1, p. K9—K14 (1986).

K. HELMING and B. RAUSCHENBACH
*Determination of Implantation-Induced Textures by Electron Diffraction
(61.14 and 68.55; 61.70; S1.2; S1.62)
(b) Vol. **138**, No. 1, p. K1—K6 (1986).

T. JUNG and W. TITEL
On the Influence of Ar^+ Ion Bombardment on the Si_{LVV} Auger Line in Silicon Nitride Films
(79.20; 68.55; S6)
(a) Vol. **98**, No. 1, p. 63—67 (1986).

R. HERGT, H. PFEIFFER, and L. FRITZSCH
*Formation of Texture in Co/Cr Films for Perpendicular Magnetic Recording
(68.55; 75.70; S1.2)
(a) Vol. **98**, No. 1, p. 69—80 (1986).

P. DANESH, I. SAVATINOVA, E. ANACHKOVA, and ST. GEORGIEV
Raman Scattering in Glow Discharge Si:H:Cl Films
(78.30; 61.40; 68.55; S5.11)
(a) Vol. **98**, No. 1, p. 127—134 (1986).

K. SOMOGYI and I. GYURÓ
*A Method for Studying Parameter Variations in VPE Technology
(68.55; S7.12)
(a) Vol. **98**, No. 1, p. K1—K4 (1986).

F. L. EDELMAN, J. HEYDENREICH, D. HOEHL, J. MATTHÄI, I. MELNIK, A. RZHANOV, M. VOELSKOV, and P. WERNER
*Structure of Annealed Polycrystalline Silicon Films. I. Recrystallization
(68.55; 64.70; S5.11)
(a) Vol. **98**, No. 2, p. 383—390 (1986).

K. NITSCH, M. MATYÁŠ, JR., and J. BOK
Validity of the Anionic Model of Rare Earth Iron Garnet Solubility in $PbO-B_2O_3$ Flux
(64.75; 68.55; S11.2)
(a) Vol. **98**, No. 2, p. 465—469 (1986).

S. NAGATA, M. OGINO, S. TANIGUCHI, and A. NOZAKI
*Indirect Observation of the Transition from an Amorphous to a Crystalline State in Pd Film Vapor-Quenched at 77 K
(68.55; 73.60; S1)
(a) Vol. **98**, No. 2, p. K105—K108 (1986).

E. VETTER, T. GESSNER, E. WIESER, and C. WEISE
Redistribution of Implanted Arsenic in a $MoSi_2$/Polysilicon Structure by Silicidation Tempering
(61.70; 68.55; S1.61)
(a) Vol. **98**, No. 2, p. K129—K132 (1986).

68.60. Physical properties of thin films; nonelectronic

Original Papers

J. Kub
*A Study of the Lattice Relaxation around Cracks in LPE Garnet Films
(61.70 and 68.60; 75.70; S11.2)
(a) Vol. **96**, No. 1, p. 249—253 (1986).

N. Laouini et F. Hila
*Diffusion dans les films minces. Cas des bicouches Au–Ni
(Diffusion in Thin Films. Case of Au–Ni Couple)
(68.60; 68.48; S1.2; S1.3)
(a) Vol. **97**, No. 1, p. 173—179 (1986).

68.70. Whiskers and dendrites: growth, structure, and nonelectronic properties

Original Paper

P. Trung Dung and M. Láznička
Origin and Formation Mechanism of Macroscopic Defects in GaAs Films Grown by Molecular Beam Epitaxy
(68.55; 68.70; S7.12)
(a) Vol. **97**, No. 1, p. 103—109 (1986).

68.90. Other topics in the structure and nonelectronic properties of surfaces and thin films

Original Papers

D. Lehmann and P. Ziesche
Hellmann-Feynman Forces on and between Jellia
(71.10; 68.90; 73.40)
(b) Vol. **135**, No. 2, p. 651—660 (1986).

R. Kaschner and P. Ziesche
*Force Sum Rules at Surfaces
(61.50 and 68.90; 71.10)
(b) Vol. **138**, No. 1, p. 65—71 (1986).

70. Condensed matter: electronic structure; electrical, magnetic, and optical properties

71. Electronic states
(see also 63. Lattice dynamics, 73. Electronic structure and electrical properties of surfaces, interfaces, and thin films)

71.10. General theories and computational techniques

Review Article

YU. N. SHUNIN and K. K. SHVARTS
Calculation of the Electronic Structure in Disordered Semiconductors
(71.25; 71.10; 71.20; S5; S5.11)
(b) Vol. **135**, No. 1, p. 15—36 (1986).

Original Papers and Short Notes

L. A. BUGAEV, I. I. GEGUSIN, V. N. DATSYUK, A. A. NOVAKOVICH, and R. V. VEDRINSKII
*Multiple Scattering Approach to the XANES Theory of Alkali Halide Crystals. I. Crystalline Potential in the X-Ray Absorption Spectra Problem
(71.10; 78.70; S9.11)
(b) Vol. **133**, No. 1, p. 195—202 (1986).

D. J. GONZÁLEZ
*Density Functional Pseudopotential Calculation of the Cohesive Properties of Ordered (CsCl-Type) Compounds of Nontransition Metals
(71.10; S2; S3; S4)
(b) Vol. **133**, No. 1, p. 277—284 (1986).

S. V. VLASOV and O. V. FARBEROVICH
*APW Research of Europium Monochalcogenides. II. Magnetic Form Factor and Insulator–Metal Transition in EuO under Pressure
(71.10; 71.25; 71.30; 75.25; S10.1)
(b) Vol. **133**, No. 2, p. 601—609 (1986).

R. SHARMA, K. S. SHARMA, and L. DASS
Linearised Screened Pseudopotential and Superconducting State Parameters of a Number of Metals
(74.20; 71.10; S1.2; S2; S3; S4)
(b) Vol. **133**, No. 2, p. 701—706 (1986).

G. GROSSO and G. PASTORI PARRAVICINI
*A Projective Procedure for Embedding a Finite Cluster in an Effective Medium
(71.10)
(b) Vol. **133**, No. 2, p. K107—K110 (1986).

D. LEHMANN and P. ZIESCHE
*A Rigorous Theorem for the Pressure of Extended Electron Systems
(71.10)
(b) Vol. **133**, No. 2, p. K115—K117 (1986).

V. ČÁPEK
*GME and the Electron or Exciton Initial Condition Term Due to Lattice Relaxation. Dimer
(71.10; 71.35; 71.38)
(b) Vol. **136**, No. 1, p. 251—259 (1986).

P. V. SMITH, J. E. SZYMANSKI, and J. A. D. MATTHEW
*A Reformulation of the LUC-CNDO Approach to the Properties of Solids. Application to Diamond and Silicon
(71.10; 63.20; S5; S5.11)
(b) Vol. **136**, No. 1, p. 261—271 (1986).

A. A. KARPUSHIN, A. N. SOROKIN, I. YU. SEMENOVA, and M. TOMÁŠEK
*Method of Generalized Bethe Lattices in Chemisorption Theory
(71.10 and 73.20; 68.45; S5.11)
(b) Vol. **136**, No. 1, p. 331—339 (1986).

Y. CHEN and B. GIL
*Simple Description of Hydrogen-Like Systems in Arbitrary Magnetic Fields
(71.10 and 75.10)
(b) Vol. **136**, No. 2, p. 629—635 (1986).

S. K. LAI
Density Functional Approach to the Surface Tension of Liquid Metals
(68.40; 71.10; S2)
(b) Vol. **136**, No. 2, p. 685—691 (1986).

A. V. FISENKO, S. A. PROSANDEEV, and V. P. SACHENKO
*Electronic Structure of Intrinsic Point Defects in Perovskite-Type Crystals
(71.10 and 71.55; S11.1)
(b) Vol. **137**, No. 1, p. 187—197 (1986).

R. KASCHNER and P. ZIESCHE
Force Sum Rules at Surfaces
(61.50 and 68.90; 71.10)
(b) Vol. **138**, No. 1, p. 65—71 (1986).

71.20. Electronic density of states determinations (including energy states of liquid semiconductors) (see also 65.40. Electronic heat capacity)

Review Article

Original Papers and Short Notes

P. Blaudeck
*Cluster Embedding in an Effective Shell and Calculation of the Density of States for Amorphous Copper
(71.20; 61.40; S1.2)
(b) Vol. **136**, No. 2, p. 673—678 (1986).

V. Bethkenhagen and K. Hübner
Cluster-Bethe-Lattice Approach to the Electronic Structure of Intrinsic Defects in v-SiO_2
(71.55; 71.20; 78.50; 78.55; S10.1)
(b) Vol. **136**, No. 2, p. 721—734 (1986).

G. D. Maksimović, D. R. Mašović, and Z. S. Popović
*Application of the Modified Veljković-Slavić Pseudopotential for Calculation of the Electronic Charge Density of Si
(71.20; S5.11)
(b) Vol. **136**, No. 2, p. K101—K104 (1986).

S. Wilke, J. Mašek, and B. Velický
*The Electronic Structure of Polysilane Alloys
(71.20; S5.11)
(b) Vol. **136**, No. 2, p. K113—K117 (1980).

B. Esser and F. Marlow
*On the Influence of Noise on the Density of States and Absorption in Disordered Molecular-Systems
(71.20 and 71.35; 78.20)
(b) Vol. **137**, No. 1, p. 169—181 (1986).

V. A. Terekhov, S. N. Trostyanskii, E. P. Domashevskaya, O. A. Golikova, M. M. Mezdrogina, K. L. Sorokina, and M. M. Kazanin
*Density of States and Photoconductivity of Hydrogenated Amorphous Silicon
(71.20; 72.40; 73.60; 78.70; S5.11)
(b) Vol. **138**, No. 2, p. 647—653 (1986).

A. Jezierski
*Electronic Density of States in Binary Non-Crystalline Alloys with Different Chemical Short Range Ordering
(71.20; S1.1)
(b) Vol. **138**, No. 2, p. K119—K123 (1986).

L. Gier, A. Scharmann, and D. Schalch
Hopping Conduction and Defect States in Reactively Sputtered Silicon Nitride Thin Films
(73.60; 71.20; S7)
(a) Vol. **98**, No. 2, p. 605—610 (1986).

71.25. Nonlocalized single-particle electronic states

Review Article

Yu. N. Shunin and K. K. Shvarts
*Calculation of the Electronic Structure in Disordered Semiconductors
(71.25; 71.10; 71.20; S5; S5.11)
(b) Vol. **135**, No. 1, p. 15—36 (1986).

Original Papers and Short Notes

S. G. Abdullaeva and N. T. Mamedov
*Band Structure of $TlGaSe_2$ Ternary Layered Crystals
(71.25; S8.16)
(b) Vol. **133**, No. 1, p. 171—177 (1986).

G. C. Fletcher
*The Band Structure of Co_3Ga and Co_2MnGa and the Electrical Conductivity of Related Alloys
(71.25 and 72.15; 75.50; S1.2)
(b) Vol. **133**, No. 1, p. 241—248 (1986).

O. P. Singh and V. P. Gupta
*Electronic Properties of Iron Chalcogenides FeS, FeS_2, $FeSe_2$, and $FeTe_2$
(71.25; 76.80; S8)
(b) Vol. **133**, No. 1, p. 249—252 (1986).

V. K. Bashenov, V. E. Gorbachev, and D. I. Marvakov
*Bond Charge Approximation for Valence Electron Density in Elemental Semiconductors
(71.25; S5; S5.11; S5.12)
(b) Vol. **133**, No. 1, p. 285—296 (1986).

B. Kumar
Carrier Drift Mobilities in Some Organic Solids. A Search for Organic Photovoltaic Materials. I. Charge Transport Calculations in 2-2′ bipyridine and pyridazino [4,5-d] pridazine
(72.40 and 72.80; 71.25; S12)
(b) Vol. **133**, No. 1, p. 395—402 (1986).

O. P. Singh and V. P. Gupta
*Modified PVV Theory and X-Ray K Absorption Edge Studies in Zinc Chalcogenides (Ionicity Determination and d Electron Effect on It)
(71.25; 78.70; S8.11; S8.12; S8.13)
(b) Vol. **133**, No. 1, p. K17—K19 (1986).

A. R. Jani and H. K. Patel
*Quantum Number Dependent Model Pseudopotential for Metallic Properties
(71.25; 72.15; S4)
(b) Vol. **133**, No. 1, p. K21—K24 (1986).

V. I. Gatalskaya
Cyclotron Resonance Anisotropy of Charge Carriers in Irradiated Germanium
(76.40; 71.25; S5.12)
(b) Vol. **133**, No. 1, p. K31—K33 (1986).

V. P. Shirokovskii, N. A. Shilkova, and N. A. Trubitsina
*Calculation of Electron Velocity in Crystal ab Initio. Results of Computation by Green's Function Method for Vanadium
(71.25; S1.2)
(b) Vol. **133**, No. 2, p. 593—599 (1986).

S. V. Vlasov and O. V. Farberovich
APW Research of Europium Monochalcogenides. II. Magnetic Form Factor and Insulator–Metal Transition in EuO under Pressure
(71.10; 71.25; 71.30; 75.25; S10.1)
(b) Vol. **123**, No. 2, p. 601—609 (1986).

V. Milanović, D. Tjapkin, and R. Šašić
The Influence of Nonparabolicity on Self-Consistent Treatment for GaAs–$Al_xGa_{1-x}As$ Heterojunction
(73.40; 71.25; S7.12; S7.15)
(b) Vol. **133**, No. 2, p. 775—784 (1986).

V. N. Datsyuk, I. I. Gegusin, and R. V. Vedrinskii
Multiple Scattering Approach to the XANES Theory of Alkali Halide Crystals. II. Band Structure Approach and Cluster Approximations
(71.10; 71.25; 78.70; S9.11)
(b) Vol. **134**, No. 1, p. 175—184 (1986).

S. I. Gubarev
*Valence Band Splittings in Wurtzite-Type Semimagnetic Semiconductors
(71.25 and 71.70; 78.20; S8.16)
(b) Vol. **134**, No. 1, p. 211—222 (1986).

B. A. Oli and J. S. Ononiwu
Effect of Non-Locality in the Calculation of Liquid Metal Resistivity of Transition Metals
(72.15; 71.25; S1.1; S1.2; S1.3)
(b) Vol. **134**, No. 1, p. 325—330 (1986).

Sh. M. Sharipov, K. M. Mukimov, and L. A. Ernazarova
The Heavy Rare-Earth Contribution to the Electronic Structure of the RFe_2 and R_2Fe_{17} Intermetallic Compounds
(78.20; 71.25; S1.1)
(b) Vol. **134**, No. 1, p. K59—K62 (1986).

H. Wiesner, W. Cordts, and J. Monecke
*On the Influence of the Non-Locality of Exchange and Correlation on the Total Energy and on the Band Gap of Silicon
(71.25; S5.11)
(b) Vol. **134**, No. 1, p. K63—K66 (1986).

C. M. van Vliet and P. Vasilopoulos
*Electrical and Optical Bandgap Narrowing Due to Heavy Doping in Silicon
(71.25 and 71.55; S5.11)
(a) Vol. **94**, No. 2, p. 635—644 (1986).

P. Enders and R. Schuchardt
*Anomalous Composition Dependence of the Dielectric Constant in Bismuth–Antimony Alloys Due to Disorder-Induced Indirect Interband Polarization Processes
(71.25; 77.20; S5)
(b) Vol. **135**, No. 1, p. 207—216 (1986).

N. N. Ablyazov, A. G. Areshkin, V. G. Melekhin, L. G. Suslina, and D. L. Fedorov
Fluctuation-Induced Broadening of Exciton Reflection Spectra in $A^{II}B^{VI}$ Solid Solutions
(71.35 and 78.40; 71.25; S8.15)
(b) Vol. **135**, No. 1, p. 217—225 (1986).

J. Petrů
Computed X-Ray Emission Spectra for Rhodium under High Pressure
(78.70; 71.25; S1)
(b) Vol. **135**, No. 1, p. 233—237 (1986).

M. Mondal and K. P. Ghatak
*Effect of Carrier Degeneracy on the Screening Length in Degenerate Tetragonal Semiconductors
(71.25; S7)
(b) Vol. **135**, No. 1, p. 239—251 (1986).

A. M. El-Dib and H. F. Hassan
*Energy Gap, Effective Mass, and Critical Phenomena in $NaNO_2$ Crystal
(71.25 and 77.80; 65.70; S11.1)
(b) Vol. **135**, No. 1, p. 253—259 (1986).

D. Genzow, M. Mocker, and E. Normantas
*The Influence of Uniaxial Strain on the Band to Band Recombination in $Pb_{1-x}Sn_xTe$
(71.25; 73.60; S8.15)
(b) Vol. **135**, No. 1, p. 261—270 (1986).

G. Gobsch, G. Paasch, and H. Übensee
*Self-Consistent Calculation of the Electronic Structure of n-Inversion Layers Adjacent to the Grain Boundary in InSb Bicrystals
(71.25 and 73.40; S7.13)
(b) Vol. **135**, No. 1, p. 283—297 (1986).

G. Valiukonis, D. A. Guseinova, G. Krivaitė, and A. Šileika
*Optical Spectra and Energy Band Structure of Layer-Type $A^{IV}B^{VI}$ Compounds
(71.25 and 78.20; S8)
(b) Vol. **135**, No. 1, p. 299—307 (1986).

S. Huant, L. Dmowski, M. Baj, and L. C. Brunel
*Pressure Dependence of the Electronic Effective Mass and Effective *g*-Factor in the Narrow Gap Semiconductor InSb
(71.25; S7.13)
(b) Vol. **135**, No. 2, p. K129—K132 (1986).

Ke-Jun Ma and W. Giriat
*Dependence of Energy Gap on Temperature and Composition in $Mn_xCd_{1-x}S$
(71.25; 72.40; 78.20; S8.15)
(a) Vol. **95**, No. 2, p. K135—K138 (1986).

L. Palermo and X. A. da Silva
Magnetic Order of Singlet–Singlet Ions Coupled to Conduction Electrons. Band Structure Effects
(75.10; 71.25; S1.4)
(b) Vol. **136**, No. 1, p. 97—102 (1986).

W. Borgiel, W. Nolting, and G. Borstel
Self-Consistent CPA Treatment of the Conduction Electron Spin Polarization of Degenerate Ferromagnetic Semiconductors
(75.10; 71.25; S1.4; S8)
(b) Vol. **136**, No. 1, p. 131—144 (1986).

L. S. Grigoryan, I. L. Kharutyunyan, A. A. Matnishyan, R. S. Nikogosyan, and E. G. Sharoyan
*Charged Elementary Excitations in Anthracene Polymers. A Systematical Electrical and EPR Study as a Function of Doping
(71.25 and 71.38; 72.80; 76.30; S12)
(b) Vol. **136**, No. 1, p. 145—151 (1986).

P. V. Geld, V. V. Singer, I. Z. Radovskii, and S. V. Konovalov
*Variable Valence of Ce and Pr in Liquid Alloys with Fe, Co, and Ni
(71.25; 61.25; 72.15; 75.30; S1.1; S1.2; S1.4)
(b) Vol. **136**, No. 1, p. 153—158 (1986).

A. M. El-Dib and H. F. Hassan
*The Electronic Energy Bands of $AgNO_2$ Crystal at Room Temperature
(71.25; S11.1)
(b) Vol. **136**, No. 1, p. 159—166 (1986).

L. M. Sandratskii
*Energy Band Structure Calculations for Crystals with Spiral Magnetic Structure
(71.25 and 75.10)
(b) Vol. **136**, No. 1, p. 167—180 (1986).

J. Pabst and H. Röll
*Induced Magnetic Form Factors for 3d Transition Metals and 3d–3d Alloys. II. Induced Spin Form Factors
(71.25 and 75.25; S1.1; S1.2)
(b) Vol. **136**, No. 1, p. 225—231 (1986).

C. Schober, G. Kurz, H. Wonn, V. V. Nemoshkalenko, and V. N. Antonov
*The Gyromagnetic Factor of Conduction Electrons in Silver, Gold, Palladium, and Platinum
(71.25; S1.3)
(b) Vol. **136**, No. 1, p. 233—239 (1986).

A. K. Abass, Z. A. Ahmed, and R. E. Tahir
*Interband Transitions of Chemically Deposited Pyrite FeS_2 Films in the Fundamental Absorption Region between 1 and 3.8 eV
(71.25 and 78.65; S8)
(a) Vol. **97**, No. 1, p. 243—247 (1986).

F. López-Aguilar and J. Costa-Quintana
*Electronic Structure of the Ce Systems. I. General Theory
(71.25; S1.4)
(b) Vol. **137**, No. 2, p. 533—539 (1986).

P. K. Mukhopadhyay and K. C. Das
*Effect of Non-Spherical Potential on the Band Structure of Metallic Dysprosium
(71.25; S1.4)
(b) Vol. **137**, No. 2, p. 541—546 (1986).

F. Lukeš, E. Schmidt, J. Humlíček, P. Dub, and F. Kosek
Absorption Edge of SnSe
(78.20; 71.25; S8)
(b) Vol. **137**, No. 2, p. 569—579 (1986).

H. Haberlandt
*Determination of Titanium CNDO Parameters and Application in Cluster Calculations of Different Titanium Compounds
(71.25; S1.2; S10.1)
(b) Vol. **137**, No. 2, p. 581—589 (1986).

N. I. Papanicolaou, N. C. Bacalis, and D. A. Papaconstantopoulos
*Calculation of Compton Profiles of Tantalum and Tungsten
(71.25; 78.70; S1)
(b) Vol. **137**, No. 2, p. 597—606 (1986).

M. Załużny and A. Łukasik
Theory of the Infrared Spectrum of Size-Quantized Bi Films
(78.30; 71.25; 78.65; S4)
(b) Vol. **137**, No. 2, p. 607—617 (1986).

A. Brezini, M. Sebbani, and L. Dahmani
Localization in Random Systems. I. Effect of Off-Diagonal Disorder
(71.50; 71.25)
(b) Vol. **137**, No. 2, p. 667—673 (1986).

O. P. Singh and V. P. Gupta
*Investigations on Uranium Chalcogenides (US, USe, and UTe)
(71.25; S1.63)
(b) Vol. **137**, No. 2, p. K153—K156 (1986).

H. Wiesner, W. Cordts, and J. Monecke
*On the Shaw Potential and Its Application to Self-Consistent Band Structure and Total Energy Calculations for Silicon
(71.25; S5.11)
(b) Vol. **137**, No. 2, p. K161—K166 (1986).

71.30. Metal–insulator transitions

Original Papers and Short Notes

N. N. BELEVICH and L. A. MAKOVETSKAYA
Semiconductor–Metal Transition and Variable Range Hopping Conductivity in $CuInSe_2$
(72.60; 71.30; S8.16)
(b) Vol. **133**, No. 1, p. K53—K56 (1986).

S. V. VLASOV and O. V. FARBEROVICH
APW Research of Europium Monochalcogenides. II. Magnetic Form Factor and Insulator–Metal Transition in EuO under Pressure
(71.10; 71.25; 71.30; 75.25; S10.1)
(b) Vol. **133**, No. 2, p. 601—609 (1986).

M. ARNOLD and P. BUSSEMER
*A Modified Bruggeman Theory of Optical Properties for Metal–Isolator Transitions
(71.30 and 77.20)
(b) Vol. **134**, No. 1, p. K43—K45 (1986).

W. KRAAK, U. SCHALLER, and R. HERRMANN
*Metal–Insulator Transition and Metastable State in High-Purity n-InSb under High Hydrostatic Pressure
(71.30; 71.55; 72.20; S7.13)
(b) Vol. **134**, No. 2, p. 779—788 (1986).

E. KOLLEY and W. KOLLEY
*Electron Localization in the Site-Percolation Limit
(71.30 and 72.10)
(b) Vol. **135**, No. 1, p. K89—K93 (1986).

B. DAS, K. S. BARTWAL, and O. N. SRIVASTAVA
*Metal–Insulator Transition in Titanium Disulphide Crystals
(71.30 and 72.60; S1.63)
(b) Vol. **136**, No. 1, p. 365—371 (1986).

R. RENTZSCH, K. J. FRIEDLAND, A. N. IONOV, M. N. MATVEEV, I. S. SHLIMAK, C. GLADUN, and H. VINZELBERG
Variable-Range Hopping in Neutron-Transmutation-Doped Gallium Arsenide
(72.20; 71.30; S7.12)
(b) Vol. **137**, No. 2, p. 691—700 (1986).

B. P. GORSHUNOV, G. V. KOZLOV, A. A. VOLKOV, J. PETZELT, and V. ŽELEZNÝ
Dielectric Response of Some Semiconducting TCNQ Salts in the Submillimeter Range
(72.20 and 77.20; 71.30; S12.1)
(b) Vol. **138**, No. 1, p. 347—355 (1986).

71.35. Excitons and related phenomena (including electron–hole drops)

Review Article

F. Henneberger
Optical Bistability at the Absorption Edge of Semiconductors
(78.20; 71.35; S7.16; S8.16)
(b) Vol. **137**, No. 2, p. 371—432 (1986).
Erratum
(b) Vol. **138**, No. 2, p. 744 (1986).

Original Papers and Short Notes

Kh. Pushkarov and M. T. Primatarova
Solitary Excitations in Molecular Chains in the Presence of Vibronic States
(71.38; 71.35; S12.1)
(b) Vol. **133**, No. 1, p. 253—265 (1986).

M. Kato and Y. H. Ohtsuki
Stopping Power of Atoms near Metal Surfaces at Grazing Incidence Condition
(61.80; 71.35; S2)
(b) Vol. **133**, No. 1, p. 267—275 (1986).

S. G. Guseinov, G. D. Guseinov, N. Z. Gasanov, and S. B. Kyazimov
*Special Features of Exciton Absorption Spectra of $A^{III}B^{III}X_2^{VI}$-Type Layer-Semiconductor Crystals
(71.35 and 78.40; S8.16)
(b) Vol. **133**, No. 1, p. K25—K30 (1986).

K. T. Stoychev
Non-Linear Optical Susceptibilities of Molecular Crystals in the Region of Vibronic Excitations. III. Non-Totally Symmetric Vibrations
(78.20; 63.20; 71.35)
(b) Vol. **133**, No. 2, p. 557—562 (1986).

V. A. Kiselev, B. V. Novikov, A. E. Cherednichenko, and E. A. Ubushiev
*Effect of Thin Space-Charge Layers on Exciton Reflectance
(71.35 and 78.40; S8.11)
(b) Vol. **133**, No. 2, p. 573—581 (1986).

H. Conzelmann, A. Hangleiter, and J. Weber
Thallium-Related Isoelectronic Bound Excitons in Silicon. A Bistable Defect at Low Temperatures
(71.70; 71.35; 78.55; S5.11)
(b) Vol. **133**, No. 2, p. 655—668 (1986).

Qu Zhe and Pan Jin-Sheng
*The Behaviour of Excitons near the Interface of a Polar Crystal
(71.35; 73.40)
(b) Vol. **133**, No. 2, p. K119—K124 (1986).

W. DUŁAK and H. MĘCZYŃSKA
Exciton Spectra of Thin CdS Films Prepared by Spray Pyrolysis
(78.65; 71.35; S8.11)
(a) Vol. **93**, No. 2, p. K171—K172 (1986).

A. J. PAL and T. N. MISRA
Delayed Luminescence of Organic Mixed Crystals. Guest Concentration Effect in the p-Terphenyl Host–Pyrene Guest System
(78.55; 71.35; S12)
(b) Vol. **134**, No. 1, p. 243—249 (1986).

A. N. GEORGOBIANI, V. B. GUTAN, I. M. TIGINYANU, V. V. URSAKI, and T. F. FILINA
Excitonic Luminescence in $CdGa_2S_4$ Single Crystals
(78.55 and 78.60; 71.35; S8.16)
(b) Vol. **134**, No. 1, p. K47—K49 (1986).

K. A. DMITRENKO, S. G. SHEVEL, L. V. TARANENKO, and A. V. MARINTCHENKO
*The Temperature Dependence (4.2 to 293 K) of the Resonance Energies of Excitonic Transitions in II–VI Compounds
(71.35 and 78.40; S8.11; S8.12; S8.13)
(b) Vol. **134**, No. 2, p. 605—613 (1986).

A. A. MANENKOV, G. N. MIKHAILOVA, S. YU. SOKOLOV, and S. G. TIKHODEEV
*Phonon Wind Generation in Ge by CO_2 Laser Induced Heating of Electron–Hole Drops
(71.35; 71.38; 72.30; 78.70; S5.12)
(b) Vol. **134**, No. 2, p. 631—639 (1986).

E. N. MYASNIKOV, A. E. DRUZHININ, and A. P. POPOV
*On Participation of Medium Oscillations in Wannier Exciton Motion
(71.35)
(b) Vol. **134**, No. 2, p. 651—658 (1986).

S. O. CHOLAKH, V. A. PUSTOVAROV, and B. V. SHULGIN
*The Effects of Electron Excitation Energy Transfer to the Impurity Centre in LiH Crystals
(71.35 and 71.55; 78.55; S2)
(b) Vol. **134**, No. 2, p. 741—744 (1986).

V. E. MASHCHENKO, V. F. KHARSIK, and S. V. BREZHNEVA
*Secondary Emission of Excitons in CuCl Polycrystals
(71.35; 78.55; S9)
(b) Vol. **135**, No. 1, p. 201—206 (1986).

N. N. ABLYAZOV, A. G. ARESHKIN, V. G. MELEKHIN, L. G. SUSLINA, and D. L. FEDOROV
*Fluctuation-Induced Broadening of Exciton Reflection Spectra in $A^{II}B^{VI}$ Solid Solutions
(71.35 and 78.40; 71.25; S8.15)
(b) Vol. **135**, No. 1, p. 217—225 (1986).

M. C. DESJARDINS-DERUELLE, J. P. LASCARAY, D. COQUILLAT, and R. TRIBOULET
*Magneto-Optical Measurements in $Zn_{1-x}Mn_xTe$ with High Manganese Concentration
(71.35 and 71.70; 75.60; 78.20; S8.15)
(b) Vol. **135**, No. 1, p. 227—232 (1986).

R. MADELON, J. HUREL, and E. PAUMIER
*Exciton Absorption Spectrum in a $ZnSiP_2$ Crystal
(71.35; 78.40; S7.16)
(b) Vol. **136**, No. 2, p. 679—684 (1986).

V. A. HAROUTUNIAN and E. M. KAZARIAN
*Feasibility of the Bose Condensation of Excitons in a Quantizing Semiconductor Sphere
(71.35)
(b) Vol. **136**, No. 2, p. K105—K108 (1986).

R. STEGMANN, G. NEUBERT, E. DOBRUININA, and M. V. CHUKHICHEV
*The Influence of the Energy Spectrum of the Nitrogen Bound Exciton on the Efficiency in $GaAs_{1-x}P_x$:N
(71.35; 78.60; S7.15)
(a) Vol. **96**, No. 2, p. 643—650 (1986).

P. REINEKER, R. G. WINKLER, and H. PORT
*Temperature Dependence of Resonance Splittings in Molecular Pairs. Comparison of Optical Experiments and Theoretical Results for the Phenazine Triplet State
(63.20 and 71.35; S12)
(b) Vol. **137**, No. 1, p. 137—147 (1986).

B. ESSER and F. MARLOW
*On the Influence of Noise on the Density of States and Absorption in Disordered Molecular Systems
(71.20 and 71.35; 78.20)
(b) Vol. **137**, No. 1, p. 169—181 (1986).

D. V. KORBUTYAK, YU. V. KRYUCHENKO, L. A. DEMCHINA, V. G. LITOVCHENKO, and V. I. SUGAKOV
Influence of Macroscopic Inclusions on Excitonic Luminescence of InSe Crystals
(78.55; 71.35; 71.36; S8)
(b) Vol. **137**, No. 1, p. 217—228 (1986).

K. THONKE, U. SCHALL, and R. SAUER
Photoluminescence Study of the Gallium Defect Spectrum at $\approx$ 0.928 eV (Ga3) in Irradiated Silicon
(71.55 and 78.55; 71.35; S5.11)
(b) Vol. **137**, No. 1, p. 291—303 (1986).

U. SCHALL, K. THONKE, und R. SAUER
Photoluminescence Study of the Gallium Defect Spectrum at $\approx$ 1.049 eV (Ga2) in Irradiated Silicon
(71.55 and 78.55; 71.35; S5.11)
(b) Vol. **137**, No. 1, p. 305—317 (1986).

D. LJ. MIRJANIĆ and Z. RAJILIĆ
*Influence of the Quasiparticle Non-Conservation on the Interaction of Exciton Currents
(71.35)
(b) Vol. **137**, No. 1, p. K69—K72 (1986).

F. R. VUKAJLOVIĆ and S. I. VINITSKII
*Adiabatic Method for Excitonic Molecules
(71.35)
(b) Vol. **138**, No. 2, p. 553—557 (1986).

J. Y. BIGOT, M. FRINDI, F. FIDORRA, R. LEVY, and B. HÖNERLAGE
Nonlinear Propagation of Nanosecond Laser Pulses through Polycrystalline CuCl Layers
(78.20; 71.35; S9)
(b) Vol. **138**, No. 2, p. 621—631 (1986).

J. GUTOWSKI, J. SCHOTT, and I. BROSER
*Resonant Brillouin Scattering at Bound-Exciton States in CdS
(71.35; 78.20; S8.11)
(b) Vol. **138**, No. 2, p. 673—683 (1986).

71.36. Polaritons (including photon–phonon and photon–magnon interactions)

Original Papers and Short Notes

L. Merten and R. Kotzott
*Elliptically Polarized Magnon-Polaritons with Longitudinal and Transverse Components in Uniaxial Ferrites
(71.36; 75.50; 78.30)
(b) Vol. **133**, No. 2, p. 547—556 (1986).

A. N. Botvich, V. F. Shabanov, A. N. Vtyurin, and T. A. Pozdnyakova
*Photon–Phonon Interaction Influence on the External Vibration Spectra of Molecular Crystals
(63.20 and 71.36; 78.30; S12)
(b) Vol. **134**, No. 2, p. 515—522 (1986).

L. Wendler and R. Pechstedt
*Effects of a Thin Curved Transition Layer on Surface Polaritons. Local States
(71.36 and 73.20)
(b) Vol. **134**, No. 2, p. 707—715 (1986).

Yu. M. Azhnyuk, V. V. Artamonov, M. Ya. Valakh, and A. P. Litvinchuk
Raman Scattering from Polaritons and Plasmaritons in 6H-SiC
(78.30; 71.36; 71.45; S6)
(b) Vol. **135**, No. 1, p. 75—84 (1986).

V. Schulz and L. Merten
Euler's Formulae and Sturm's Sequence and the Stimulating Regions for Stimulated Raman Effect in Simple Piezoelectric Cubic Crystals
(71.10 and 78.30; 63.20; 71:36; 78.45)
(b) Vol. **135**, No. 1, p. 105—118 (1986).

H. Stolz
A Kinetic Equation for Polaritons
(71.36)
(b) Vol. **135**, No. 2, p. 639—650 (1986).

R. Pechstedt and L. Leine
*Effects of a Thin Transition Layer on the Dispersion Relation of Surface Polaritons on a Randomly Rough Surface
(71.36 and 73.20; 68.30; 78.30; S4)
(b) Vol. **135**, No. 2, p. 749—758 (1986).

L. Wendler
*S-Polarized Nonlinear Surface Polaritons. Effects of a Transition Layer
(71.36 and 73.20; 68.30; 78.20)
(b) Vol. **135**, No. 2, p. 759—774 (1986).

V. M. Fomin and E. P. Pokatilov
*Optical Properties of Multi-Layer Structures. I. Polaritons
(71.36; 73.40; 78.20; S7.12; S7.15)
(b) Vol. **136**, No. 1, p. 187—199 (1986).

V. M. Fomin and E. P. Pokatilov
*Optical Properties of Multi-Layer Structures. II. Reflection and Transmission
(71.36; 73.40; 78.20; S7.12; S7.15)
(b) Vol. **136**, No. 2, p. 593—602 (1986).

D. V. Korbutyak, Yu. V. Kryuchenko, L. A. Demchina, V. G. Litovchenko, and V. I. Sugakov
Influence of Macroscopic Inclusions on Excitonic Luminescence of InSe Crystals
(78.55; 71.35; 71.36; S8)
(b) Vol. **137**, No. 1, p. 217—228 (1986).

L. Wendler and R. Haupt
*Damping of Polaritons in Bilayer Systems. Long-Range Surface Phonon-Polaritons
(68.30 and 71.36; S7.11; S7.12; S9; S9.12)
(b) Vol. **137**, No. 1, p. 269—290 (1986).

J. Barnaś
*Non-Reciprocal Character of Wave Propagation in Magnetic Systems with Spontaneous Electric Polarization
(71.36 and 75.30; 78.20; S9.16)
(b) Vol. **137**, No. 1, p. K43—K47 (1986).

C. Gourdon and P. Lavallard
*Comments on "Transmission and Damping of Excitonic Polaritons in CdS" by I. Broser, K.-H. Pantke, and M. Rosenzweig
(71.35 and 71.36; S8.11)
(b) Vol. **138**, No. 1, p. K29—K30 (1986).

I. Broser, K.-H. Pantke, and M. Rosenzweig
*Response to "Comments on Transmission and Damping of Excitonic Polaritons in CdS" by C. Gourdon and P. Lavallard
(71.35 and 71.36; S8.11)
(b) Vol. **138**, No. 1, p. K31—K32 (1986).

71.38. Polarons and electron–phonon interactions (see also 63.20. phonon–electron interactions in lattices)

Original Papers and Short Notes

F. M. Gashimzade and E. V. Tagirov
*Selection Rules for Matrix Elements of Electron–Phonon Scattering Processes in Crystals with TlSe Structure
(71.38; S8)
(b) Vol. **133**, No. 1, p. 179—182 (1986).

Wu Xiaoguang, F. M. Peeters, and J. T. Devreese
*Screening of the Electron–Phonon Interaction in GaAs Heterostructures
(71.38; 73.40; S7.12; S7.15)
(b) Vol. **133**, No. 1, p. 229—240 (1986).

Kh. I. Pushkarov and M. T. Primatarova
*Solitary Excitations in Molecular Chains in the Presence of Vibronic States
(71.38; 71.35; S12.1)
(b) Vol. **133**, No. 1, p. 253—265 (1986).

V. V. Krasnogolovets and P. M. Tomchuk
*Erratum to: The Coulomb Correlations and the Electric Field Local Heterogeneities on the Proton Conductivity of Hydrogen Bonded Chains
(71.38 and 77.30; S12.1)
(b) Vol. **133**, No. 1, p. 431 (1986).

H.-J. Fitting and A. von Czarnowski
Electron Breakdown in Polar Insulating and Semiconducting Layers
(73.60; 71.38; 79.70; S8.11; S10.1)
(a) Vol. **93**, No. 1, p. 385—396 (1986).

Yu. L. Shitikov, N. A. Tulina, and M. G. Zemlyanov
*Special Features of Lattice Dynamics and Electron–Phonon Interaction in Re
(63.20 and 71.38; S1)
(b) Vol. **133**, No. 2, p. 469—474 (1986).

C. Trallero Giner and M. Antón
Quantum Theory of Free-Carrier Absorption in Quasi-Two-Dimensional Semiconducting Structures
(78.20; 71.38; 78.65; S5.12; S7.12)
(b) Vol. **133**, No. 2, p. 563—572 (1986).

H. J. Fischbeck
*The Incommensurate CDW Phase of a Quasi-One-Dimensional Peierls-Fröhlich System with a Nearly Quarter-Filled Band. II. Numerical Results
(64.70 and 71.38)
(b) Vol. **134**, No. 1, p. 223—238 (1986).

A. A. MANENKOV, G. N. MIKHAILOVA, S. YU. SOKOLOV, and S. G. TRIKHODEEV
Phonon Wind Generation in Ge by CO_2 Laser Induced Heating of Electron–Hole Drops
(71.35; 71.38; 72.30; 78.70; S5.12)
(b) Vol. **134**, No. 2, p. 631—639 (1986).

M. I. BELINSKII, V. YA. GAMURAR, and B. S. TSUKERBLAT
Heisenberg-Type and Double Exchange in Mixed-Valence Clusters. II. Electron–Vibrational Interactions
(71.70 and 75.30; 71.38)
(b) Vol. **135**, No. 2, p. 555—569 (1986).

MAO ER-WANG, ZHANG HUNG-REI, WANG WEI-YUAN, YE LIANG, and FANG GUO-PING
Inelastic Scattering of Electrons at the InP–SiO_2 Interface
(73.40; 71.38; S7.10; S10.1)
(b) Vol. **135**, No. 2, p. 825—830 (1986).

L. S. GRIGORYAN, I. L. KHARUTYUNYAN, A. A. MATNISHYAN, R. S. NIKOGOSYAN, and E. G. SHAROYAN
*Charged Elementary Excitations in Anthracene Polymers. A Systematical Electrical and EPR Study as a Function of Doping
(71.25 and 71.38; 72.80; 76.30; S12)
(b) Vol. **136**, No. 1, p. 145—151 (1986).

V. L. BONCH-BRUEVICH
*On a Certain Model Leading to a Fractal Dimension in the Schrödinger Problem
(71.10 and 71.38)
(b) Vol. **136**, No. 1, p. 207—210 (1986).

V. ČÁPEK
GME and the Electron or Exciton Initial Condition Term Due to Lattice Relaxation. Dimer
(71.10; 71.35; 71.38)
(b) Vol. **136**, No. 1, p. 251—259 (1986).

G. P. TRIBERIS
The Effect of Temperature and Coupling Strength on the Thermoelectric Power in the Small-Polaron Hopping Regime
(72.10; 71.38)
(b) Vol. **136**, No. 1, p. 393—398 (1986).

E. PIOTROWSKI and J. ŁUCZKA
Simple Derivation of the Direct Spin–Phonon Interaction
(75.10; 71.38)
(b) Vol. **136**, No. 1, p. K27—K31 (1986).

P. ENDERS
*On the Chatterjee-Mitra Transformation for the Electron–Phonon Interaction Problem
(71.38)
(b) Vol. **136**, No. 1, p. K37—K39 (1986).

L. E. Reznik
Electron–Phonon Interaction and Structure of Molecular Luminescence Center in Crystals
(63.20 and 71.70; 71.38; 78.50; 78.55; S11.1)
(b) Vol. **136**, No. 2, p. 417—424 (1986).

Yu. E. Perlin, Sh. N. Gifeisman, and V. P. Koropchanu
*Free Carriers in Ionic Semiconductors with a Degenerate Valence Band Edge
(71.38)
(b) Vol. **136**, No. 2, p. K119—K122 (1986).

A. Ausmees, M. Elango, A. Kikas, and J. Pruulmann
*Monte Carlo Simulation of Electron–Phonon Scattering in the XUV-Induced Electron Emission of NaCl
(71.38 and 79.60; S9.11)
(b) Vol. **137**, No. 2, p. 495—500 (1986).

P. A. Apanasevich, V. I. Kruglov, and I. E. Tralle
*Interaction of Crystal Impurity Centres with a Coherent Electromagnetic Field
(71.38 and 71.55; 78.55)
(b) Vol. **137**, No. 2, p. 641—653 (1986).

L. Wendler and R. Pechstedt
*Dynamical Screening, Collective Excitations, and Electron–Phonon Interaction in Heterostructures and Semiconductor Quantum Wells. General Theory
(71.38 and 71.45; 63.20; 73.40; 77.30; 78.20)
(b) Vol. **138**, No. 1, p. 197—217 (1986).

F. Comas, C. Trallero Giner, and H. León
Quantum Theory of Free-Carrier Absorption in Quasi-Two-Dimensional Semiconducting Structures. Degenerate Carrier Case
(78.20; 71.38)
(b) Vol. **138**, No. 1, p. 219—227 (1986).

Yu. D. Arbuzov and M. Yu. Kolenkin
On the Interband Absorption in Disordered Semiconductors
(78.20; 71.38)
(b) Vol. **138**, No. 2, p. 547—551 (1986).

Yan-Min Li and K. A. Chao
*Dissociation and Thermodynamic Properties of Bipolarons
(71.38; 65.50)
(b) Vol. **138**, No. 2, p. 577—587 (1986).

J. Monecke, W. Cordts, G. Irmer, B. H. Bairamov, and V. V. Toporov
LO-Phonon Modes Bound to Neutral Impurities in Polar Semiconductors. I. General Theory
(63.20 and 71.55; 71.38)
(b) Vol. **138**, No. 2, p. 685—691 (1986).

V. V. KRASNOGOLOVETS and P. M. TOMCHUK
The Influence of External Effects on the Proton Conductivity of an Hydrogen-Bonded Chain
(66.30; 62.65; 71.38; S12.1)
(b) Vol. **138**, No. 2, p. 727—733 (1986).

MAO-ER-WANG, ZHANG HUNG-REI, WANG WEI-YUAN, YE LIANG, and FANG GUO-PING
Localization Effect of Electrons in the InP–SiO_2 Interface
(73.40; 71.38; S7.11)
(b) Vol. **138**, No. 2, p. 735—743 (1986).

71.45. Collective effects

Original Papers and Short Notes

T. Nattermann
Dielectric Tails in K_2ZnCl_4: No Evidence for an Intrinsic Chaotic State
(77.20; 71.45; S9.16)
(b) Vol. **133**, No. 1, p. 65—70 (1986).

D. K. Papademetriou, D. Katsanos, and A. G. Doukas
*Plasmon Bands in X-Ray Spectra of Polycrystalline Beryllium
(71.45 and 78.70; S3)
(b) Vol. **133**, No. 1, p. 223—228 (1986).

Tran Thoai Duy Bao
*Static Polarizability and Dipolar Surface Plasmon of Small Metal Particles. II. Including the Lattice Structure
(71.45 and 73.20; S2)
(b) Vol. **133**, No. 1, p. 329—332 (1986).

Tran Thoai Duy Bao
*Linear Density Response at the Metal Surface in the Hydrodynamic Approximation
(71.45 and 73.20)
(b) Vol. **133**, No. 2, p. 633—644 (1986).

G. V. Ionova, Yu. N. Kostrubov, and A. V. Nikolaev
*Charge Ordering in Intermediate-Band Crystals
(71.45)
(b) Vol. **134**, No. 1, p. 239—242 (1986).

A. M. Brodskii
*Nonlinear Optics of Metal Surface
(71.45 and 73.20; 78.20)
(b) Vol. **134**, No. 1, p. 251—256 (1986).

H. Wiesner, W. Cordts, and J. Monecke
*Self-Consistent Calculation of the Equilibrium Lattice Constant and of the Bulk Modulus of Silicon with Local and Non-Local Exchange-Correlation Functionals
(61.50 and 71.45; S5.11)
(b) Vol. **134**, No. 2, p. K93—K95 (1986).

Yu. M. Azhnyuk, V. V. Artamonov, M. Ya. Valakh, and A. P. Litvinchuk
Raman Scattering from Polaritons and Plasmaritons in 6H-SiC
(78.30; 71.36; 71.45; S6)
(b) Vol. **135**, No. 1, p. 75—84 (1986).

K. M. Sultanov, Sh. A. Kuliev, and N. G. Abdullaev
*Coupled Waves in Conducting Spin Glasses
(71.45 and 75.30; 76.50)
(b) Vol. **135**, No. 1, p. 163—171 (1986).

H. K. Sy
*Effects of Disorder on Peierls Distortion
(71.45)
(b) Vol. **135**, No. 1, p. K17—K20 (1986).

A. Dittmar-Wituski, M. Naparty, and J. Skonieczny
*Total Current Spectroscopy of Vapour Deposited Layers of Ag, Sn, and Ag_xSn_{1-x} Alloys
(68.55 and 71.45; S1.3)
(a) Vol. **95**, No. 1, p. 255—260 (1986).

Tran Thoai Duy Bao
*Induced Field at a Dipole near the Metal Surface in the Hydrodynamic Approximation
(71.45 and 73.20; 78.20)
(b) Vol. **135**, No. 2, p. 775—781 (1986).

H. K. Sy and L. M. Song
*Exchange Potential and Plasma Modes in Coupled Two-Dimensional Electron Systems
(71.45)
(b) Vol. **136**, No. 1, p. 201—206 (1986).

Lj. Ristovski, G. S. Davidović-Ristovski, and V. Ristić
*Bisolitons in a Molecular Polymer Chain
(71.45; S12.1)
(b) Vol. **136**, No. 2, p. 615—622 (1986).

P. Kleinert and M. Giehler
Theory of Free-Carrier Infrared Absorption in GaAs
(72.10; 63.20; 71.45; 78.30; S7.12)
(b) Vol. **136**, No. 2, p. 763—777 (1986).

T. Rother and W. D. Kraeft
*On the Thermodynamics of Strongly Degenerate Two-Component Plasmas
(65.50 and 71.45; S5.12; S7.12)
(b) Vol. **137**, No. 1, p. 103—108 (1986).

F. Bechstedt and R. Enderlein
*Interface Plasmon Modes as Collective Electronic Excitations of Semiconductor Superlattices
(71.45 and 73.40)
(b) Vol. **137**, No. 1, p. 109—115 (1986).

P. Kleinert
*Ionized-Impurity-Mediated Free-Carrier Absorption in n-GaAs Slabs
(63.20 and 71.45; 78.30; S7.12)
(b) Vol. **137**, No. 1, p. 125—135 (1986).

M. Rösler and W. Stolzmann
*Thermodynamic Functions and Self-Energy of Interacting Coulombic Systems. Numerical Results
(65.50 and 71.45)
(b) Vol. **137**, No. 1, p. 149—167 (1986).

71.50. Localized single-particle electronic states (excluding impurities)

Original Papers and Short Notes

B. Viswanathan, S. Madhavan, and C. S. Swamy
Charge Transfer Satellites in X-Ray Photoelectron Spectra of La_2CuO_4
(79.60; 71.50; S10.15)
(b) Vol. **133**, No. 2, p. 629—632 (1986).

A. Brezini and M. Sebbani
*Power Law Localization in One-Dimensional Random Potentials
(71.50 and 72.10)
(b) Vol. **133**, No. 2, p. K129—K132 (1986).

U. Pietsch
*Bond Charges and Electronic Charge Transfer in Ternary Semiconductors
(61.50 and 71.50; S7.1; S7.15; S7.16)
(b) Vol. **134**, No. 1, p. 21—27 (1986).

M. Kasner and W. Weller
*Localization in a One-Dimensional Random Potential with Spatial Correlation
(71.50)
(b) Vol. **134**, No. 2, p. 731—739 (1986).

C.-M. Liegener, G. Biczó, and A. Godzik
*On the Concept of Intermediate One-Particle States in Finite Chains. A Self-Consistent Field Study
(71.50; 73.20; S12.1)
(b) Vol. **136**, No. 1, p. 281—289 (1986).

K. B. Garg, D. Norman, and R. B. Beeken
An X-Ray Absorption Study of Intermediate Valence in $Sm_{1-x}Sc_xS$ Alloys
(78.70; 71.50; S1.63)
(a) Vol. **96**, No. 1, p. 301—309 (1986).

A. Brezini, M. Sebbani, and L. Dahmani
*Localization in Random Systems. I. Effect of Off-Diagonal Disorder
(71.50; 71.25)
(b) Vol. **137**, No. 2, p. 667—673 (1986).

A. Brezini, M. Sebbani, and F. Behilil
*Power-Law Localization in One-Dimensional Random Potentials
(71.50; 72.10)
(b) Vol. **138**, No. 2, p. K137—K142 (1986).

71.55. Impurity and defect levels

Review Article

Original Papers and Short Notes

D. Wolf and S. Kemmler-Sack
The Luminescence of Nd^{3+} in the Polymorphic Perovskites $Ba_6B_2W_3O_{18}$ (B = Gd, Lu, Y). A Case with a Low-Lying Charge Transfer Transition
(78.55; 71.55; 78.40; S11)
(a) Vol. **93**, No. 1, p. 271—275 (1986).

O. Goede, W. Heimbrodt, and F. Sittel
Photoacoustic Spectroscopy of Isovalent Impurities in II–VI Semiconductors
(78.20; 71.55; 78.50; S8.11)
(a) Vol. **93**, No. 1, p. 277—282 (1986).

S. G. Gagarin, A. V. Mudryi, A. L. Pushkarchuk, A. G. Ulyashin, and Yu. A. Teterin
*Theoretical Analysis of the Chemical Activity of Noble Gas Atoms in Silicon
(61.70 and 71.55; S5.11)
(b) Vol. **133**, No. 2, p. 499—503 (1986).

V. A. Radtsig and A. A. Bobyshev
*Twofold Coordinated Si and Ge Atoms and "Hydrogen" Paramagnetic Centers in Amorphous SiO_2 and GeO_2
(71.55; 61.70; 76.30; S10.1)
(b) Vol. **133**, No. 2, p. 621—627 (1986).

V. M. Nabutovskii and D. A. Romanov
*Electron States in Periodic Dislocation Networks in Semiconductors
(71.55; 72.20; 72.30)
(b) Vol. **133**, No. 2, p. 645—653 (1986).

D. S. Domanevskii, V. V. Krasovskii, M. V. Prokopenya, V. A. Vilkotskii, and S. V. Zhokhovets
*The Origin of the Minority Impurity States in Heavily Doped Semiconductors
(71.55; 78.55; S7.11; S7.12)
(b) Vol. **133**, No. 2, p. 693—700 (1986).

S. Wilke, J. Mašek, and B. Velický
*A CBLM Approach to the Electronic Structure of Transition Metal Impurities in Silicon
(71.55; S5.11)
(b) Vol. **133**, No. 2, p. K111—K114 (1986).

K. D. Glinchuk, N. M. Litovchenko, and V. Yu. Ptitsin
A Variety of Oxygen-Induced Recombination Centres in 450 to 600 °C Heat-Treated Silicon
(61.70; 71.55; S5.11)
(a) Vol. **93**, No. 2, p. 565—571 (1986).

H. Indusekhar, V. Kumar, and D. Sengupta
*Investigation of Deep Defects Due to α-Particle Irradiation in n-Silicon
(71.55; 61.80; S5.11)
(a) Vol. **93**, No. 2, p. 645—653 (1986).

K. MASUDA-JINDO
On the Grain Boundary Segregation of sp-Valence Impurities in B.C.C. Transition Metal
(64.75; 71.55; S1.1)
(b) Vol. **134**, No. 2, p. 545—550 (1986).

J. MYCIELSKI and A. M. WITOWSKI
*Effect of Chemical and Spin Disorder on Shallow Impurity States in Semiconductor Alloys
(71.55; 71.70; 75.30; S8.15)
(b) Vol. **134**, No. 2, p. 675—689 (1986).

S. O. CHOLAKH, V. A. PUSTOVAROV, and B. V. SHULGIN
*The Effects of Electron Excitation Energy Transfer to the Impurity Centre in LiH Crystals
(71.35 and 71.55; 78.55; S2)
(b) Vol. **134**, No. 2, p. 741—744 (1986).

W. KRAAK, U. SCHALLER, and R. HERRMANN
Metal–Insulator Transition and Metastable State in High-Purity n-InSb under High Hydrostatic Pressure
(71.30; 71.55; 72.20; S7.13)
(b) Vol. **134**, No. 2, p. 779—788 (1986).

E. I. DMITRIEVA, N. N. KRISTOFFEL, I. A. PARFIANOVICH, L. G. POPOV, and V. N. SALOMATOV
*The $F_H(F^-)$ Center Energy Structure in KCl
(71.55; S9.11)
(b) Vol. **134**, No. 2, p. K125—K128 (1986).

C. M. VAN VLIET and P. VASILOPOULOS
*Electrical and Optical Bandgap Narrowing Due to Heavy Doping in Silicon
(71.25 and 71.55; S5.11)
(a) Vol. **94**, No. 2, p. 635—644 (1986).

A. B. AHMED and R. K. GARTIA
Trapping Levels in $KAlSi_3O_8$
(78.60; 71.55; S11)
(a) Vol. **94**, No. 2, p. 645—651 (1986).

G. D. GUSEINOV, V. A. ALIEV, and E. F. BAGIRZADE
*Recombination Levels in Non-Stoichiometric $TlGaSe_2$
(71.55; 72.40; S8.16)
(a) Vol. **94**, No. 2, p. 653—657 (1986).

L. DÓZSA
*On the Reliability of Minority Carrier Injection DLTS Spectra
(71.55; 73.40)
(a) Vol. **94**, No. 2, p. 735—743 (1986).

P. A. PUNDUR, J. G. SHAVALGIN, and V. A. GRITSENKO
*On the Nature of Deep Centres Responsible for the Memory Effect and Luminescence of a-SiN_x with $x \leqq 4/3$
(71.55; 78.55; 78.60; S7)
(a) Vol. **94**, No. 2, p. K107—K112 (1986).

W. Hoerstel and T. Otto
Density of Trap Tunnelling States in Diffused $PbS_{1-x}Se_x$ p–n Junctions
(73.40; 71.55; S8.15)
(a) Vol. **95**, No. 1, p. 361—366 (1986).

K. Schmalz, V. V. Bolotov, H. Richter, and K. Tittelbach-Helmrich
DLTS Investigation on the Interaction of Radiation Defects with Thermal Defects in Czochralski p-Si
(61.70 and 61.80; 71.55; S5.11)
(a) Vol. **95**, No. 1, p. K15—K20 (1986).

N. A. Yakusheva, V. Ya. Prinz, and Yu. B. Bolkhovityanov
*Discovery of Electron Traps in LPE GaAs Grown from a Bismuth Melt
(71.55; S7.12)
(a) Vol. **95**, No. 1, p. K43—K46 (1986).

D. Crandles and E. Fortin
*The Trap Distribution and the Luminescence Band in $ZnIn_2S_4$
(71.55; 78.55; S8.16)
(a) Vol. **95**, No. 1, p. K47—K51 (1986).

L. Tulach, H. Frank, B. Pína, and M. Janů
*Gold as Residual Impurity in Silicon Devices
(71.55; S5.11)
(a) Vol. **95**, No. 1, p. K87—K89 (1986).

W. Heimbrodt and O. Goede
*Energy Transfer Processes between Te_n Centres in ZnS:Te and CdS:Te
(71.35 and 71.55; 78.55; S8.11; S8.15)
(b) Vol. **135**, No. 2, p. 795—804 (1986).

K. Chetouane, J. Hugel, and J. C. Parlebas
Optical Absorption in Binary Semiconductors Containing Substitutional Impurities
(78.50; 71.55)
(b) Vol. **135**, No. 2, p. 805—812 (1986).

A. Lodder, I. Mertig, R. Zeller, E. Mrosan, and P. H. Dederichs
*Influence of Charge Transfer on ab initio Calculated Impurity Resistivities in Noble Metals
(71.55 and 72.15; S1.2; S1.3)
(b) Vol. **135**, No. 2, p. 831—836 (1986).

M. V. Chukhichev, V. I. Petrov, and R. Stegmann
Observation of the $N_{\Gamma-x}$-Line in $GaAs_{0.42}P_{0.58}$:N
(78.60; 71.35; 71.55; S7.15)
(b) Vol. **135**, No. 2, p. K139—K142 (1986).

A. V. Dvurechenskii, V. A. Dravin, I. A. Ryazantsev, A. Kh. Antonenko, and I. G. Landochkin
*Transport Phenomena in Amorphous Silicon Doped by Ion Implantation of 3d Metals
(71.55; 72.80; S5.11)
(a) Vol. **95**, No. 2, p. 635—640 (1986).

B. C. GRABMAIER and R. OBERSCHMID
*Properties of Pure and Doped $Bi_{12}GeO_{20}$ and $Bi_{12}SiO_{20}$ Crystals
(61.70 and 71.55; 72.40; 78.50; 78.55; S10)
(a) Vol. **96**, No. 1, p. 199—210 (1986).

L. A. KAZAKEVICH, V. I. KUZNETSOV, and P. F. LUGAKOV
The Effect of Dislocations on the Activation Energy of Radiation Defect Thermal Ionization
(61.70 and 61.80; 71.55; 72.80; S5.11)
(a) Vol. **96**, No. 1, p. 243—247 (1986).

P. N. METLINSKII, S. I. RADAUTSAN, I. M. TIGINYANU, V. G. TYRZIU, and V. V. URSAKI
*Deep Levels in $HgGa_2Se_4$ Single Crystals
(71.55; 72.40; 78.55; S8.16)
(a) Vol. **96**, No. 1, p. 295—299 (1986).

T. WOSINSKI and O. BREITENSTEIN
*Spatial Distribution of Dominant Electron and Hole Traps in Plastically Deformed GaAs
(71.55; 61.70; S7.12)
(a) Vol. **96**, No. 1, p. 311—315 (1986).

V. B. ORLETSKII, V. Z. YURTSENYUK, and K. D. TOVSTYUK
Localized Defect States in $PbSe_xTe_{1-x}$
(73.60; 71.55; S8.15)
(a) Vol. **96**, No. 1, p. K61—K63 (1986).

P. G. BARANOV, V. A. VETROV, N. G. ROMANOV, and V. TOPA
*ODMR in Triplet Excited States of ns^2 Ions in Alkali Halide Crystals
(71.55 and 76.70; 78.55; S9.11)
(b) Vol. **136**, No. 2, p. 699—707 (1986).

P. K. KHOWASH, V. A. SINGH, and D. C. KHAN
*A Theoretical Study of the Semiinsulating Behaviour of Fe in InSb
(71.55; 76.80; S7.11)
(b) Vol. **136**, No. 2, p. 715—719 (1986).

V. BETHKENHAGEN and K. HÜBNER
*Cluster-Bethe-Lattice Approach to the Electronic Structure of Intrinsic Defects in v-SiO_2
(71.55; 71.20; 78.50; 78.55; S10.1)
(b) Vol. **136**, No. 1, p. 721—734 (1986).

E. RADZHABOV
*Optical Transitions of Chalcogen-Vacancy Centers in Ionic Crystals
(71.55; 78.50; 78.55; S9.11)
(b) Vol. **136**, No. 2, p. K139—K143 (1986).

I. M. FILIPPOV, L. A. KAZAKEVICH, P. F. LUGAKOV, and V. V. SHUSHA
*The Charge Carrier Recombination Processes in Dislocated Silicon
(61.70 and 71.55; 72.80; S5.11)
(a) Vol. **96**, No. 2, p. 527—531 (1986).

N. S. Bogdanyuk and S. S. Ostapenko
Structure Peculiarities and Annealing Mechanism of the Red Flash-Like Luminescent Centre in CdS Single Crystals Irradiated with 1.2 MeV Electron Beam
(78.60; 61.80; 71.55; S8.11)
(a) Vol. **96**, No. 2, p. 621—628 (1986).

C. A. Londos
*Charge-Dependent Defect Traces in the DLTS and MCTS Spectra of Silicon
(71.55; 61.80; S5.11)
(a) Vol. **96**, No. 2, p. 637—642 (1986).

M. Kittler, W. Seifert, K. Schmalz, and K. Tittelbach-Helmrich
*Comparison of EBIC and DLTS Measurements on Boron-Doped CZ Silicon Contamined with Iron
(61.70 and 71.55; S5.11)
(a) Vol. **96**, No. 2, p. K133—K137 (1986).

V. P. Markevich and L. I. Murin
*The Nature of Intrinsic Defects in n-Type Silicon γ-Irradiated at 77 K
(61.70 and 71.55; 61.80; S5.11)
(a) Vol. **96**, No. 2, p. K151—K155 (1986).

M. Tomak and V. E. Godwin
*Electron Energy Levels in $GaAs/Ga_{1-x}Al_xAs$ Heterojunctions (I)
(71.55 and 73.40; S7.12; S7.15)
(b) Vol. **137**, No. 1, p. 183—186 (1986).

A. V. Fisenko, S. A. Prosandeev, and V. P. Sachenko
*Electronic Structure of Intrinsic Point Defects in Perovskite-Type Crystals
(71.10 and 71.55; S11.1)
(b) Vol. **137**, No. 1, p. 187—197 (1986).

H. A. Kassim
*The F^+-Centre in Alkali Oxides of Antifluorite Structure. A Model Potential Numerical Calculation
(71.55; 61.70; S10.1)
(b) Vol. **137**, No. 1, p. 199—206 (1986).

L. N. Kantorovich
*Multipole Theory of the Polarization of Solids by Point Defects. III. Transition Energy Calculations, the Approximation of Non-Point Charges
(71.55 and 77.30; S9.11)
(b) Vol. **137**, No. 1, p. 229—240 (1986).

K. Thonke, U. Schall, and R. Sauer
*Photoluminescence Study of the Gallium Defect Spectrum at $\approx$ 0.928 eV (Ga3) in Irradiated Silicon
(71.55 and 78.55; 71.35; S5.11)
(b) Vol. **137**, No. 1, p. 291—303 (1986).

U. Schall, K. Thonke, and R. Sauer
*Photoluminescence Study of the Gallium Defect Spectrum at ≈ 1.049 eV (Ga2) in Irradiated Silicon
(71.55 and 78.55; 71.35; S5.11)
(b) Vol. **137**, No. 1, p. 305—317 (1986).

W. K. Choi, J. J. Delima, and A. E. Owen
A Model for the Variations in the Field-Dependent Behaviour of the Poole-Frenkel Effect
(72.20; 71.55)
(b) Vol. **137**, No. 1, p. 345—351 (1986).

M. Georgiev and A. Manov
*Non-Radiative Charge Transfer Rate at the E_1' Center in α-Quartz
(71.55; S10.1)
(b) Vol. **137**, No. 1, p. K83—K87 (1986).

R. K. Gartia and A. B. Ahmed
Low-Temperature Thermally Stimulated Luminescence in $KAlSi_3O_8$
(78.60; 71.55; S11)
(a) Vol. **97**, No. 1, p. 257—264 (1986).

M. Matyáš, Jr.
*Deep Levels and Processes in the Course of Degradation of GaP:N LED's
(71.55; 73.40; 78.60; S7.11)
(a) Vol. **97**, No. 1, p. 297—305 (1986).

A. S. Popov and D. B. Kushev
*Deep Levels in Fe-Doped InP and GaP Formed by Iron Complexes
(71.55; S7.11)
(a) Vol. **97**, No. 1, p. K63—K67 (1980).

V. E. Godwin and M. Tomak
*Electron Energy Levels in $GaAs/Ga_{1-x}Al_xAs$ Heterojunctions. II. Optical Properties
(71.55 and 73.40; 78.50; S7.12; S7.15)
(b) Vol. **137**, No. 2, p. 619—622 (1986).

P. A. Apanasevich, V. I. Kruglov, and I. E. Tralle
*Interaction of Crystal Impurity Centres with a Coherent Electromagnetic Field
(71.38 and 71.55; 78.55)
(b) Vol. **137**, No. 2, p. 641—653 (1986).

S. V. Voitikov and V. P. Gribkovskii
*On the Theory of Crystal Excitation by Electric Discharges. II. Ionization of Defects
(71.55 and 72.90)
(b) Vol. **137**, No. 2, p. 655—665 (1986).

Uma Subramanian and M. L. Mukherjee
Thermoluminescence Study of CaF_2:Nd Single Crystals from LNT to 400 °C
(78.60; 71.55; S9.12)
(b) Vol. **137**, No. 2, p. 675—681 (1986).

M. Höhne
*A Silicon–Gold Pair in Silicon Investigated by EPR
(71.55; 76.30; S5.11)
(b) Vol. **138**, No. 1, p. 337—346 (1986).

J. F. Boas
Electron Spin Resonance Studies of Light Sensitive Centers with Tetragonal Symmetry in CaO
(61.70 and 76.30; 71.55; S10.1)
(b) Vol. **138**, No. 2, p. 501—507 (1986).

J. Monecke, W. Cordts, G. Irmer, B. H. Bairamov, and V. V. Toporov
*LO-Phonon Modes Bound to Neutral Impurities in Polar Semiconductors. I. General Theory
(63.20 and 71.55; 71.38)
(b) Vol. **138**, No. 2, p. 685—691 (1986).

Fat Duen Ho
*Further Calculations of the F^+-Center Wave Function for Lithium Niobate
(71.55; S11.1)
(b) Vol. **138**, No. 2, p. 693—697 (1986).

D. Wolf and S. Kemmler-Sack
Cathodo- and Photoluminescence in Rare Earth Activated Garnets of Type $Ca_3M_2Ge_{3-x}Si_xO_{12}$ (M = Al, Ga, Sc, Y)
(78.55 and 78.60; 71.55; S11.2)
(a) Vol. **98**, No. 2, p. 567—575 (1986).

A. Popov and R. Atanasova
The Origin of Nonradiative Centers in Degraded LEDs from $GaAs_{0.6}P_{0.4}$
(73.40 and 78.60; 71.55; S7.15)
(a) Vol. **98**, No. 2, p. 629—632 (1986).

G. J. Papaioannou, P. C. Euthymiou, C. Carabatos, B. Lepley, and A. Bath
*On the Electron Irradiation Induced Defects in GaP:Zn
(61.80 and 71.55; S7.11)
(a) Vol. **98**, No. 2, p. K125—K127 (1986).

71.70. Level splitting and interactions
(see 75.10. — in magnetic phenomena, 75.30. Exchange and superexchange interactions, 73.20. Electronic surface states)

Original Papers and Short Notes

V. P. SETH, A. YADAV, and R. S. BANSAL
Solution of the Spin-Hamiltonian with Orthorhombic HF and g-Tensors ($S = \frac{1}{2}$).,II. Application to VO^{2+} Centres in Oxide Glasses
(76.30; 71.70; S11.1)
(b) Vol. **133**, No. 1, p. 297—300 (1986).

A. K. PETROSYAN and A. A. MIRZAKHANYAN
*Zero-Field Splitting and g-Values of d^8 Ions in a Trigonal Crystal Field
(71.70; S10; S10.1)
(b) Vol. **133**, No. 1, p. 315—319 (1986).

S. K. HOFFMANN, J. GOSLAR, and L. S. SZCZEPANIAK
Low-Symmetry Doped Cu(II) Complexes in $(gly)_2CaCl_2 \cdot 4\,H_2O$ Single Crystals: EPR Studies
(76.30; 71.70; S12)
(b) Vol. **133**, No. 1, p. 321—328 (1986).

C. K. JAYASANKAR, E. ANTIC-FIDANCEV, M. LEMAITRE BLAISE, and P. PORCHER
*Optical Properties of EuF_3 and GdF_3 and Paramagnetic Susceptibility of EuF_3
(71.70; 75.20; 78.55; S9)
(b) Vol. **133**, No. 1, p. 345—353 (1986).

M. KRUPSKI, A. A. MIRZAKHANYAN, W. HILCZER, and E. G. SHAROYAN
Effect of Pressure and Temperature on the EPR Spectrum of Cr^{3+} Ions in α-$LiIO_3$ Single Crystals
(76.30; 71.70; S11.1)
(b) Vol. **133**, No. 1, p. 355—361 (1986).

B. POUMELLEC, F. LAGNEL, J. F. MARUCCO, and B. TOUZELIN
X-Ray Absorption K-Edge Spectra of Titanium in $(Ti, Nb)O_2$ Rutile Solid Solutions
(78.70; 71.70; S10.15)
(b) Vol. **133**, No. 1, p. 371—377 (1986).

I. B. ERMOLOVICH and V. V. MILENIN
Nature of Deep Luminescence Centres in CdSe and $CdSe_xTe_{1-x}$
(78.55; 71.70; S8.12; S8.15)
(b) Vol. **133**, No. 2, p. 611—620 (1986).

H. CONZELMANN, A. HANGLEITER, and J. WEBER
*Thallium-Related Isoelectronic Bound Excitons in Silicon. A Bistable Defect at Low Temperatures
(71.70; 71.35; 78.55; S5.11)
(b) Vol. **133**, No. 2, p. 655—668 (1986).

R. Lacroix
*The Ligand Field on d^n Ions in a Strained Corundum Crystal
(71.70; 62.20; S10.1)
(b) Vol. **133**, No. 2, p. 669—682 (1986).

A. P. Dudka, A. A. Kaminskii, and V. I. Simonov
Refinement of $NaGdGeO_4$, $NaYGeO_4$, and $NaLuGeO_4$ Single-Crystal Structures
(61.55; 71.70; 78.45; 78.55; S11.1)
(a) Vol. **93**, No. 2, p. 495—502 (1986).

S. I. Gubarev
*Valence Band Splittings in Wurtzite-Type Semimagnetic Semiconductors
(71.25 and 71.70; 78.20; S8.16)
(b) Vol. **134**, No. 1, p. 211—222 (1986).

L. I. Levin
*Semiphenomenological Theory of the Gd^{3+} S-State Splitting in Low-Symmetry Crystals
(71.70)
(b) Vol. **134**, No. 1, p. 275—280 (1986).

S. M. Ryabchenko and Yu. G. Semenov
*EPR and Spin-Flip Raman Scattering in the n-Type Magnetically Mixed Semiconductors
(71.70; 76.30; 78.30)
(b) Vol. **134**, No. 1, p. 281—288 (1986).

K. K. P. Srivastava and S. N. Choudhary
*Orbit–Lattice Interaction and Mössbauer Quadrupole Splitting of Fe(II) in $(NMe_4)_2FeCl_4$
(71.70; 63.20; S11)
(b) Vol. **134**, No. 1, p. 289—296 (1986).

S. M. Pillai and C. P. G. Vallabhan
A Study on the Electroluminescence Spectrum of CaS:Sm and CaS:Dy
(78.60; 71.70; S8)
(b) Vol. **134**, No. 1, p. 383—390 (1986).

B. Ng and D. J. Newman
*Tensorial Coupling and the Correlation Crystal Field
(71.70)
(b) Vol. **134**, No. 1, p. K51—K53 (1986).

J. A. Tuszyński and J. M. Dixon
*Comments on Tensorial Coupling and Matrix Elements of the Correlation Crystal Field
(71.70)
(b) Vol. **134**, No. 1, p. K55—K57 (1986).

J. Mycielski and A. M. Witowski
Effect of Chemical and Spin Disorder on Shallow Impurity States in Semiconductor Alloys
(71.55; 71.70; 75.30; S8.15)
(b) Vol. **134**, No. 2, p. 675—689 (1986).

A. A. Kaminskii, A. A. Kornienko, and M. I. Chertanov
*Parametrization of Electric-Dipole Intensities in f^N Systems Due to Electron-Correlation Effects
(71.70; 78.50; S9; S11)
(b) Vol. **134**, No. 2, p. 717—729 (1986).

G. Irmer and J. Monecke
*On the Magnitude of the Jahn-Teller Effect on Shallow Donors in Many-Valley Compound Semiconductors
(71.70; S7.11)
(b) Vol. **134**, No. 2, p. K129—K132 (1986).

M. I. Belinskii, V. Ya. Gamurar, and B. S. Tsukerblat
*Heisenberg-Type and Double Exchange in Mixed-Valence Clusters. I. The Microscopic Theory of Transition Metal Dimers
(71.70 and 75.30)
(b) Vol. **135**, No. 1, p. 189—199 (1986).

M. C. Desjardins-Deruelle, J. P. Lascaray, D. Coquillat, and R. Triboulet
*Magneto-Optical Measurements in $Zn_{1-x}Mn_xTe$ with High Manganese Concentration
(71.35 and 71.70; 75.60; 78.20; S8.15)
(b) Vol. **135**, No. 1, p. 227—232 (1986).

H. R. Asatryan, A. S. Kuzanyan, A. G. Petrosyan, A. K. Petrosyan, and E. G. Sharoyan
*Single Crystal Growth and Investigation of Optical and ESR Absorption Spectra of Zirconium-Doped YAG
(71.70; 76.30; 78.50; S11.2)
(b) Vol. **135**, No. 1, p. 343—352 (1986).

L. A. Yuguryan and A. G. Makhanek
*Coherent Four-Photon Spectroscopy of Two-Photon Resonances of Trivalent Rare-Earth Ions
(71.70; 78.20)
(b) Vol. **135**, No. 1, p. 353—367 (1986).

K. Schmitt
Absorption and Uniaxial Stress Experiments in the B, C, D′ Bands of Pb^{2+} Centres in KBr, RbBr, KI, and RbI
(78.50; 71.70; S9.11)
(b) Vol. **135**, No. 1, p. 389—398 (1986).

P. A. Nagaraja, A. S. Jacob, and B. J. Reddy
*Electronic Spectra of Cu^{2+} and VO_4^{3-} in Vanadinite
(71.70; 78.50; S11)
(b) Vol. **135**, No. 1, p. K39—K42 (1986).

E. J. K. B. Banda
*Optical Absorption of $CoPS_3$ in the Energy Range 0.73 to 5.40 eV
(71.70 and 78.50; S11)
(b) Vol. **135**, No. 1, p. K43—K47 (1986).

W. PASTUSIAK and J. KURIATA
*On the Microscopic Interpretation of Tetragonal Scond-Order Spin-Lattice Coefficients of $4f^7$ Ions in the Dynamic Relativistic Crystal Field Approach
(71.70; S9.12)
(b) Vol. **135**, No. 1, p. K69—K73 (1986).

M. I. BELINSKII, V. YA. GAMURAR, and B. S. TSUKERBLAT
*Heisenberg-Type and Double Exchange in Mixed-Valence Clusters. II. Electron-Vibrational Interactions
(71.70 and 75.30; 71.38)
(b) Vol. **135**, No. 2, p. 555—569 (1986).

H. M. A. SCHEJPEN and F. A. P. BLOM
*Anisotropy in the Shubnikov-de Haas Effect in $Hg_{1-x}Mn_xSe$
(71.25 and 71.70; 72.20; 75.30; S8.15)
(b) Vol. **135**, No. 2, p. 605—614 (1986).

M. J. PONNAMBALAM
*Electric Field Gradient in Ag Alloys
(71.70; S1.3)
(b) Vol. **135**, No. 2, p. 715—721 (1986).

M. E. FOGLIO and P. SCHLOTTMANN
*Intermediate Valence Model of the Hyperfine Field and Relaxation Rate of Cerium in Gadolinium Metal
(71.70; 75.10; 76.20; S1.4)
(b) Vol. **136**, No. 1, p. 273—280 (1986).

J. ULNER
*Zur Theorie des durch das äußere Magnetfeld induzierten Jahn-Teller-Effektes in orthorhombischen $J = 1$-Paramagneten
(On the Theory of the Jahn-Teller Effect Induced by an External Magnetic Field in Orthorhombic $J = 1$ Paramagnets)
(71.70; 75.80)
(b) Vol. **136**, No. 1, p. 299—308 (1986).

Y. KAZUMATA, Y. NAKANO, S. YUGO, and T. KIMURA
Exchange Coupling Model in Neutron-Irradiated Graphite
(76.30; 61.80; 71.70; S5)
(b) Vol. **136**, No. 1, p. 325—329 (1986).

L. E. REZNIK
*Electron-Phonon Interaction and Structure of Molecular Luminescence Center in Crystals
(63.20 and 71.70; 71.38; 78.50; 78.55; S11.1)
(b) Vol. **136**, No. 2, p. 417—424 (1986).

M. WÖHLECKE and A. JOVANOVIĆ
*Invariant Harmonic Oscillator Functions. II. An Alternative Method for the Groups T_d, O, and O_h
(63.20 and 71.70)
(b) Vol. **136**, No. 2, p. 467—473 (1986).

A. V. Malakhovskii, G. G. Vasilev, and T. P. Morozova
*Temperature Dependence of $^6A_{1g} \to {}^4A_{1g}, {}^4E_g(^4G)$ and $^6A_{1g} \to {}^4E_g(^4D)$ Absorption Bands in $RbMnF_3$ in Paramagnetic State
(71.70 and 78.20; S9.16)
(b) Vol. **138**, No. 1, p. 285—293 (1986).

M. Souiri, A. Goltzene, and C. Schwab
*Correlations between Electron Paramagnetic Resonance and Bonding Parameters. Evidence of a Lattice Relaxation
(71.45 and 71.70; 76.30; S7; S8; S9; S10)
(b) Vol. **138**, No. 1, p. 321—330 (1986).

C. Weber, E. Sigmund, and M. Wagner
*An Unconventional Jahn-Teller Phenomenon in a Valency-Mixing Surrounding
(71.70; 76.30; S1.4)
(b) Vol. **138**, No. 2, p. 661—671 (1986).

71.90. Other topics in electron states

Short Note

72. Electronic transport in condensed matter (for surfaces, interfaces, and thin films, see 73)

72.10. Theory of electronic transport; scattering mechanism

Original Papers and Short Notes

A. L. VARTANIAN and A. A. KIRAKOSIAN
*Calculation of the Charge Carrier Mobility of a Size Quantized Semiconductor Film by the Energy-Loss Method
(72.10; 73.60)
(b) Vol. **133**, No. 1, p. 389—394 (1986).

O. ZIEP, A. G. MIRONOV, and R. KEIPER
*Nonlinear High-Electric-Field Transport in the Presence of Strong Laser Radiation
(72.10 and 72.20; 71.20)
(b) Vol. **133**, No. 1, p. 409—416 (1986).

R. LENK
*The Density Matrix of an Infinite Muffin-Tin System and Mutually Avoiding Scattering Paths
(72.10)
(b) Vol. **133**, No. 2, p. 745—754 (1986).

H. BURGHARDT, T. FRAUENHEIM, and C. HAMANN
*The Influence of an Electric Field on the Mobility in Semiconductors
(72.10; 72.20; S7.12)
(b) Vol. **133**, No. 2, p. 755—767 (1986).

R. KEIPER and O. ZIEP
*Theory of Quantum Hall Effect
(72.10; 73.40)
(b) Vol. **133**, No. 2, p. 769—773 (1986).

A. BREZINI and M. SEBBANI
*Power Law Localization in One-Dimensional Random Potentials
(71.50 and 72.10)
(b) Vol. **133**, No. 2, p. K129—K132 (1986).

J. L. FARVACQUE
*Improvement of Debye-Hückel Evaluation of Free Carrier Screening Effects in Semiconductors
(72.10; S7.12; S7.13)
(b) Vol. **134**, No. 1, p. 343—349 (1986).

F. VÖLKLEIN and E. KESSLER
*Temperature and Thickness Dependence of Electrical and Thermal Transport Coefficients of $Bi_{1-x}Sb_x$ Films in an Anisotropic, Non-Degenerate Two-Band Model
(72.10 and 73.60; S4)
(b) Vol. **134**, No. 1, p. 351—362 (1986).

E. Heiner
*A Dynamical Theory for a Disordered Hubbard Chain far from Equilibrium. III. The High-Field Case
(71.25 and 72.10)
(b) Vol. **136**, No. 2, p. 669—672 (1986).

B. M. Askerov, B. I. Kuliev, S. R. Figarova, and R. F. Eminov
*Non-Monotonous Behaviour of the Nernst-Ettingshausen Effect in Conductive Films in a Strong Magnetic Field
(72.10 and 73.60)
(b) Vol. **136**, No. 2, p. 743—748 (1986).

P. Kleinert and M. Giehler
*Theory of Free-Carrier Infrared Absorption in GaAs
(72.10; 63.20; 71.45; 78.30; S7.12)
(b) Vol. **136**, No. 2, p. 763—777 (1986).

P. Kopčanský
*The Kondo Effect in Amorphous Alloys
(72.10)
(b) Vol. **137**, No. 2, p. K171—K174 (1986).

R. Lenk
*Diffusion of Non-Interacting Quantum Particles in a Disordered System. The Transport Length in the Classical Limit
(72.10)
(b) Vol. **138**, No. 1, p. 365—371 (1986).

E. Kolley and W. Kolley
*Numerical Study of the Localization-Affected Conductivity in Random Hubbard Alloys
(72.10; 72.15)
(b) Vol. **138**, No. 1, p. K65—K69 (1986).

L. Reggiani, R. Brunetti and E. Normantas
*Comparison between Monte-Carlo and Davydov Approaches to Hot-Carrier Transport in Semiconductors
(72.10; S5.11; S5.12)
(b) Vol. **138**, No. 2, p. 717—725 (1986).

A. Brezini, M. Sebbani, and F. Behilil
Power-Law Localization in One-Dimensional Random Potentials
(71.50; 72.10)
(b) Vol. **138**, No. 2, p. K137—K142 (1986).

S. Balaska, A. Brezini, and F. Behilil
*Quantum Transport Properties in a Dynamically Stochastic Medium
(72.10)
(b) Vol. **138**, No. 2, p. K143—K147 (1986).

72.15. Electronic conduction in metals and alloys

Original Papers and Short Notes

G. C. FLETCHER
*The Band Structure of Co_3Ga and Co_2MnGa and the Electrical Conductivity of Related Alloys
(71.25 and 72.15; 75.50; S1.2)
(b) Vol. **133**, No. 1, p. 241—248 (1986).

A. R. JANI and H. K. PATEL
Quantum Number Dependent Model Pseudopotential for Metallic Properties
(71.25; 72.15; S4)
(b) Vol. **133**, No. 1, p. K21—K24 (1986).

S. GARG, H. C. GUPTA, and B. B. TRIPATHI
*The Width of the Scattering Resonance and the Electrical Transport Properties in Partially Filled d-Band Transition Metals
(72.15; S1.2)
(b) Vol. **133**, No. 1, p. K73—K76 (1986).

J. KADLECOVÁ, A. HANDSTEIN, T. ZEMČÍK, and K. ZÁVĚTA
Electrical and Magnetic Properties of Annealed Amorphous $Fe_8Ni_{72}P_{10}B_{10}$ at Low Temperatures
(75.50; 72.15; 76.80; S1.1)
(a) Vol. **93**, No. 1, p. 213—220 (1986).

J. FILIPECKI, E. JAKUBCZYK, B. CAŁUSIŃSKI, and D. GAUDYN
Electron Phenomena Connected with Hydriding of the Metallic Glass $Fe_{81}B_{13.5}Si_{3.5}C_2$
(61.70; 72.15; 78.70; S1.1)
(a) Vol. **93**, No. 1, p. K25—K28 (1986).

A. N. VOLOSHINSKII, N. V. RYZHANOVA, L. YU. VISHNEKOV, and A. G. OBUKHOV
*Off-Diagonal Conductivity and Magnetization of PdFe, PdCo, and PdNi Alloys in Coherent Potential Approximation Calculated on the Basis of Realistic Models of Component Densities of States
(72.15; 71.20; 75.60; S1.1; S1.2)
(b) Vol. **133**, No. 2, p. 735—743 (1986).

A. ZENTKO, P. DUNHAJ, M. TIMKO, and V. KAVEČANSKÝ
*The Effect of Neutron Irradiation on the Electrical Resistivity of Amorphous $Co_{47}Ni_{25}B_{18}Si_{10}$ Alloys
(72.15; 61.40; 61.80; S1.2)
(a) Vol. **93**, No. 2, p. 685—693 (1986).

J. M. PELLETIER, F. FOUQUET, and J. HILLAIRET
*Evolution of the Electrical Properties of Amorphous $Ni_{24}Zr_{76}$ and $Ni_{35}Ti_{65}$ as a Result of Structural Relaxation and Crystallization
(72.15; S1.2)
(a) Vol. **93**, No. 2, p. 703—710 (1986).

A. Rojek, G. Sułkowski, A. Zygmunt, and G. Kozłowski
Low-Temperature Properties of Antiferromagnetic $GdRh_xSn_y$ Compound
(75.50; 72.15; S1.4)
(b) Vol. **134**, No. 1, p. 125—129 (1986).

T. M. A. Khajil and M. Tomak
*The Resistivity of Liquid Metals with Finite Mean-Free Path
(72.15; S4)
(b) Vol. **134**, No. 1, p. 321—324 (1986).

B. A. Oli and J. S. Ononiwu
*Effect of Non-Locality in the Calculation on Liquid Metal Resistivity of Transition Metals
(72.15; 71.25; S1.1; S1.2; S1.3)
(b) Vol. **134**, No. 1, p. 325—330 (1986).

P. L. Rossiter and R. H. Mair
*The Electrical Properties of Rapidly Quenched Cu–Ni Alloys
(72.15; 75.90; S1.2)
(b) Vol. **134**, No. 1, p. 335—342 (1986).

G. Schubert and V. Christoph
*Phonon Drag Contribution to the Electrical Resistivity of Alkali and Noble Metals
(72.15; S1.3; S2)
(b) Vol. **134**, No. 2, p. 771—777 (1986).

R. Ducastel and G. Pitsi
*The Lorenz Number of Some Ag–Cr Alloys
(72.15; S1.3)
(b) Vol. **134**, No. 2, p. 799—803 (1986).

A. Rojek and C. Sułkowski
*Electric, Magnetic, and Superconducting Properties of Type-II Stannides $ErRh_{1.1}Sn_{3.6}$ and $LuRh_{1.2}Sn_{4.0}$
(72.15 and 74.70; 75.50; S1.2; S1.4)
(b) Vol. **134**, No. 2, p. K133—K137 (1986).

P. Muth
*On the Anisotropy of the Magnetoresistance of Nickel- and Iron-Based Alloys
(72.15; S1.1; S1.2)
(b) Vol. **135**, No. 1, p. K75—K78 (1986).

S. Takamura and M. Kobiyama
Dislocation Pinning in Al and Ag Alloy after Low-Temperature Deformation
(61.70; 62.20; 72.15; S1.3; S4)
(a) Vol. **95**, No. 1, p. 165—172 (1986).

A. Raoufi and S. Arajs
*Effects of Magnetic Field and Tensile Stress on the Electrical Resistivity of Some Amorphous Ferromagnets
(72.15 and 75.80; S1.1)
(a) Vol. **95**, No. 1, p. 193—196 (1986).

T. Luciński and B. Szymański
*Hall Effect and Magnetic Anisotropy in Amorphous $Fe_{1-x}Si_x$ Films
(72.15 and 75.70; S1.1)
(a) Vol. **96**, No. 1, p. K53—K56 (1986).

F. M. Muntyanu
*Quantum Oscillations of Magnetoresistance and Thermomagnetic Power and the Fermi Surface of As–Sb Alloys
(72.15; 71.25; S5)
(b) Vol. **136**, No. 2, p. 749—756 (1986).

A. B. Kaiser
*Low Temperature Resistivity Anomaly in Glassy Metals. Two-Level Systems and Incipient Localization versus Correlations
(72.15; S1.2; S3)
(b) Vol. **136**, No. 2, p. 779—784 (1986).

G. Schubert and V. Christoph
*Phonon Drag Contribution to the Thermopower of Simple Metals
(72.15)
(b) Vol. **136**, No. 2, p. K151—K153 (1986).

G. L. Whittle, A. M. Stewart, and A. B. Kaiser
Magnetic and Electrical Properties of $(FeTM)_{85}B_{15}$ (TM = Cr, Ni) Metallic Glasses
(75.50; 72.15; 76.80; S1.1)
(a) Vol. **97**, No. 1, p. 199—211 (1986).

J. Mucha, A. V. Krasnokutskii, V. A. Merisov, N. Azumai, and J. Rafałowicz
*Thermal Conductivity Minimum of 6N Purity Aluminium Monocrystal
(72.15; S4)
(a) Vol. **97**, No. 1, p. 307—312 (1986).

M. A. Kenawy, M. R. Nagy, M. S. Sakr, and M. T. Mostafa
*The Effect of Precipitation on the Electrical Resistivity in Al–2.5 wt% Cu
(72.15; 64.75; S4)
(a) Vol. **97**, No. 1, p. K73—K78 (1986).

A. Raoufi, S. Arajs, and K. V. Rao
*Effects of Short-Range Order on the Electrical Resistivity of Plastically Deformed Pd–Ag and Pd–Ag–Fe Alloys
(72.15; S1.1; S1.3)
(a) Vol. **97**, No. 2, p. 571—577 (1986).

M. Ohta, T. Kanadani, and A. Sakakibara
*Effect of Quenching Conditions on the Growth of GP Zones in Al–Zn Alloys
(61.50 and 72.15; S4)
(a) Vol. **97**, No. 2, p. K193—K196 (1986).

72.20. Conductivity phenomena in semiconductors and isulators (for nonelectronic conduction, see 66.70)

Review Article

K. Lischka
Deep Level Defects in Narrow Gap Semiconductors
(71.55; 72.20; 72.40; S8; S8.15)
(b) Vol. **133**, No. 1, p. 17—46 (1986).

Original Papers and Short Notes

O. Ziep, A. G. Mironov, and R. Keiper
*Nonlinear High-Electric-Field Transport in the Presence of Strong Laser Radiation
(72.10 and 72.20; 71.20)
(b) Vol. **133**, No. 1, p. 409—416 (1986).

P. J. S. Foot and B. A. Nevett
*Electronic Conduction in p- and n-Type $NiPS_3$
(72.20; 78.30; S8)
(a) Vol. **93**, No. 1, p. 283—292 (1986).

M. Kittler, W. Seifert, and K.-W. Schröder
*Temperature-Dependent EBIC Diffusion-Length Measurements in Silicon
(72.20; S5.11)
(a) Vol. **93**, No. 1, p. K101—K104 (1986).

V. M. Nabutovskii and D. A. Romanov
Electron States in Periodic Dislocation Networks in Semiconductors
(71.55; 72.20; 72.30)
(b) Vol. **133**, No. 2, p. 645—653 (1986).

H. Burghardt, T. Frauenheim, and C. Hamann
The Influence of an Electric Field on the Mobility in Semiconductors
(72.10; 72.20; S7.12)
(b) Vol. **133**, No. 2, p. 755—767 (1986).

V. I. Gatalskaya
*On the Mechanism of Charge Carrier Scattering in Germanium with Point Defects under Cyclotron Resonance
(72.20; 76.40; S5.12)
(b) Vol. **133**, No. 2, p. K137—K141 (1986).

H. Böttger, P. Szyler, and D. Wegener
*Numerical Study of Non-Ohmic R–ε Hopping Transport
(72.20)
(b) Vol. **133**, No. 2, p. K143—K147 (1986).

B. MANSOUR, F. MUKHTAR, and G. G. BARAKATI
*Electrical and Thermoelectric Properties of Copper Tellurides
(72.20; S1.63)
(a) Vol. **95**, No. 2, p. 703—707 (1986).

V. A. KASIYAN, D. D. NEDEOGLO, A. V. SIMASHKEVICH, and I. N. TIMCHENKO
*Electron Mobility in Heavily Doped Strongly Compensated ZnSe Crystals
(72.20; S8.12)
(b) Vol. **136**, No. 1, p. 341—347 (1986).

J. T. LENKKERI
*Nonlinear Effects in the Piezoresistivity of p-Type Silicon
(72.20; S5.11)
(b) Vol. **136**, No. 1, p. 373—385 (1986).

O. ZIEP and R. KEIPER
*On the Theory of Quantum Hall Effect
(72.20 and 73.40)
(b) Vol. **136**, No. 2, p. 785—794 (1986).

I. G. KIRNAS, P. G. LITOVCHENKO, E. E. PETROSYAN, and N. N. PASHCHUK
*Peculiar Features of the Electric Field Distribution in One-Open-Face Coaxial Germanium Detectors of Nuclear Radiation
(72.20; 73.40; S5.12)
(a) Vol. **96**, No. 2, p. 657—662 (1986).

M. DRESSEL and H. W. HELBERG
*AC Conductivity of Deformed Germanium Single Crystals at $T = 4.2$ K
(72.20; 61.70; S5.12)
(a) Vol. **96**, No. 2, p. K199—K202 (1986).

A. DIMOULAS
*Negative Momentum Relaxation Rate and Transport in Polar Semiconductors
(72.20)
(b) Vol. **137**, No. 1, p. 319—329 (1986).

H. HAVLOVÁ and L. SMRČKA
*Quantum Oscillations of Thermogalvanomagnetic Coefficients in Two-Dimensional Electron Systems
(72.20 and 73.40; S5.11; S7.15)
(b) Vol. **137**, No. 1, p. 331—343 (1986).

W. K. CHOI, J. J. DELIMA, and A. E. OWEN
*A Model for the Variations in the Field-Dependent Behaviour of the Poole-Frenkel Effect
(72.20; 71.55)
(b) Vol. **137**, No. 1, p. 345—351 (1986).

P. ENDERS
Photon Recycling in Double Heterostructures. I. The Case of Perfect Optical Confinement
(78.55; 72.20; 72.90)
(b) Vol. **137**, No. 1, p. 353—360 (1986).

A. P. MEDVID, J. B. LITAUNIEKS, A. P. KRIVICH, and V. E. PRUDENTS
*Magnetoconcentration Effect in Inhomogeneous Electric Field
(72.20; S5.12)
(a) Vol. **97**, No. 1, p. K91—K94 (1986).

S. S. Kubakaddi, B. G. Mulimani, and V. M. Jali
*Thermopower in Quasi-Two-Dimensional Semiconductor Quantum Well Structures
(72.20; 73.40; S7.15)
(b) Vol. **137**, No. 2, p. 683—689 (1986).

R. Rentzsch, K. J. Friedland, A. N. Ionov, M. N. Matveev, I. S. Shlimak, C. Gladun, and H. Vinzelberg
*Variable-Range Hopping in Neutron-Transmutation-Doped Gallium Arsenide
(72.20; 71.30; S7.12)
(b) Vol. **137**, No. 2, p. 691—700 (1986).

P. Enders
Photon Recycling in Double Heterostructures. II. The Case of Non-Perfect Optical Confinement
(78.55; 72.20; 72.90)
(b) Vol. **137**, No. 2, p. 701—708 (1986).

S. N. Biswas and K. P. Ghatak
*On the Modification of the Einstein Relation in Quasi-One-Dimensional Systems of Bismuth
(72.20; S4)
(b) Vol. **137**, No. 2, p. K175—K180 (1986).

K. K. Kulshrestha, A. Pandya, P. K. Goyal, and P. C. Mathur
*DC Galvanomagnetic Properties in a Single Crystal of n-Type CdSe Optimally Annealed in Molten Cd
(72.20; S8.12)
(a) Vol. **97**, No. 2, p. 557—563 (1986).

C. Julien, E. Hatzikraniotis, and K. Kambas
*Electrical Transport Properties of Impurity-Doped In_2Se_3
(72.20; 71.55; S8)
(a) Vol. **97**, No. 2, p. 579—585 (1986).

B. P. Gorshunov, G. V. Kozlov, A. A. Volkov, J. Petzelt, and V. Železný
*Dielectric Response of Some Semiconducting TCNQ Salts in the Submillimeter Range
(72.20 and 77.20; 71.30; S12.1)
(b) Vol. **138**, No. 1, p. 347—355 (1986).

M. Koyano, H. Negishi, Y. Ueda, M. Sasaki, and M. Inoue
*Electrical Resistivity and Thermoelectric Power of Intercalation Compounds M_xTiS_2 (M = Mn, Fe, Co, and Ni)
(72.20; S1.63)
(b) Vol. **138**, No. 1, p. 357—363 (1986).

L. Wendler and M. I. Kaganov
*Quantized Undamping of Helicon Waves in Semiconductor Superlattices
(72.20 and 73.40; 72.30)
(b) Vol. **138**, No. 1, p. K33—K38 (1986).

E. Caselli, R. Cabanillas, R. Wainschenker, and M. Cutella
*High Temperature Drift Mobilities in High Resistivity Silicon
(72.20; S5.11)
(a) Vol. **98**, No. 2, p. K179—K183 (1986).

72.30. High-frequency effects; plasma effects

Original Papers and Short Notes

K. P. MAHESHWARI and G. TAREY
*Resonant Excitation of Helicon Waves by Two Microwave Beams in a Solid State Plasma
(72.30; S7.13)
(b) Vol. **133**, No. 1, p. 417—423 (1986).

V. M. NABUTOVSKII and D. A. ROMANOV
Electron States in Periodic Dislocation Networks in Semiconductors
(71.55; 72.20; 72.30)
(b) Vol. **133**, No. 2, p. 645—653 (1986).

I. I. SASOVSKAYA and V. P. KORABEL
Optical Properties of α- and β-CuZn Brasses in the Region of Quantum Absorption
(78.20; 64.70; 71.25; 72.30; S1.2)
(b) Vol. **134**, No. 2, p. 621—630 (1986).

A. A. MANENKOV, G. N. MIKHAILOVA, S. YU. SOKOLOV, and S. G. TIKHODEEV
Phonon Wind Generation in Ge by CO_2 Laser Induced Heating of Electron–Hole Drops
(71.35; 71.38; 72.30; 78.70; S5.12)
(b) Vol. **134**, No. 2, p. 631—639 (1986).

MANVIR S. KUSHWAHA
*Helicon Wave Propagation in Semiconductor Superlattices
(72.30)
(b) Vol. **136**, No. 2, p. 757—762 (1986).

G. JUNGK and A. RÖSELER
*Optical Properties of Metals: Au
(72.30 and 78.20; 78.65; S1.3)
(b) Vol. **137**, No. 1, p. 117—123 (1986).

B. P. GORSHUNOV, G. V. KOZLOV, A. A. VOLKOV, H. OZAKI, J. PETZELT, and V. ŽELEZNÝ
*Submillimetre Conductivity and Dielectric Function of 1T-TaS_2
(72.30 and 77.20; S1.63)
(b) Vol. **137**, No. 1, p. K89—K93 (1986).

L. WENDLER and M. I. KAGANOV
Quantized Undamping of Helicon Waves in Semiconductor Superlattices
(72.20 and 73.40; 72.30)
(b) Vol. **138**, No. 1, p. K33—K38 (1986).

72.40. Photoconduction and photovoltaic effects; photodielectric effects

Review Articles

Original Papers and Short Notes

S. I. RADAUTSAN and O. V. KULIKOVA
Photocapacitance Effect in $Zn_xCd_{1-x}Te$-InSb Structures
(73.40; 72.40; S7.13; S8.15)
(a) Vol. **93**, No. 2, p. K201—K203 (1986).

NGUYEN HONG SHON and VO HONG ANH
*Photovoltaic Effect in Semiconductors Stimulated by an Additional Electromagnetic Wave
(72.40)
(b) Vol. **134**, No. 1, p. 363—366 (1986).

K. P. MÖLLMANN and M. HAPP
*Inhomogeneous Electric-Field Quenching of Persistent Photoconductivity in Strongly Indium Doped Lead Tin Telluride
(72.40; S8.15)
(b) Vol. **134**, No. 1, p. K75—K78 (1986).

F. A. BENKO and F. P. KOFFYBERG
The Optical Interband Transitions of the Semiconductor $CuGaO_2$
(71.25; 72.40; S10)
(a) Vol. **94**, No. 1, p. 231—234 (1986).

J. HLÁVKA
Excess Carrier Recombination Centre Distribution in Implanted and Annealed Silicon
(61.70; 72.40; S5.11)
(a) Vol. **94**, No. 1, p. 353—355 (1986).

MA KE-JUN
*Photoconductivity in $Mn_xCd_{1-x}Se$
(72.40; S8.15)
(a) Vol. **94**, No. 1, p. K75—K78 (1986).

G. D. GUSEINOV, V. A. ALIEV, and E. F. BAGIRZADE
Recombination Levels in Non-Stoichiometric $TlGaSe_2$
(71.55; 72.40; S8.16)
(a) Vol. **94**, No. 2, p. 653—657 (1986).

H. BERGNER, V. BRÜCKNER, F. KERSTAN, L. LEINE, M. SCHULZE, and W. NOWICK
*Thermally and Carrier Induced Changes in the Photoconductivity in a-Si in the Subnanosecond Time Domain
(72.40; S5.11)
(a) Vol. **94**, No. 2, p. K137—K141 (1986).

S. KUŹMIŃSKI and A. T. SZAYNOK
Surface Photovoltage Spectroscopy of $Cd_{0.85}Mn_{0.15}Te$
(73.20 and 73.25; 72.40; S8.15)
(a) Vol. **95**, No. 1, p. 279—282 (1986).

G. D. GUSEINOV, E. F. BAGIRZADE, S. N. MUSTAFAEVA, N. Z. GASANOV, and S. G. GUSEINOV
*Ionic and Electronic Relaxation of the Conductivity of Intercalated $TlGaSe_2$ Single Crystals
(72.40; 72.80; S8.16)
(a) Vol. **95**, No. 1, p. K71—K74 (1986).

N. V. UNNIKRISHNAN and R. D. SINGH
*Temperature Dependence of Lux–Ampere Characteristics of CdI_2 Single Crystals
(72.40; S9)
(a) Vol. **96**, No. 2, p. K219—K222 (1986).

T. WILSON and P. D. PESTER
*Optical Beam Induced Minority Carrier Distribution in Semiconductors
(72.40 and 73.25; S5.11)
(a) Vol. **97**, No. 1, p. 323—329 (1986).

YA. D. LEBEDEV and M. D. SHAGALOV
*Electrophysical Properties of Gallium Nitride in the Model of a Nonhomogeneous Semiconductor
(72.40 and 78.55; 71.55; S7.1)
(a) Vol. **97**, No. 2, p. 539—548 (1986).

V. MUNOZ, J. GONZALEZ, and A. CHEVY
*Effect of Surface Damage on the Photoconductivity Response of Zinc Phosphide
(72.40; 73.20; S7)
(a) Vol. **97**, No. 2, p. 587—591 (1986).

J. MENDOLIA and D. LEMOINE
Electrical and Optical Properties of Evaporated Thin Films of a-CdTe
(73.60; 61.40; 72.40; S8.13)
(a) Vol. **97**, No. 2, p. 601—607 (1986).

J. PASTRŇÁK, F. KAREL, J. OSWALD, and O. PETŘÍČEK
*The Electric Field Dependence of Photoconductivity Spectra in AlGaAs/GaAs DH Laser Diodes in Relation with Laser Parameters
(72.40 and 73.40; 78.45; S7.12; S7.15)
(a) Vol. **97**, No. 2, p. 657—665 (1986).

CHANG YOUMING, LÜ SHENGQUING, and QIAN RENYUAN
Polymorphism and Electronic Properties of 3-Ethyl-5-[2-(3-Ethyl-2-Benzothiazolinylidene)-Ethylidene]-Rhodanine
(61.65; 72.40; 78.30; S12)
(a) Vol. **98**, No. 1, p. 37—42 (1986).

D. A. ARONOV and B. R. MAMATKULOV
*The Field and Spectrum Dependences of Impurity Photoconductivity with Exclusion in Compensated Semiconductors
(72.40; 73.40; S5.11; S5.12)
(a) Vol. **98**, No. 1, p. 263—277 (1986).

V. A. TEREKHOV, S. N. TROSTYANSKII, E. P. DOMASHEVSKAYA, O. A. GOLIKOVA, M. M. MEZDROGINA, K. L. SOROKINA, and M. M. KAZANIN
Density of States and Photoconductivity of Hydrogenated Amorphous Silicon
(71.20; 72.40; 73.60; 78.70; S5.11)
(b) Vol. **138**, No. 2, p. 647—653 (1986).

72.50. Acoustoelectric effects

Original Paper

72.60. Mixed conductivity and conductivity transitions

Original Papers and Short Notes

N. N. BELEVICH and L. A. MAKOVETSKAYA
*Semiconductor–Metal Transition and Variable Range Hopping Conductivity in $CuInSe_2$
(72.60; 71.30; S8.16)
(b) Vol. **133**, No. 1, p. K53—K56 (1986).

P. DORDOR, S. JOIRET, J. P. DOUMERC, J. C. LAUNAY, J. CLAVERIE, and P. HAGENMÜLLER
Electrical Properties of Lead-Doped $LaCoO_3$ Single Crystals
(72.80; 72.60; S11.1)
(a) Vol. **93**, No. 1, p. 321—328 (1986).

B. DAS, K. S. BARTWAL, and O. N. SRIVASTAVA
*Metal–Insulator Transition in Titanium Disulphide Crystals
(71.30 and 72.60; S1.63)
(b) Vol. **136**, No. 1, p. 365—371 (1986).

G. V. LOSEVA and L. I. RYABINKINA
*Electrical and Magnetic Properties of the Antiferromagnetic Semiconductor α-Mn_xS
(72.60 and 75.50; S8)
(a) Vol. **96**, No. 2, p. K195—K197 (1986).

P. SIMONNIN, R. TOURBOT, B. BOUCHER, M. PERRIN, and J. VANHAUTE
Thermal, Electrical, and Magnetic Properties of the Amorphous Alloys RE_xSi_{1-x} (RE = Gd or Tb; $x = 0.87, 0.59, 0.18$)
(75.50; 61.40; 64.75; 72.60; S1.4; S1.61)
(a) Vol. **98**, No. 2, p. 551—566 (1986).

72.70. Noise processes and phenomena

Original Paper

YU MING-KANG
*A Discussion on the "Magic Number" in Hooge's Formula
(72.70)
(b) Vol. **134**, No. 2, p. 757—762 (1986).

72.80. Conductivity of specific semiconductors and insulators

Original Papers and Short Notes

B. Kumar
*Carrier Drift Mobilities in Some Organic Solids. A Search for Organic Photovoltaic Materials. I. Charge Transport Calculations in 2-2′ bipyridine and pyridazino [4,5-d] pridazine
(72.40 and 72.80; 71.25; S12)
(b) Vol. **133**, No. 1, p. 395—402 (1986).

K. I. Andronik, V. F. Banar, V. G. Kantser, and A. S. Sidorenko
*Kondo-Type Behaviour of Conductivity in Tl-Doped PbTe Single Crystals
(72.80; S8)
(b) Vol. **133**, No. 1, p. K61—K65 (1986).

V. N. Brudnyi, V. A. Charchenko, N. G. Kolin, V. A. Novikov, A. D. Pogrebnyak, and Sh. M. Ruzimov
Electrical Properties and Positron Annihilation in Neutron-Irradiated n-InP
(61.80; 72.80; 78.70; S7.11)
(a) Vol. **93**, No. 1, p. 195—200 (1986).

S. K. Mendiratta
*Electrical Conduction Mechanism in Lead Borate and Lead Silicate Glasses Containing Fe Ions
(72.80; S10.15)
(a) Vol. **93**, No. 1, p. 293—298 (1986).

P. Dordor, S. Joiret, J. P. Doumerc, J. C. Launay, J. Claverie, and P. Hagenmüller
*Electrical Properties of Lead-Doped $LaCoO_3$ Single Crystals
(72.80; 72.60; S11.1)
(a) Vol. **93**, No. 1, p. 321—328 (1986).

A. Kuczkowski and J. Liwo
*Change in Electrical Conductivity of Polymers with Conjugated Bonds by Means of H_2^+ Ion Implantation
(72.80; 61.70; S12)
(a) Vol. **93**, No. 1, p. K69—K73 (1986).

T. Y. Ayyub and W. Bogusz
*On Electrical Conductivity of $Li_{16-2x}Zn_x(SiO_4)_4$
(72.80; S11.1)
(a) Vol. **93**, No. 1, p. K97—K100 (1986).

S. Guimarães, R. Kögler, E. Landi, S. Solmi, and E. Wieser
Anomalous Enhanced Diffusion and Electrical Activation of Boron in Silicon after Rapid Isothermal Annealing
(66.30; 72.80; S5.11)
(a) Vol. **93**, No. 2, p. 549—557 (1986).

M. M. EL-ZAIDIA, A. A. AMMAR, and A. A. EL-RHMANY
Isothermal Crystallization and Kinetics of Amorphous $As_{34.}Ge_xTe_{65.8-x}$
(64.70; 72.80; S8.15)
(a) Vol. **93**, No. 2, p. K111—K116 (1986).

N. I. AKULOVICH, V. M. MAKSIMENKO, and V. I. UTENKO
On Localization of Phosphorus in Neutron-Doped Silicon
(61.70; 72.80; S5.11)
(a) Vol. **93**, No. 2, p. K157—K159 (1986).

J. BETKO and K. MĚŘÍNSKÝ
*Temperature Dependences of Electrical Parameters of Semi-Insulating GaAs Determined from Galvanomagnetic Measurements
(72.80; S7.12)
(a) Vol. **93**, No. 2, p. K205—K209 (1986).

S. BUMELIENĖ, J. POŽELA, and A. TAMAŠEVIČIUS
*Period Multiplying and Chaoting Response in Driven n-Ge with Repulsive Defect Centres
(72.80; S5.12)
(b) Vol. **134**, No. 1, p. K71—K74 (1986).

K. MOCHIZUKI and M. TAKAKUSAKI
Growth of In-Doped ZnS_xSe_{1-x} Single Crystals and Their Photoluminescence
(78.55; 61.70; 72.80; S8.15)
(a) Vol. **94**, No. 1, p. 243—250 (1986).

M. PRASAD and S. K. GUPTA
*Electrical Conductivity of $CuI_{1.0016}$ and $CuI_{1.0046}$ at Different Pressure
(72.80; S9)
(a) Vol. **94**, No. 1, p. 283—288 (1986).

M. PRASAD and S. K. GUPTA
*Hall Effect Measurement of a $CuI_{1.0046}$ Pellet
(72.80; S9)
(a) Vol. **94**, No. 1, p. K65—K68 (1986).

V. P. POPOV, A. V. DVURECHENSKII, B. P. KASHNIKOV, and A. I. POPOV
Defects in Supersaturated Solid Solution of Arsenic in Silicon at Rapid Thermal Annealing
(61.70; 72.80; S5.11)
(a) Vol. **94**, No. 2, p. 569—572 (1986).

K. SIERAŃSKI and J. SZATKOWSKI
*Electrical Conductivity of Zn_3P_2
(72.80; S7)
(a) Vol. **94**, No. 2, p. K133—K135 (1986).

K. R. ALLAKHVERDIEV, SH. G. GASYMOV, and T. G. MAMEDOV
*Effect of Hydrostatic Pressure on the Electrical Conductivity and Hall Coefficient of $TlGaTe_2$ Crystals
(72.80; S8.16)
(a) Vol. **94**, No. 2, p. K143—K146 (1986).

D. I. PISKUNOV, YU. A. DANILOV, S. K. MAKSIMOV, and E. A. PITIRIMOVA
Nonstoichiometry in Ion Implantation of Gallium Arsenide. Two Competing Mechanisms
(61.70; 61.16; 72.80; S7.12)
(a) Vol. **96**, No. 1, p. 141—153 (1986).

L. A. KAZAKEVICH, V. I. KUZNETSOV, and P. F. LUGAKOV
The Effect of Dislocations on the Activation Energy of Radiation Defect Thermal Ionization
(61.70 and 61.80; 71.55; 72.80; S5.11)
(a) Vol. **96**, No. 1, p. 243—247 (1986).

OM PARKASH, DEVENDRA KUMAR, D. K. GANGOPADHAYAY, and D. BAHADUR
*Electrical Properties of La_2TiMO_6 (M = Fe, Ni, Cu, and Zn)
(72.80; S10)
(a) Vol. **96**, No. 1, p. K79—K83 (1986).

E. CULEA, C. GHEORGHIU, and AL. NICULA
*Electrical Conductivity of Vitreous $75\%V_2O_5$–$25\%(As_2O_3 \cdot B_2O_3)$
(72.80; S10.15)
(a) Vol. **96**, No. 1, p. K85—K88 (1986).

I. M. FILIPPOV, L. A. KAZAKEVICH, P. F. LUGAKOV, and V. V. SHUSHA
The Charge Carrier Recombination Processes in Dislocated Silicon
(61.70 and 71.55; 72.80; S5.11)
(a) Vol. **96**, No. 2, p. 527—531 (1986).

I. M. CHAPNIK
*On the Relation between the Energy Gap Data of Semiconducting Compounds and the Position of the Components of the Compounds in the Periodic Table
(72.80; S5; S6; S7; S8; S10.1)
(b) Vol. **137**, No. 1, p. K95—K100 (1986).

V. D. VULCHEV, L. D. BORISOVA, and S. K. DIMITROVA
*Preparation and Properties of the $Pb_{1-x}Cr_xTe$ System
(72.80; 64.75; S8.15)
(a) Vol. **97**, No. 1, p. K79—K82 (1986).

S. SHIGETOMI and T. MATSUMORI
*Electrical Properties of p-Type Layer in Si-Implanted GaAs
(72.80; 61.70; S7.12)
(a) Vol. **97**, No. 1, p. K83—K86 (1986).

D. GERTHSEN
Hall Effect of Plastically Deformed GaAs
(71.55; 72.80; S7.12)
(a) Vol. **97**, No. 2, p. 527—537 (1986).

A. A. BUGAI, V. A. GIRII, V. M. MAKSIMENKO, YU. V. POMOZOV, and V. I. SHAKHOVTSOV
*Investigation of the Production of Quenching Thermodefects in Silicon
(71.55 and 72.80; 76.30; S5.11)
(a) Vol. **97**, No. 2, p. 549—555 (1986).

72.90. Other topics in electronic transport in condensed matter

Original Papers

P. ENDERS
Photon Recycling in Double Heterostructures. I. The Case of Perfect Optical Confinement
(78.55; 72.20; 72.90)
(b) Vol. **137**, No. 1, p. 353—360 (1986).

S. V. VOITIKOV and V. P. GRIBKOVSKII
*On the Theory of Crystal Excitation by Electric Discharges. II. Ionization of Defects
(71.55 and 72.90)
(b) Vol. **137**, No. 2, p. 655—665 (1986).

P. ENDERS
Photon Recycling in Double Heterostructures. II. The Case of Non-Perfect Optical Confinement
(78.55; 72.20; 72.90)
(b) Vol. **137**, No. 2, p. 701—708 (1986).

73. Electronic structure and electrical properties of surfaces, interfaces, and thin films

73.20. Electronic surface states (for emission and impact phenomena, see 79)

Review Article

T. Terzibaschian and R. Enderlein
*The Irreducible Representations of the Two-Dimensional Space Groups of Crystal Surfaces. Theory and Applications
(68.20 and 73.20)
(b) Vol. **133**, No. 2, p. 443—461 (1986).

Original Papers and Short Notes

Tran Thoai Duy Bao
*Static Polarizability and Dipolar Surface Plasmon of Small Metal Particles. II. Including the Lattice Structure
(71.45 and 73.20; S2)
(b) Vol. **133**, No. 1, p. 329—332 (1986).

Tran Thoai Duy Bao
*Linear Density Response at the Metal Surface in the Hydrodynamic Approximation
(71.45 and 73.20)
(b) Vol. **133**, No. 2, p. 633—644 (1986).

Jian-Wei He and P. J. Møller
*On the Energy Dependence of the Very Low Energy Reflectance from Clean and Manganese-Doped MgO (100) Surfaces
(73.20; 79.20; S10.1)
(b) Vol. **133**, No. 2, p. 687—691 (1986).

A. M. Brodskii
*Nonlinear Optics of Metal Surface
(71.45 and 73.20; 78.20)
(b) Vol. **134**, No. 1, p. 251—256 (1986).

R. A. Vardanyan and L. B. Khovakimyan
*Capture of Majority Carriers by a Semiconductor Surface
(73.20)
(b) Vol. **134**, No. 1, p. 307—312 (1986).

V. V. Ustinov and D. Z. Khusainov
*Boundary Conditions at Multichannel Specular Reflection of the Conduction Electrons on the Surface
(73.20 and 73.25)
(b) Vol. **134**, No. 1, p. 313—319 (1986).

O. V. Romanov and A. M. Yafyasov
*Manifestation of Structure and Occupation of Energy Bands in Surface Effects of Semiconductors
(73.20; 73.25; S7.1; S8.15)
(a) Vol. **94**, No. 1, p. 235—242 (1986).

B. Stankiewicz
*Surface States — an Effect of Overlap Integrals
(71.25 and 73.20)
(b) Vol. **134**, No. 2, p. 691—697 (1986).

L. Wendler and R. Pechstedt
*Effects of a Thin Curved Transition Layer on Surface Polaritons. Local States
(71.36 and 73.20)
(b) Vol. **134**, No. 2, p. 707—715 (1986).

S. Kuźmiński and A. T. Szaynok
*Surface Photovoltage Spectroscopy of $Cd_{0.85}Mn_{0.15}Te$
(73.20 and 73.25; 72.40; S8.15)
(a) Vol. **95**, No. 1, p. 279—282 (1986).

R. Pechstedt and L. Leine
*Effects of a Thin Transition Layer on the Dispersion Relation of Surface Polaritons on a Randomly Rough Surface
(71.36 and 73.20; 68.30; 78.30; S4)
(b) Vol. **135**, No. 2, p. 749—758 (1986).

L. Wendler
*S-Polarized Nonlinear Surface Polaritons. Effect of a Transition Layer
(71.36 and 73.20; 68.30; 78.20)
(b) Vol. **135**, No. 2, p. 759—774 (1986).

Tran Thoai Duy Bao
*Induced Field at a Dipole near the Metal Surface in the Hydrodynamic Approximation
(71.45 and 73.20; 78.20)
(b) Vol. **135**, No. 2, p. 775—781 (1986).

C.-M. Liegener, G. Biczó, and A. Godzik
On the Concept of Intermediate One-Particle States in Finite Chains. A Self-Consistent Field Study
(71.50; 73.20; S12.1)
(b) Vol. **136**, No. 1, p. 281—289 (1986).

Tran Thoai Duy Bao
*Surface Excitations on Small Metal Particles by Fast Electrons
(61.80 and 73.20; S4)
(b) Vol. **136**, No. 1, p. 291—298 (1986).

A. A. Karpushin, A. N. Sorokin, I. Yu. Semenova, and M. Tomášek
*Method of Generalized Bethe Lattices in Chemisorption Theory
(71.10 and 73.20; 68.45; S5.11)
(b) Vol. **136**, No. 1, p. 331—339 (1986).

V. P. Parkhutik, V. A. Labunov, and G. G. Chigir
Kinetics and Mechanisms of Transient Thermal Oxidation of Silicon
(68.55; 73.20; 73.40; 79.60; S5.11; S10.1)
(a) Vol. **96**, No. 1, p. 11—18 (1986).

G. V. Gadiyak, A. A. Karpushin, I. V. Korolenko, Yu. N. Morokov, I. Yu. Semenova, A. N. Sorokin, and M. Tomášek
*Calculation of the Electronic Structure of Disordered Hydrogen Adsorption on the Si(111) Surface
(73.20; 73.40; S5.11)
(b) Vol. **137**, No. 2, p. 633—639 (1986).

V. Munoz, J. Gonzalez, and A. Chevy
Effect of Surface Damage on the Photoconductivity Response of Zinc Phosphide
(72.40; 73.20; S7)
(a) Vol. **97**, No. 2, p. 587—591 (1986).

A. Beckmann
*Modelling of Angle Resolved Photoemission Spectra at Low Energy
(73.20 and 79.60; S1.2)
(b) Vol. **138**, No. 1, p. 295—308 (1986).

V. N. Davydov and A. S. Petrov
*Electric Field Effect on Characteristics of Indium Antimonide MOS Structures
(73.20 and 73.40; S7.13)
(a) Vol. **98**, No. 1, p. 253—261 (1986).

73.25. Surface conductivity

Original Papers and Short Notes

V. V. USTINOV and D. Z. KHUSAINOV
*Boundary Conditions at Multichannel Specular Reflection of the Conduction Electrons on the Surface
(73.20 and 73.25)
(b) Vol. **134**, No. 1, p. 313—319 (1986).

O. V. ROMANOV and A. M. YAFYASOV
Manifestation of Structure and Occupation of Energy Bands in Surface Effects of Semiconductors
(73.20; 73.25; S7.1; S8.15)
(a) Vol. **94**, No. 1, p. 235—242 (1986).

R. COELHO, L. LEVY, and D. SARRAIL
*On the Natural Decay of Corona Charged Insulating Sheets
(73.25 and 77.50; 77.30; S.12)
(a) Vol. **94**, No. 1, p. 289—298 (1986).

S. KUŹMIŃSKI and A. T. SZAYNOK
*Surface Photovoltage Spectroscopy of $Cd_{0.85}Mn_{0.15}Te$
(73.20 and 73.25; 72.40; S8.15)
(a) Vol. **95**, No. 1, p. 279—282 (1986).

T. WILSON and P. D. PESTER
*Optical Beam Induced Minority Carrier Distribution in Semiconductors
(72.40 and 73.25; S5.11)
(a) Vol. **97**, No. 1, p. 323—329 (1986).

73.30. Surface double layers, Schottky barriers, and work functions

Original Papers and Short Notes

N. T. Tam and Tran Chot
*Experimental Richardson Constant of Metal–Semiconductor Schottky Barrier Contacts
(73.30; S5.11; S7.11; S7.12; S8.11; S8.12)
(a) Vol. **93**, No. 1, p. K91—K95 (1986).

D. A. Aronov, D. K. Isamukhamedova, and Yu. M. Yuabov
On the Theory of Majority-Carrier Silicon Solar-Cell MTIS Structures with a Schottky Barrier
(73.40; 73.30; S5.11)
(a) Vol. **93**, No. 2, p. K219—K221 (1986).

D. E. Ioannou, Y. J. Huang, P. K. McLarty, and S. M. Johnson
*On the Barrier Height of Al/p-Si Schottky Diodes
(73.30; S5.11)
(a) Vol. **93**, No. 2, p. K223—K226 (1986).

I. Bryłowska, K. Paprocki, and M. Subotowicz
Obtaining Hafnium Silicides by the Mixing Method
(61.80 and 68.55; 73.30; S1.61)
(a) Vol. **94**, No. 1, p. 423—426 (1986).

H. Unlu and A. Nussbaum
Position-Dependent Theory of Heterojunctions
(73.40; 73.30)
(a) Vol. **94**, No. 2, p. 687—691 (1986).

Zs. J. Horváth, I. Gyúró, M. Németh-Sallay, B. Szentpáli, and K. Kazi
GaAs Schottky Varactors for Linear Frequency Tuning in X-Band
(73.40; 73.30; S7.12)
(a) Vol. **94**, No. 2, p. 719—726 (1986).

D. K. Rao, J. J. B. Prasad, D. Sridevi, K. V. Reddy, and J. Sobhanadri
*Properties of p-$CuInSe_2$/Al Schottky Devices
(73.30; S8.16)
(a) Vol. **94**, No. 2, p. K153—K158 (1986).

V. A. Kiselev, B. V. Novikov, A. S. Batyrev, E. A. Ubushiev, and A. E. Cherednichenko
Localization of Excitons in Space-Charge Layers
(71.35; 73.30; 78.40; S8.11; S8.12)
(b) Vol. **135**, No. 2, p. 597—604 (1986).

R. J. Iwanowski and D. Trivich
*Enhancement of the Photovoltaic Conversion Efficiency in Cu/Cu_2O Schottky Barrier Solar Cells by H^+ Ion Irradiation
(73.30 and 73.40; S10.1)
(a) Vol. **95**, No. 2, p. 735—741 (1986).

H. MENNIGER, M. BEER, and H. LANGE
*Surface Imperfection Induced Electrical Leakage Paths in Silicide Schottky Diodes Investigated by SEM/EBIC and AES
(61.70 and 73.30; 61.14; 79.20; S1.61)
(a) Vol. **96**, No. 1, p. 353—362 (1986).

H. LEMKE
*Eine analytische Approximation für die Ortsabhängigkeit der Ladungsträgerkonzentration in Schottky-Randschichten
(An Analytical Approximation for the Spatial Dependence of the Charge Carrier Concentration in Schottky Layers)
(73.30)
(a) Vol. **96**, No. 1, p. K65—K68 (1986).

R. S. GUPTA, G. S. CHILANA, and G. P. SRIVASTAVA
*A Simplified Approach to Reduce the Effective Barrier Height of Triangular Barrier Structures
(73.30)
(a) Vol. **96**, No. 1, p. K89—K93 (1986).

ZHANG JI-KANG, G. N. OKEKE, and C. H. B. MEE
*Work Function Measurements on a Silver–Oxygen–Caesium Photocathode Prepared on an Ag(111) Single Crystal
(73.30 and 79.60; S1.3)
(a) Vol. **96**, No. 2, p. 691—696 (1986).

H. LEMKE
*Eine analytische Approximation für den Potentialverlauf bei Schottky-Randschichten
(An Analytical Approximation for the Potential Distribution in Schottky Layers)
(73.30)
(a) Vol. **96**, No. 2, p. K213—K217 (1986).

J. BOLLMANN, H. KLOSE, and A. MERTENS
*Formation of Schottky Junctions in Silicon by Ion Implantation
(73.30; 61.70; S5.11)
(a) Vol. **97**, No. 1, p. K95—K99 (1986).

TRAN CHOT and N. T. TAM
*Schottky Barriers on Annealed GaAs
(73.30; S7.12)
(a) Vol. **97**, No. 1, p. K101—K105 (1986).

H. POHLACK
*Optically Enhanced Schottky Barrier Photodetectors for IR Image Sensor Application
(73.30; 78.90; S1.61)
(a) Vol. **97**, No. 2, p. K211—K214 (1986).

H. LEMKE
*Eine analytische Approximation für die Ortsabhängigkeit des Potentials bei Schottky-Randschichten
(An Analytic Approximation for the Spatial Dependence of the Potential in Schottky Barriers)
(73.30)
(a) Vol. **98**, No. 2, p. K185—K190 (1986).

J. TOUŠKOVA, D. KINDL, and J. TOUŠEK
*Schottky Photocells on Electrodeposited CdTe Layers
(73.30; S8.13)
(a) Vol. **98**, No. 2, p. K197—K199 (1986).

73.40. Interfaces

Review Article

V. V. Rossin and V. G. Sidorov
*Reabsorption of Recombination Radiation in Semiconductors with High Internal Quantum Efficiency
(73.40 and 78.60; 72.40; 78.20; S7.12)
(a) Vol. **95**, No. 1, p. 15—40 (1986).

Original Papers and Short Notes

A. M. Gabovich and A. I. Voitenko
*Electron Potential Energy near the Ferromagnetic Metal–Vacuum Interface
(73.40 and 75.10)
(b) Vol. **133**, No. 1, p. 135—142 (1986).

Wu Xiaoguang, F. M. Peeters, and J. T. Devreese
Screening of the Electron–Phonon Interaction in GaAs Heterostructures
(71.38; 73.40; S7.12; S7.15)
(b) Vol. **133**, No. 1, p. 229—240 (1986).

W. Kraak, G. Nachtwei, and R. Herrmann
*Quantum Hall Effect in the Inversion Layer of p-Type InSb Bicrystals under High Hydrostatic Pressure
(73.40; S7.13)
(b) Vol. **133**, No. 1, p. 403—408 (1986).

R. Herrmann, U. Preppernau, and M. Gliński
*Thermopower Measurements on the Two-Dimensional Electron Gas of a Grain Boundary in an InSb Bicrystal
(73.40; S7.13)
(b) Vol. **133**, No. 1, p. K57—K59 (1986).

V. A. Gurtov and A. I. Nazarov
*Isothermal Annealing of Space Charge in MNOS Structures
(73.40; S5.11)
(a) Vol. **93**, No. 1, p. 347—351 (1986).

P. R. Vaya, J. Majhi, B. S. V. Gopalam, and C. Dattatreyan
*Study on n-PbTe/p-Si Heterostructures
(73.40; S5.11; S8)
(a) Vol. **93**, No. 1, p. 353—360 (1986).

S. V. Bogdanov, A. B. Kravchenko, A. F. Plotnikov, and V. E. Shubin
*Model of the Avalanche Multiplication in MIS Structures
(73.40; S5.11)
(a) Vol. **93**, No. 1, p. 361—367 (1986).

A. B. Kravchenko, A. F. Plotnikov, A. G. Solonko, and V. E. Shubin
*Avalanche Spectroscopy of MIS Structure Generation Centres
(73.40; S5.11)
(a) Vol. **93**, No. 1, p. 369—376 (1986).

M. Mondal and K. P. Ghatak
*Effect of a Quantizing Magnetic Field on the Gate Capacitance of MOS Structures of Kane-Type Semiconductors
(73.40; S7.12)
(a) Vol. **93**, No. 1, p. 377—383 (1986).

A. A. Akhmedov, S. Kh. Khaililov, A. G. Kyazymzade, and Ya. A. Bairamov
*Prolonged Relaxation and Residual Conductivity in Sn_2O–GaSe Heterotransitions
(73.40; S8; S10.1)
(a) Vol. **93**, No. 1, p. K79—K83 (1986).

O. G. Balev and I. I. Boiko
*Conductivity Tensor of a Two-Dimensional Electron Gas in a Strong Magnetic Field in Higher Orders of Perturbation Theory. Quantum Hall Effect
(73.40)
(b) Vol. **133**, No. 2, p. 721—734 (1986).

R. Keiper and O. Ziep
Theory of Quantum Hall Effect
(72.10; 73.40)
(b) Vol. **133**, No. 2, p. 769—773 (1986).

V. Milanović, D. Tjapkin, and R. Šašić
*The Influence of Nonparabolicity on Self-Consistent Treatment for GaAs–$Al_xGa_{1-x}As$ Heterojunction
(73.40; 71.25; S7.12; S7.15)
(b) Vol. **133**, No. 2, p. 775—784 (1986).

Qu Zhe and Pan Jin-Sheng
The Behaviour of Excitons near the Interface of a Polar Crystal
(71.35; 73.40)
(b) Vol. **133**, No. 2, p. K119—K124 (1986).

W. Braune, N. Kubicki, N. Pruss, and R. Herrmann
*Subband Characteristics of the Two-Dimensional Inversion Carrier System in p-InSb Bicrystals
(73.40; S7.13)
(b) Vol. **133**, No. 2, p. K149—K151 (1986).

I. Petr
*Activation Energy of Trapping Centres in SiO_2 Gate Insulator
(73.40; 71.55; S10.1)
(a) Vol. **93**, No. 2, p. 711—716 (1986).

P. Chakrabarti and P. B. Pal
*Computer Simulation of Ion-Implanted DAR Impatt around 94 GHz under Steady State Condition
(73.40; S5.11; S7.11; S7.12)
(a) Vol. **94**, No. 1, p. 305—313 (1986).

F. Cembali, M. Servidori, E. Landi, and S. Solmi
*Influence of Damage Depth Profile on the Characteristics of Shallow p^+/n Implanted Junctions
(73.40; 61.70; S5.11)
(a) Vol. **94**, No. 1, p. 315—319 (1986).

H. Bachert, W. Pittroff, I. Rechenberg, S. Stoeff, and K. Vogel
The Influence of Mg Doping on the Optical and Electrical Properties of GaAs/(AlGa)As Injection Lasers
(66.30 and 68.55; 73.40; 78.45; S7.12; S7.15).
(a) Vol. **94**, No. 1, p. 321—327 (1986).

P. Krispin and J. Maege
*The Ge-Related DX Level in Sn/Ge-Doped $Al_xGe_{1-x}As$ Heterojunctions Grown by LPE
(73.40; 71.55; S7.15)
(a) Vol. **94**, No. 1, p. 329—335 (1986).

P. Enders
*Interface Recombination Velocity for a Barrier at the p–n-Heterojunction
(73.40)
(a) Vol. **94**, No. 1, p. K85—K88 (1986).
Erratum
(a) Vol. **96**, No. 1, p. K103 (1986).

F. Berg, H. Pagnia, and N. Sotnik
*Influence of Ion Bombardment on the Electrical Behaviour of Discontinuous Metal Film Diodes
(73.40; 61.80; S10.1)
(a) Vol. **94**, No. 1, p. K89—K92 (1986).

J. Pultorak
*Current Induced Shift of the Maximum Radiation Plane in LE GaAs:Si Diodes Obtained by LPE Technique
(73.40 and 78.60; S7.12)
(a) Vol. **94**, No. 1, p. K93—K96 (1986).

O. Ziep, M. Suhrke, and R. Keiper
*Theory of the Carrier Mobility in Quantum Wells Including Multi-Subband Coupling
(73.40)
(b) Vol. **134**, No. 2, p. 789—798 (1986).

G. Paasch, T. Fiedler, and I. Bartoš
*Subband Energies at Semiconductor Heterojunctions. An Approximate Quantization Formula
(73.40; 71.25)
(b) Vol. **134**, No. 2, p. 825—835 (1986).

G. Gobsch, G. Paasch, and H. Übensee
*Self-Consistent Calculation of the Electronic Structure of n-Inversion Layers Adjacent to the Grain Boundary in InSb Bicrystals
(71.25 and 73.40; S7.13)
(b) Vol. **135**, No. 1, p. 283—297 (1986).

P. Streitenberger
*On the Force between High-Density Jellium-Metal Surfaces at Small Separation
(68.40 and 73.40)
(b) Vol. **135**, No. 1, p. 321—333 (1986).

D. Chattopadhyay
*Piezoelectric and Deformation Potential Acoustic Scattering Mobility of a Two-Dimensional Electron Gas in Quantum Wells
(73.40; S7.12)
(b) Vol. **135**, No. 1, p. 409—413 (1986).

R. Herrmann, W. Kraak, G. Nachtwei, and Th. Schurig
*Quantum Properties of the Two-Dimensional Electron Gas in the n-Inversion Layers of InSb Grain Boundaries under High Hydrostatic Pressure
(73.40; S7.13)
(b) Vol. **135**, No. 1, p. 423—435 (1986).

H. Indusekhar and V. Kumar
Properties of Iron Related Quenched-In Levels in p-Silicon
(71.55; 61.70; 73.40; S5.11)
(a) Vol. **95**, No. 1, p. 269—278 (1986).

V. I. Turchanikov, V. S. Lysenko, and V. A. Gusev
Isothermal DLTS Method Using Sampling Time Scanning
(71.55; 73.40; S5.11)
(a) Vol. **95**, No. 1, p. 283—289 (1986).

I. L. Baginskii and E. G. Kostsov
*Experimental Determination of the Mechanism of Transient Single-Injection Current Flow in Dielectrics
(73.40 and 77.55; S7)
(a) Vol. **95**, No. 1, p. 291—301 (1986).

M. K. Agarwal, P. K. Garg, and L. T. Talele
*Energy Band Location and Redox Analysis of Molybdenum Sulphoselenides (MoS_xSe_{2-x}, $0 \leqq x \leqq 2$) Photoelectrodes
(73.40; S8.15)
(a) Vol. **95**, No. 1, p. 303—307 (1986).

G. Ghibaudo
*An Analytical Model of Conductance and Transconductance for Enhanced-Mode MOSFETs
(73.40)
(a) Vol. **95**, No. 1, p. 323—335 (1986).

Mao Er-wang, Zhang Hung-rei, Wang Wei-yuan, Ye Liang, and Fang Guo-ping
*Inelastic Scattering of Electrons at the InP–SiO_2 Interface
(73.40; 71.38; S7.11; S10.1)
(b) Vol. **135**, No. 2, p. 825—830 (1986).

N. V. Rumak and V. V. Khatko
Stimulated Crystallization of Thermal Silica Films. I. Heat Treatment
(68.55; 68.48; 73.40; S10.1)
(a) Vol. **95**, No. 2, p. 391—395 (1986).

O. Engström, H. Pettersson, and B. Sernelius
Photoelectric Yield Spectra of Metal–Semiconductor Structures
(79.60; 73.40; S1; S1.3; S1.61; S5.11)
(a) Vol. **95**, No. 2, p. 691—701 (1986).

R. J. Iwanowski and D. Trivich
*Enhancement of the Photovoltaic Conversion Efficiency in Cu/Cu_2O Schottky Barrier Solar Cells by H^+ Ion Irradiation
(73.30 and 73.40; S10.1)
(a) Vol. **95**, No. 2, p. 735—741 (1986).

A. M. Narsale and B. M. Arora
*C–V–ω and G–V–ω Characteristics of Native Oxide–GaAs Interfaces of ⟨$\bar{1}\bar{1}\bar{1}$⟩ and ⟨100⟩ Orientations
(73.40; S7.12)
(a) Vol. **95**, No. 2, p. 743—748 (1986).

B. I. Sysoev, N. N. Bezryadin, and Yu. K. Shlyk
*Electrophysical Properties of In_2Te_3–InAs Heterojunctions
(73.40; S7.12; S8)
(a) Vol. **95**, No. 2, p. K169—K173 (1986).

F. El Akkad, A. Felimban, and F. Sallam
*Indium Surface Barriers on $Mg_xZn_{1-x}Te$
(73.40; S8.15)
(a) Vol. **95**, No. 2, p. K183—K186 (1986).

T. Gengenbach, H. Pagnia, and N. Sotnik
*Influence of Different Electrode Materials on the I–U Characteristics of Planar MIM Diodes
(73.40)
(a) Vol. **95**, No. 2, p. K199—K202 (1986).

A. Balasiński, A. Jakubowski, T. Brożek, and K. Iniewski
*On γ-Radiation and Electron Beam Dose and Gate Oxide Thickness Dependence of Radiation Defects in MOS Capacitors
(61.80 and 73.40; S5.11)
(a) Vol. **95**, No. 2, p. K203—K206 (1986).

V. M. Fomin and E. P. Pokatilov
Optical Properties of Multi-Layer Structures. II. Reflection and Transmission
(71.36; 73.40; 78.20; S7.12; S7.15)
(b) Vol. **136**, No. 2, p. 593—602 (1986).

V. Milanović
Influence of Nonparabolicity on the Superlattice Band Structure
(71.25; 73.40; S7.12; S7.15)
(b) Vol. **136**, No. 2, p. 661—667 (1986).

O. Ziep and R. Keiper
*On the Theory of Quantum Hall Effect
(72.20 and 73.40)
(b) Vol. **136**, No. 2, p. 785—794 (1986).

M. Zvára, P. Hlídek, P. Höschl, V. Koubele, Z. Kučera, V. Prosser, and V. Szöcs
Landau Levels and Phonon Oscillations in Photovoltage Response of $Cd_xHg_{1-x}Te$ Diodes
(72.40; 73.40; S8.15)
(b) Vol. **136**, No. 2, p. K145—K150 (1986).

I. G. Kirnas, P. G. Litovchenko, E. E. Petrosyan, and N. N. Pashchuk
Peculiar Features of the Electric Field Distribution in One-Open-Face Coaxial Germanium Detectors of Nuclear Radiation
(72.20; 73.40; S5.12)
(a) Vol. **96**, No. 2, p. 657—662 (1986).

E. Gehrke, K. Wandel, and H. Weinert
*Experimental Investigations on the Influence of Surface and Interface Recombination on the Minority Carrier Lifetime in LPE-GaP:N, Si, VPE-GaP:N, Te, and VPE-$GaAs_{0.12}P_{0.88}$:N, Te
(73.40 and 78.55; S7.11; S7.15)
(a) Vol. **96**, No. 2, p. 683—689 (1986).

V. Belenkii, V. G. Veselago, M. S. Gilenko, A. Yu. Leiderman, L. P. Maksimov, V. D. Ovsyannikov, and P. M. Karageorgy-Alkalaev
*On the Enhanced Magnetosensitivity of Semiconductor Structures with Isotype Junction
(73.40; S5.11)
(a) Vol. **96**, No. 2, p. K223—K228 (1986).

F. Bechstedt and R. Enderlein
*Interface Plasmon Modes as Collective Electronic Excitations of Semiconductor Superlattices
(71.45 and 73.40)
(b) Vol. **137**, No. 1, p. 109—115 (1986).

M. Tomak and V. E. Godwin
*Electron Energy Levels in $GaAs/Ga_{1-x}Al_xAs$ Heterojunctions (I)
(71.55 and 73.40; S7.12; S7.15)
(b) Vol. **137**, No. 1, p. 183—186 (1986).

G. HOLZ and CH. SCHNITTLER
*Calculation of the Electronic Properties of a High Electron Mobility Transistor. I. Equilibrium Conditions
(73.40; 71.25; S7.12; S7.15)
(a) Vol. **97**, No. 2, p. 635—644 (1986).

J. PASTRŇÁK, F. KAREL, J. OSWALD, and O. PETŘÍČEK
*The Electric Field Dependence of Photoconductivity Spectra in AlGaAs/GaAs DH Laser Diodes in Relation with Laser Parameters
(72.40 and 73.40; 78.45; S7.12; S7.15)
(a) Vol. **97**, No. 2, p. 657—665 (1986).

V. F. GREMENOK, N. N. KOREN, V. V. KINDYAK, V. V. MOISEENKO, and K. P. GRIGOREV
*Properties of Heterojunctions Based on CdS
(73.40; S8.11)
(a) Vol. **97**, No. 2, p. K205—K209 (1986).

L. WENDLER and R. PECHSTEDT
Dynamical Screening, Collective Excitations, and Electron–Phonon Interaction in Heterostructures and Semiconductor Quantum Wells. General Theory
(71.38 and 71.45; 63.20; 73.40; 77.30; 78.20)
(b) Vol. **138**, No. 1, p. 197—217 (1986).

L. WENDLER and M. I. KAGANOV
*Quantized Undamping of Helicon Waves in Semiconductor Superlattices
(72.20 and 73.40; 72.30)
(b) Vol. **138**, No. 1, p. K33—K38 (1986).

H. YAMAMOTO
*Resonant Tunneling through Amorphous Double-Barrier Structures
(73.40; S5.11; S6)
(b) Vol. **138**, No. 1, p. K71—K73 (1986).

V. N. DAVYDOV and A. S. PETROV
*Electric Field Effect on Characteristics of Indium Antimonide MOS Structures
(73.20 and 73.40; S7.13)
(a) Vol. **98**, No. 1, p. 253—261 (1986).

D. A. ARONOV and B. R. MAMATKULOV
The Field and Spectrum Dependences of Impurity Photoconductivity with Exclusion in Compensated Semiconductors
(72.40; 73.40; S5.11; S5.12)
(a) Vol. **98**, No. 1, p. 263—277 (1986).

F. P. KORSHUNOV, I. G. MARCHENKO, and V. T. TROSHCHINSKII
*Peculiarities of Minority Carrier Lifetime Variations in p–n–p–n Structures Fabricated upon Neutron-Transmutation Doped Silicon and Irradiated by Fast Electrons
(61.80 and 73.40; S5.11)
(a) Vol. **98**, No. 1, p. K39—K42 (1986).

73.60. Electronic properties of thin films

Original Papers and Short Notes

A. L. Vartanian and A. A. Kirakosian
Calculation of the Charge Carrier Mobility of a Size Quantized Semiconductor Film by the Energy-Loss Method
(72.10; 73.60)
(b) Vol. **133**, No. 1, p. 389—394 (1986).

M. Mondal and K. P. Ghatak
*On the Generalized Einstein Relation in Ultrathin Films of Ternary Chalcopyrite Semiconductors in the Presence of Crossed Electric and Magnetic Fields
(73.60; S7.16)
(b) Vol. **133**, No. 1, p. K67—K72 (1986).

V. M. Koleshko, V. F. Belitskii, and A. A. Khodin
Electrical Conductance and Phase Transformations in Dy and Tb Thin Films on Si Substrates Subjected to Heat Treatment
(68.55; 73.60; S1.4; S1.61)
(a) Vol. **93**, No. 1, p. 29—33 (1986).

J. Szczyrbowski, K. Schmalzbauer, and H. Hoffmann
*Optical Properties and Dynamic Current Density for Polycrystalline Metallic Films
(73.60 and 78.20; 78.65; S1.3)
(a) Vol. **93**, No. 1, p. 329—339 (1986).

N. Saito and N. Tanaka
*Preparation and Properties of Magnetron Sputtered Fluorinated and Hydrogenated Amorphous Si–C Films
(73.60 and 78.65; 68.55; S6)
(a) Vol. **93**, No. 1, p. 341—345 (1986).

H.-J. Fitting and A. von Czarnowski
*Electron Breakdown in Polar Insulating and Semiconducting Layers
(73.60; 71.38; 79.70; S8.11; S10.1)
(a) Vol. **93**, No. 1, p. 385—396 (1986).

S. K. Sharma, A. G. Gupta, and P. K. C. Pillai
*Studies on the Temperature Dependence of the DC Conductivity of Thermally Sensitized Poly(Acrylonitrile Butadiene Styrene)
(73.60; S12)
(a) Vol. **93**, No. 1, p. K75—K78 (1986).

P. G. di Marco, G. Giro, and R. Degli Esposti
*The Photoconductivity of Polycrystalline Films of Thionaphthenindol
(72.40 and 73.60; S12)
(a) Vol. **93**, No. 1, p. K85—K89 (1986).

W. Wesch and G. Götz
Rapid Annealing of Ion-Implanted GaAs
(61.70; 61.80; 73.60; 78.20; S7.12)
(a) Vol. **94**, No. 2, p. 745—766 (1986).

B. I. Sysoev, N. N. Bezryadin, Yu. V. Synorov, and B. L. Agapov
*Investigation of Gallium Selenide Films, Grown by the Hot Wall Method, on Silicon Substrates
(73.60; 73.40; S8)
(a) Vol. **94**, No. 2, p. K129—K132 (1986).

J. V. Kornev, D. I. Sementsov und V. V. Sidorenkov
*Hochfrequenzeigenschaften von Kobalt-Dünnschichten mit Flächenanisotropie der elektrischen Leitfähigkeit
(High-Frequency Properties of Cobalt Thin Films with Surface Anisotropy of the Electrical Conductivity)
(73.60; S1.2)
(a) Vol. **94**, No. 2, p. K147—K152 (1986).

D. Genzow, M. Mocker, and E. Normantas
The Influence of Uniaxial Strain on the Band to Band Recombination in $Pb_{1-x}Sn_xTe$
(71.25; 73.60; S8.15)
(b) Vol. **135**, No. 1, p. 261—270 (1986).

F. El Akkad, M. Abdel Naby, and A. Felimban
*Some Electrical Properties of RF Sputtered CdS Thin Films
(73.60; S8.11)
(a) Vol. **95**, No. 1, p. K63—K65 (1986).

A. M. Karguppikar and A. G. Vedeshwar
*Transport Properties of Thin Iron Pyrite Films
(73.60; S8)
(a) Vol. **95**, No. 2, p. 717—720 (1986).

P. Burgaud, J. C. Bernede, G. Safoula, and A. Ameziane
*Effect of Chlorine on Some Properties of Chlorine-Doped Selenium Films
(73.60; 68.55; S5)
(a) Vol. **95**, No. 2, p. 721—733 (1986).

M. V. Ramana Reddy and K. Narasimha Reddy
*Characterization and Electrical Behaviour of $NaYF_4$ Thin Films
(73.60; S9.16)
(a) Vol. **95**, No. 2, p. K193—K198 (1986).

T. Jung and M. Schmidt
*Properties of AlN_xO_y/GaAs Structures Made by Plasma Nitridation of Evaporated Aluminium Films
(68.55 and 73.60; 79.20; S7.1; S7.12)
(a) Vol. **96**, No. 1, p. 75—81 (1986).

S. A. Azimov, T. M. Razykov, V. I. Vyaly, and B. Kh. Kadyrov
Physical Properties of $(ZnSe)_x(CdTe)_{1-x}$ Multicomponent System Films Fabricated by CVD in Hydrogen Flow
(68.55 and 71.25; 73.60; S8.15)
(a) Vol. **96**, No. 1, p. 281—284 (1986).

A. N. Y. Samaan, S. M. Wasim, A. E. Hill, D. G. Armour, and R. D. Tomlinson
*Preparation and Properties of $CuInS_2$ Thin Films Deposited by RF Sputtering
(68.55 and 73.60; S8.16)
(a) Vol. **96**, No. 1, p. 317—324 (1986).

Rodica M. Cândea, Rodica Turcu, P. Mărgineanu, and D. Dădârlat
*The Dynamic Behavior of the Electrical Conductivity of CAD-PbSe Films
(73.60; S8)
(a) Vol. **96**, No. 1, p. 337—343 (1986).

B. Diegner, J. Kugler, and H. Weinert
*On the Correlations between Electrical Properties and Growth Conditions for VPE GaAs Grown in the Hydride System
(73.60; 68.55; S7.12)
(a) Vol. **96**, No. 1, p. 345—351 (1986).

V. B. Orletskii, V. Z. Yurtsenyuk, and K. D. Tovstyuk
*Localized Defect States in $PbSe_xTe_{1-x}$
(73.60; 71.55; S8.15)
(a) Vol. **96**, No. 1, p. K61—K63 (1986).

B. M. Askerov, B. I. Kuliev, S. R. Figarova, and R. F. Eminov
*Non-Monotonous Behaviour of the Nernst-Ettinghausen Effect in Conductive Films in a Strong Magnetic Field
(72.10 and 73.60)
(b) Vol. **136**, No. 2, p. 743—748 (1986).

R. Herzog, G. Schramm, and T. Flade
*Impurity Analysis of Silicon Epitaxial Layers by the 4.2 K Photoluminescence Spectroscopy
(73.60 and 78.55; 78.65; S5.11)
(a) Vol. **96**, No. 2, p. 629—636 (1986).

M. S. Rahman Khan
*Electrical Size Effects in Polycrystalline Rhodium Films
(73.60; S1)
(a) Vol. **96**, No. 2, p. 651—655 (1986).

B. A. Lombos, M. F. Lawrence, J. P. Dodelet, D. Côté, and M. Averous
*Vacancy Doping of GaAs
(73.60; 68.55; S7.12)
(a) Vol. **96**, No. 2, p. 663—675 (1986).

V. D. DAS, M. S. M. SASTRY, and M. PATTABI
*Aging and Field Effect Studies of Cu and Cu/Ag Composite Discontinuous Thin Films
(73.60; S1.2; S1.3)
(a) Vol. **96**, No. 2, p. 677—681 (1986).

S. RAGHUNATH REDDY, A. K. MALLIK, and S. R. JAWALEKAR
*Electrical and Optical Properties of Chemically Vapour Deposited Fluorine Doped SnO_2 Films
(73.60 and 78.65; S10.1)
(a) Vol. **96**, No. 2, p. K191—K194 (1986).

B. DIEGNER, H. WEINERT, R. PICKENHAIN, and W. HÖRIG
*An Experimental Study on Deep Level Incorporation and Background Doping in Hydride VPE GaAs
(73.60; 68.55; S7.12)
(a) Vol. **97**, No. 1, p. 313—321 (1986).

P. BONASEWICZ, W. HIRSCHWALD, and G. NEUMANN
*The Investigation of the Pressure and Temperature Dependence of the Electrical Conductivity of Thin Zinc Oxide Films with High Resistances
(73.60; S10.1)
(a) Vol. **97**, No. 2, p. 593—599 (1986).

J. MENDOLIA and D. LEMOINE
*Electrical and Optical Properties of Evaporated Thin Films of a-CdTe
(73.60; 61.40; 72.40; S8.13)
(a) Vol. **97**, No. 2, p. 601—607 (1986).

G. KAMPFRATH, D. DUSCHL, and C. HAMANN
*Doped Polymer Films Prepared by Glow Discharge Polymerization
(61.40 and 73.60; 61.65; S12)
(a) Vol. **97**, No. 2, p. 619—626 (1986).

T. CHERNOBELSKAYA, E. HARNIK, and S. KOVNOVICH
*Temperature Dependence of Charge-Carrier Mobility in Discontinuous Gold Films
(73.60; S1.3)
(a) Vol. **97**, No. 2, p. K187—K191 (1986).

A. URBANIAK-KUCHARCZYK
*The Magnetic Contribution to the Electrical Resistivity in Thin Films of Diluted Ferromagnetic Alloys
(73.60; 75.70)
(a) Vol. **97**, No. 2, p. K197—K200 (1986).

D. DĂDÂRLAT, M. CHIRTOC, and R. M. CÂNDEA
*A Simple Detection Method in Photothermal Deflection Measurements on Thin-Film Semiconductors
(73.60; S8)
(a) Vol. **98**, No. 1, p. 279—283 (1986).

J. SHIRAFUJI, S. NAGATA, and K. SHIRAKAWA
*γ-Irradiation Effect on Electronic Properties in Hydrogenated Amorphous Silicon
(61.40 and 73.60; 61.80; S5.11)
(a) Vol. **98**, No. 1, p. 311—318 (1986).

KH. A. MADY, M. M. EL-NAHAS, A. M. FARID, and H. S. SOLIMAN
*Electrical Resistivity of Amorphous Antimony Trisulphide Films
(73.60; S8)
(a) Vol. **98**, No. 1, p. K83—K86 (1986).

W. G. TEICH and G. MAHLER
Plasmon Excitations in Thin Metal Films. Transition from Three-Dimensional to Two-Dimensional Behaviour
(71.45; 73.60)
(b) Vol. **138**, No. 2, p. 607—620 (1986).

V. A. TEREKHOV, S. N. TROSTYANSKII, E. P. DOMASHEVSKAYA, O. A. GOLIKOVA, M. M. MEZDROGINA, K. L. SOROKINA, and M. M. KAZANIN
Density of States and Photoconductivity of Hydrogenated Amorphous Silicon
(71.20; 72.40; 73.60; 78.70; S5.11)
(b) Vol. **138**, No. 2, p. 647—653 (1986).

S. K. SHARMA and R. S. SRIVASTAVA
*Anomalous Photovoltaic Effect in CdTe Films Illuminated with Monochromatic Light
(72.40 and 73.60; S8.13)
(a) Vol. **98**, No. 2, p. 589—593 (1986).

V. K. GANDOTRA, K. V. FERDINAND, C. JAGADISH, A. KUMAR, and P. C. MATHUR
*Effect of Excess Copper on the Electrical Properties of Polycrystalline Thin Films of $CuInSe_2$
(73.60; S8.16)
(a) Vol. **98**, No. 2, p. 595—603 (1986).

L. GIER, A. SCHARMANN, and D. SCHALCH
*Hopping Conduction and Defect States in Reactively Sputtered Silicon Nitride Thin Films
(73.60; 71.20; S7)
(a) Vol. **98**, No. 2, p. 605—610 (1986).

C. A. HOGARTH and M. ZOR
*Observations of the Anomalous Poole-Frenkel Effect in Al–Polyethylene–Al Thin Film Structures
(73.60; 73.40; S12)
(a) Vol. **98**, No. 2, p. 611—616 (1986).

A. WILSON and R. A. COLLINS
*Gas–Surface Reactions on Phthalocyanine Thin Films
(68.45 and 73.60; 79.60; S12)
(a) Vol. **98**, No. 2, p. 633—644 (1986).

S. NAGATA, M. OGINO, S. TANIGUCHI, and A. NOZAKI
Indirect Observation of the Transition from an Amorphous to a Crystalline State in Pd Film Vapor-Quenched at 77 K
(68.55; 73.60; S1)
(a) Vol. **98**, No. 2, p. K105—K108 (1986).

73.90. Other topics in electrical properties of surfaces, interfaces, and thin films

Original Paper

P. Kleinert
*A Quantum Field Theoretical View on the Surface Plasmon
(71.45 and 73.90)
(b) Vol. **138**, No. 2, p. 599—605 (1986).

74. Superconductivity

74.10. Occurrence, critical temperature

Original Papers

S. L. KAKANI und U. N. UPADHYAYA
*Frequency Dependent Susceptibility of Rare Earth Ternary Superconductors
(74.10 and 75.30; S1.4)
(b) Vol. **136**, No. 1, p. 115—121 (1986).

D. SCHMELTZER
*Coexistence of Superconductivity and Sinusoidal (Helical) Magnetic Order
(74.10 and 75.30)
(b) Vol. **136**, No. 1, p. 349—364 (1986).

T. K. KOPEĆ and J. KLAMUT
*Effect of the Temperature and Concentration of Magnetic Atoms on Phase Diagrams and Order Parameters in Ferromagnetic Superconductors
(74.10 and 75.30; S1.4)
(b) Vol. **137**, No. 1, p. 73—79 (1986).

74.20. Theory

Original Papers and Short Notes

R. SHARMA, K. S. SHARMA, and L. DASS
*Linearised Screened Pseudopotential and Superconducting State Parameters of a Number of Metals
(74.20; 71.10; S1.2; S2; S3; S4)
(b) Vol. **133**, No. 2, p. 701—706 (1986).

T. KOSTYRKO
*Collective Excitations in the Hubbard Model with Intrasite Attraction
(74.20)
(b) Vol. **135**, No. 1, p. K79—K84 (1986).

I. BOSE, A. R. JANA, and S. CHATTERJEE
*A BCS Theory of Superconductivity in Heavy Fermion Systems
(74.20)
(b) Vol. **136**, No. 1, p. 387—392 (1986).

YAN-MIN LI and LI-YUAM ZHANG
*Derivation of the Ginzburg-Landau Equation in the Triplet Local-Electron-Pair System
(74.20)
(b) Vol. **138**, No. 2, p. 709—715 (1986).

74.50. Tunneling phenomena, Josephson effect, and proximity effects

Short Notes

P. Kuivalainen
*Josephson Superlattice
(74.50)
(b) Vol. **136**, No. 2, p. K155—K158 (1986).

P. Seidel, H.-J. Köhler, F. Müller, U. Moldenhauer, and U.-J. Tietz
*The Pseudo Inverse AC Josephson Effect
(74.50; S1)
(a) Vol. **96**, No. 2, p. K203—K205 (1986).

F. Schmidl, M. Siegel, P. Seidel, and H.-J. Köhler
*Some Aspects of the AC Josephson Effect at Nb/Amorphous Si/Nb Structures
(74.50; S1; S5.11)
(a) Vol. **97**, No. 1, p. K87—K89 (1986).

74.55. Type-I superconductivity

Short Notes

A. Sikora and B. Makiej
*Enhancement of Critical Currents in Superconducting Cylindrical Samples by Circular Magnetization
(74.55 and 74.60; S4)
(a) Vol. **96**, No. 1, p. K69—K72 (1986).

74.60. Type-II superconductivity

Original Papers and Short Notes

L. N. SHEHATA and T. F. REFAI
*Curved Flux Line Effects in a Type-II Superconducting Hollow Cylinder
(74.60)
(b) Vol. **133**, No. 2, p. 707—712 (1986).

P. WRÓBEL and P. TEKIEL
*Longitudinal Electric Current Distribution in the Mixed State of Type-II Superconductors
(74.60)
(b) Vol. **134**, No. 2, p. K139—K143 (1986).

L. N. SHEHATA and A. G. SAIF
*Flux Spiral Instability in a Transport Current Carrying Type-II Superconductor
(74.60)
(b) Vol. **135**, No. 1, p. 399—407 (1986).

W. ZACHARKO and R. KUBIAK
*Anisotropy Properties of H_{c2} in a Superconductor with Uniaxial Symmetry — In_3Sn
(74.60)
(b) Vol. **135**, No. 2, p. K159—K162 (1986).

A. SIKORA and B. MAKIEJ
*Enhancement of Critical Currents in Superconducting Cylindrical Samples by Circular Magnetization
(74.55 and 74.60; S4)
(a) Vol. **96**, No. 1, p. K69—K72 (1986).

B. CENDLEWSKA, J. SZYMASZEK, and J. J. WNUK
*Upper Critical Field Deduced from Critical Current Data
(74.60; S8.16)
(a) Vol. **98**, No. 1, p. 305—309 (1986).

A. G. SAIF
*Structure of the Surface Barrier in a Critical State of an Edged Type-II Superconductor
(74.60)
(a) Vol. **98**, No. 2, p. 577—588 (1986).

74.70. Superconducting materials

Original Papers and Short Notes

A. ROJEK and C. SUŁKOWSKI
*Electric, Magnetic, and Superconducting Properties of Type-II Stannides $ErRh_{1.1}Sn_{3.6}$ and $LuRh_{1.2}Sn_{4.0}$
(72.15 and 74.70; 75.50; S1.2; S1.4)
(b) Vol. **134**, No. 2, p. K133—K137 (1986).

WANG RONG-YAO
*A Further Study of Superconducting Critical Temperature of Ll_2 Type Superconductors
(61.55 and 74.70; S1.2; S1.3; S1.4; S4)
(a) Vol. **94**, No. 2, p. 445—452 (1986).

C. SUŁKOWSKI, Z. WOKULSKI, and K. WOKULSKA
*Superconducting Properties of TiN Whiskers
(74.70; S1.62)
(a) Vol. **95**, No. 1, p. K67—K70 (1986).

WANG RONG-YAO and LUO QI-GUANG
Structure of A15 Compounds
(61.55; 74.70; S1; S4)
(a) Vol. **96**, No. 2, p. 397—406 (1986).

S. MENDELSON
*Mechanisms for Martensic Transformations in A15 Superconductors
(64.60 and 74.70)
(a) Vol. **98**, No. 1, p. 133—144 (1986).

75. Magnetic properties and materials

75.10. General theory and models of magnetic ordering
(see also 71.25. Nonlocalized single-particle electronic states, 71.70. Level splitting and interactions)

Original Papers and Short Notes

A. M. GABOVICH and A. I. VOITENKO
*Electron Potential Energy near the Ferromagnetic Metal–Vacuum Interface
(73.40 and 75.10)
(b) Vol. **133**, No. 1, p. 135—142 (1986).

JIAN ZHONG, KAIYI ZHANG, K. A. CHAO, R. MICNAS, and S. ROBASZKIEWICZ
*CDW–SDW Transition in the Extended Hubbard Model
(75.10)
(b) Vol. **133**, No. 1, p. 143—150 (1986).

J. ŠLECHTA
*On SCCF Theory of Magnetic Properties of 1-D Binary Alloys (Magnon Spectrum, Magnetization)
(75.10)
(b) Vol. **133**, No. 1, p. 151—156 (1986).

G. FISCHER and A. HERR
*Detailed Low-Field Magnetic Properties of Rare-Earth Ion States in Cubic Crystal Fields
(75.10; S1.4)
(b) Vol. **133**, No. 1, p. 157—166 (1986).

A. S. T. PIRES and S. L. TALIM
*Soliton Excitations in a One-Dimensional Antiferromagnet with Two Anisotropies
(75.10)
(b) Vol. **133**, No. 1, p. K5—K9 (1986).

ZHENG SIMING
*The Combined Effect of Crystal-Field Anisotropy and Biquadratic Exchange in a Heisenberg Ferromagnet
(75.10)
(b) Vol. **133**, No. 1, p. K11—K16 (1986).

M. A. YURISHCHEV and A. M. STERLIN
*Three-Chain Ising Strip in a Nonuniform Field
(75.10; S12.1)
(b) Vol. **133**, No. 2, p. 539—546 (1986).

A. V. DE CARVALHO and N. P. SILVA
*Critical Temperature of the Two-Dimensional Ising Model with Alternated Interactions along One Direction
(75.10)
(b) Vol. **133**, No. 2, p. K95—K98 (1986).

D. Hajduković and Lj. Maurer
*About the Critical Condition of the Ising and Potts Model on the Pentagon Lattice
(75.10)
(b) Vol. **133**, No. 2, p. K99—K101 (1986).

V. Janiš
*A Self-Consistent Approximation for the Random-Field Ising Model. Phase Diagram in $d \geqq 3$
(75.10)
(b) Vol. **134**, No. 1, p. 141—146 (1986).

E. Kolley and W. Kolley
*Fokker-Planck Equation in the Spin-Glass Dynamics
(75.10)
(b) Vol. **134**, No. 1, p. 147—152 (1986).

A. Freudenhammer
*Time-Dependent Solitary Solutions of the Quantum-Mechanical Anisotropic Heisenberg Chain
(75.10)
(b) Vol. **134**, No. 1, p. 153—163 (1986).

S. L. Talim and A. S. T. Pires
*Discreteness Effects in a Classical One-Dimensional Antiferromagnet
(75.10)
(b) Vol. **134**, No. 1, p. 165—173 (1986).

A. Zagórski, T. Rżysko, and J. Żebrowski
*Multi-Spin Correlations and Magnetic Susceptibilities of Finite Frustrated Ising Chains
(75.10; 75.30)
(b) Vol. **134**, No. 1, p. K35—K38 (1986).

J. M. Wesselinowa
*The s–d Model Applied to $CdCr_2Se_4$ and $HgCr_2Se_4$
(75.10 and 75.50; S8.16)
(b) Vol. **134**, No. 1, p. K39—K42 (1986).

Bach Thanh Cong
*The Functional Integral Method in the Theory of Ferromagnetism for a Magnetic System Containing Itinerant Electrons Interacting with Localized Spins
(75.10)
(b) Vol. **134**, No. 2, p. 569—579 (1986).

K. Świderczak, A. Zagórski, and J. Król
*Amorphous Ferromagnets with Two Magnetic Components. A BPW· Approximation
(75.10 and 75.50)
(b) Vol. **134**, No. 2, p. 581—585 (1986).

T. Iwashita and N. Uryû
*The Curie Temperature of an Anisotropic Heisenberg Ferromagnet with Mixed Spins
(75.10 and 75.50)
(b) Vol. **134**, No. 2, p. 587—595 (1986).

B. A. Shiwai and M. G. Cottam
*Green Function Theory of a Semi-Infinite Ising Ferromagnet in a Transverse Field
(75.10 and 75.30)
(b) Vol. **134**, No. 2, p. 597—604 (1986).

T. Iwashita and N. Uryû
*Spin-One Ising Ferromagnet with Biquadratic Exchange and Uniaxial Anisotropy
(75.10)
(b) Vol. **134**, No. 2, p. K111—K114 (1986).

M. Matlak and K. Skrobiś
*Curie Temperature of a Doped Ferromagnetic Semiconductor
(75.10 and 75.50; S8; S10.1)
(b) Vol. **135**, No. 1, p. 155—161 (1986).

A. Bobák and M. Jaščur
*A New Type of Effective Field Theory for Ising Model with Spin —1/2
(75.10)
(b) Vol. **135**, No. 1, p. K9—K12 (1986).

L. A. B. Bernardes, O. N. Borges, and L. A. Amarante Ribeiro
*Spectral Density Method Applied to the Study of the Energy Spectrum of the XY Model in a Transverse Field
(75.10)
(b) Vol. **135**, No. 2, p. 581—586 (1986).

L. Palermo and X. A. da Silva
*Magnetic Order of Singlet–Singlet Ions Coupled to Conduction Electrons. Band Structure Effects
(75.10; 71.25; S1.4)
(b) Vol. **136**, No. 1, p. 97—102 (1986).

G. Fischer and A. Herr
*Magnetic Moments and Initial Susceptibilities of f-Electron States Experiencing a Crystal Field of Cubic Symmetry
(75.10; S1.4)
(b) Vol. **136**, No. 1, p. 103—114 (1986).

W. Borgiel, W. Nolting, and G. Borstel
*Self-Consistent CPA Treatment of the Conduction Electron Spin Polarization of Degenerate Ferromagnetic Semiconductors
(75.10; 71.25; S1.4; S8)
(b) Vol. **136**, No. 1, p. 131—144 (1986).

L. M. Sandratskii
*Energy Band Structure Calculations for Crystals with Spiral Magnetic Structure
(71.25 and 75.10)
(b) Vol. **136**, No. 1, p. 167—180 (1986).

M. E. Foglio and P. Schlottmann
Intermediate Valence Model of the Hyperfine Field and Relaxation Rate of Cerium in Gadolinium Metal
(71.70; 75.10; 76.20; S1.4)
(b) Vol. **136**, No. 1, p. 273—280 (1986).

E. Piotrowski and J. Łuczka
*Simple Derivation of the Direct Spin–Phonon Interaction
(75.10; 71.38)
(b) Vol. **136**, No. 1, p. K27—K31 (1986).

V. Urumov
*Non-Universal Behaviour on Union Jack Lattice with Three-Spin Interactions
(75.10)
(b) Vol. **136**, No. 1, p. K33—K36 (1986).

P. Wosicki
*On the Functional Integral Method to the Hubbard Model. Application to Nickel
(75.10; S1.2)
(b) Vol. **136**, No. 2, p. 567—574 (1986).

C. Z. Yang and Z. Y. Li
*Surface Magnetic Properties of the Ising Ferromagnet with a Site-Bond Diluted Anisotropic Free Surface
(75.10 and 75.30)
(b) Vol. **136**, No. 2, p. 575—583 (1986).

Y. Chen and B. Gil
*Simple Description of Hydrogen-Like Systems in Arbitrary Magnetic Fields
(71.10 and 75.10)
(b) Vol. **136**, No. 2, p. 629—635 (1986).

T. Iwashita and N. Uryû
*Ising Ferromagnet with Biquadratic Exchange Interaction and Uniaxial Anisotropy
(75.10; 75.30)
(b) Vol. **137**, No. 1, p. 65—72 (1986).

A. Kühnel
*On the Diagram Technique for Green Functions of the Heisenberg Ferromagnet
(75.10)
(b) Vol. **137**, No. 1, p. 81—89 (1986).

H. A. Brown
*The Three-Spin Box Product Interaction
(65.50 and 75.10)
(b) Vol. **137**, No. 1, p. 91—95 (1986).

H. A. Brown
*The Spin-1 Anisotropic Heisenberg Ferromagnet with Biquadratic Exchange
(75.10)
(b) Vol. **137**, No. 1, p. K21—K23 (1986).

L. R. Evangelista and V. K. Saxena
*Critical Properties of Bond- and Site-Diluted Triangular Lattice Ising Model
(75.10)
(b) Vol. **137**, No. 1, p. K31—K36 (1986).

S. Tabor and R. Micnas
*Dynamic Properties of the Heisenberg Model with Crystal Field Anisotropy: $T = \infty$ Limit
(75.10)
(b) Vol. **137**, No. 1, p. K49—K54 (1986).

S. Ichinose
*Bound Domain Wall Pairs in Wolffram's Red Salt
(75.10 and 75.60; S12.1)
(b) Vol. **137**, No. 2, p. 479—485 (1986).

P. R. Silva
*Internal Energy Expansion for the Transverse Ising Model in a Random Parallel Field at Zero Temperature
(75.10)
(b) Vol. **137**, No. 2, p. K137—K141 (1986).

V. Damjanović
*On the Critical Condition of the Ising Model on the Hypercubic Lattice
(75.10)
(b) Vol. **137**, No. 2, p. K143—K145 (1986).

O. V. Gritsenko and G. M. Zhidomirov
*A New Correlation Energy Interpolation Functional for Many-Electron Systems in the Homogeneous Electron Gas Model. The Ratio of Paramagnetic and Ferromagnetic State Energies versus Gas Density
(75.10; 75.20; 75.50; S2)
(b) Vol. **138**, No. 1, p. 161—169 (1986).

G. Wiatrowski, T. Balcerzak, L. Wojtczak, and J. Mielnicki
*The Application of Differential Operator Method to the Calculation of Magnetization in Thin Ferromagnetic Films
(75.10 and 75.70)
(b) Vol. **138**, No. 1, p. 189—195 (1986).

V. Damjanović and I. M. Savić
*On the Ground-State Properties of the Ising Model with General Spin S
(75.10)
(b) Vol. **138**, No. 1, p. K21—K24 (1986).

P. Pawlicki
*Magnetoelastic Model of the Temperature-Induced Metamagnetism for Lattices with Axial Symmetry
(75.10)
(b) Vol. **138**, No. 2, p. 531—537 (1986).

V. JANIŠ
*The Effect of Higher-Order Cumulants in the Random-Field Ising Model
(75.10)
(b) Vol. **138**, No. 2, p. 539—545 (1986).

I. M. SAVIĆ and D. HAJDUKOVIĆ
*An Approximation of the Ising Antiferromagnets in $d \geqq 2$ Dimensions with One-Dimensional Many-Neighboured Ising Antiferromagnet
(75.10)
(b) Vol. **138**, No. 2, p. K101—K105 (1986).

ZHEN-YA LI, CHUAN-ZHANG YANG, and JIA-LIN ZHONG
*Surface Phase Diagram of a Layered Antiferromagnetic Ising System with a Site-Diluted Free Surface
(75.10 and 75.40)
(b) Vol. **138**, No. 2, p. K107—K112 (1986).

75.20. Diamagnetism and paramagnetism

Original Papers and Short Notes

C. K. JAYASANKAR, E. ANTIC-FIDANCEV, M. LEMAITRE BLAISE, and P. PORCHER
Optical Properties of EuF_3 and GdF_3 and Paramagnetic Susceptibility of EuF_3
(71.70; 75.20; 78.55; S9)
(b) Vol. **133**, No. 1, p. 345—353 (1986).

W. BAŻELA, A. SZYTUŁA, and J. LECIEJEWICZ
*Magnetic Properties of ZrMnSi, ZrMnGe, and HfMnSi Intermetallics
(75.20; 61.55; S1.61)
(a) Vol. **94**, No. 1, p. 207—211 (1986).

K. P. GHATAK and M. MONDAL
Theoretical Analysis of the Magnetic Susceptibility of the Electrons in Ultrathin Films of Bi under Magnetic Quantization
(75.70; 75.20; S4)
(b) Vol. **135**, No. 2, p. K119—K124 (1986).

J. PABST and H. RÖLL
*Induced Magnetic Form Factors for 3d Transition Metals and 3d–3d Alloys. III. Diamagnetic Form Factors
(71.25 and 75.20; S1.1; S1.2)
(b) Vol. **136**, No. 2, p. 637—642 (1986).

J. ULNER
*Field-Induced Symmetry Change in the Free Energy of a Deformed Paramagnetic Tetragonal System
(75.20)
(b) Vol. **137**, No. 1, p. K25—K30 (1986).

I. OKAMOTO, T. MIYAZAKI, and M. TAKAHASHI
Spin-Glass in Rapidly Quenched $Cu_{100-x}Fe_x$ ($0 \leqq x \leqq 9.0$) Alloys
(75.50; 75.20; S1.1; S1.2)
(a) Vol. **97**, No. 1, p. 223—230 (1986).

M. KRAUSE, C. MICHALK, B. LIPPOLD, M. BENEDIX, and J. SUWALSKI
*Slow Paramagnetic Relaxation in Oxalate (2,2'-Bipyridine) Iron
(75.20; 61.40; 76.80; S1.1)
(a) Vol. **97**, No. 2, p. 521—526 (1986).

O. V. GRITSENKO and G. M. ZHIDOMIROV
A New Correlation Energy Interpolation Functional for Many-Electron Systems in the Homogeneous Electron Gas Model. The Ratio of Paramagnetic and Ferromagnetic State Energies versus Gas Density
(75.10; 75.20; 75.50; S2)
(b) Vol. **138**, No. 1, p. 161—169 (1986).

A. G. RUSTAMOV, R. Z. SADYKHOV, L. M. VALIEV, and A. O. ISMAILOV
*Paramagnetic Susceptibility of the $Me_{1-x}Fe_xCr_2S_4$ (Me = Zn, Co) Composition System
(75.20; S1.1; S1.2)
(a) Vol. **98**, No. 2, p. K153—K156 (1986).

75.25. Spin arrangements in magnetically ordered materials (neutron studies, etc.)

Original Papers and Short Notes

S. V. Vlasov and O. V. Farberovich
APW Research of Europium Monochalcogenides. II. Magnetic Form Factor and Insulator–Metal Transition in EuO under Pressure
(71.10; 71.25; 71.30; 75.25; S10.1)
(b) Vol. **133**, No. 2, p. 601—609 (1986).

H. Röll and J. Pabst
*Induced Magnetic Form Factors for 3d Transition Metals and 3d–3d Alloys. I. Results of Band Structure Calculations
(71.25 and 75.25; S1.1; S1.2)
(b) Vol. **135**, No. 2, p. 691—695 (1986).

J. Pabst and H. Röll
*Induced Magnetic Form Factors for 3d Transition Metals and 3d–3d Alloys. II. Induced Spin Form Factors
(71.25 and 75.25; S1.1; S1.2)
(b) Vol. **136**, No. 1, p. 225—231 (1986).

V. V. Lomonosov, V. D. Sedykh, and T. Zemčík
*Thermally Activated Time Fluctuations of Nuclear Spin Orientation in Amorphous $Fe_{81}B_{19}$ Alloy
(75.25; 75.50; 76.80; S1.1)
(a) Vol. **96**, No. 1, p. 271—280 (1986).

Th. Grieb, K. Wacker, E. Naumann, W. Treutmann, G. Heger, and R. Geick
*Magnetic Properties of $Rb_2Mn_xCr_{1-x}Cl_4$ Mixed Crystals
(75.25 and 75.60; 75.30; S9.16)
(b) Vol. **136**, No. 2, p. 585—591 (1986).

H. Grimm and J. Slak
*Observation of Gradual Localisation of Protons in the Mixed System $Rb_{1-x}(NH_4)_xH_2(P, As)O_4$ by Incoherent Neutron Scattering
(75.25; 77.80; S9.16)
(b) Vol. **137**, No. 1, p. K37—K41 (1986).

S. Ligenza
*On the Dynamical Nature of Superexchange Interaction in YIG Studied by Neutron Spectroscopy
(75.25; S11.2)
(b) Vol. **137**, No. 2, p. K131—K135 (1986).

S. K. Paranjpe, R. Chakravarthy, H. S. Bisht, and D. K. Kulkarni
*Neutron Diffraction Study of Some Mixed Li-Ferrites
(75.25; S11.2)
(a) Vol. **98**, No. 2, p. K157—K160 (1986).

75.30. Magnetically ordered materials, other intrinsic properties (for critical point effects, see 75.40)

Original Papers and Short Notes

H. Lass, J. Pebler, M. Pękala, and A. Polaczek
*Magnetic and Mössbauer Effect Study of the Cluster Distribution Function in the Spin-Glass Phase θ-$Me_{0.35}V_2O_5$ (Me = Fe, Al)
(75.30; 76.80; S10)
(a) Vol. **93**, No. 1, p. 233—240 (1986).

J. Pozo, R. M. Chorbadjian, and R. Ferrer
*Metamagnetic Transitions in a Random Anisotropy Model with Mixed Temperature Dependent Exchange Coupling
(75.30)
(b) Vol. **133**, No. 2, p. K103—K106 (1986).

A. A. Aly, I. H. Ibrahim, A. M. Abdeen, and M. M. Abdel-Gawad
*Variances in Molecular Magnetic Susceptibilities of Monoclinic Crystals
(75.30; S12)
(b) Vol. **134**, No. 1, p. 119—124 (1986).

A. Zagórski, T. Rżysko, and J. Żebrowski
Multi-Spin Correlations and Magnetic Susceptibilities of Finite Frustrated Ising Chains
(75.10; 75.30)
(b) Vol. **134**, No. 1, p. K35—K38 (1986).

B. A. Shiwai and M. G. Cottam
*Green Function Theory of a Semi-Infinite Ising Ferromagnet in a Transverse Field
(75.10 and 75.30)
(b) Vol. **134**, No. 2, p. 597—604 (1986).

J. Mycielski and A. M. Witowski
Effect of Chemical and Spin Disorder on Shallow Impurity States in Semiconductor Alloys
(71.55; 71.70; 75.30; S8.15)
(b) Vol. **134**, No. 2, p. 675—689 (1986).

K. M. Sultanov, Sh. A. Kuliev, and N. G. Abdullaev
*Coupled Waves in Conducting Spin Glasses
(71.45 and 75.30; 76.50)
(b) Vol. **135**, No. 1, p. 163—171 (1986).

A. I. Mitsek, K. Yu. Guslienko, and S. V. Pavlovskii
*Electron Structure and Phase Diagrams of Low-Temperature Rare-Earth Magnets
(75.30; S1.4)
(b) Vol. **135**, No. 1, p. 173—188 (1986).

M. I. Belinskii, V. Ya. Gamurar, and B. S. Tsukerblat
*Heisenberg Type and Double Exchange in Mixed-Valence Clusters. I. The Microscopic Theory of Transition Metal Dimers
(71.70 and 75.30)
(b) Vol. **135**, No. 1, p. 189—199 (1986).

A. V. Deryabin, A. V. Tkov, V. K. Kazantsev, Yu. A. Chirkov, and I. V. Zakharov
*Concentration Magnetic Transitions in Alloys with Mixed Exchange Interaction
(75.30; 65.40; 72.15; S1.1)
(a) Vol. **95**, No. 1, p. 229—237 (1986).

M. I. Belinskii, V. Ya. Gamurar, and B. S. Tsukerblat
*Heisenberg-Type and Double Exchange in Mixed-Valence Clusters. II. Electron–Vibrational Interactions
(71.70 and 75.30; 71.38)
(b) Vol. **135**, No. 2, p. 555—569 (1986).

C. E. Paraskevaidis and C. T. Papatriantafillou
*Bethe-Peierls Study of Magnetic Field Induced Phase Transitions
(75.30)
(b) Vol. **135**, No. 2, p. 571—579 (1986).

H. M. A. Schleijpen and F. A. P. Blom
Anisotropy in the Shubnikov-de Haas Effect in $Hg_{1-x}Mn_xSe$
(71.25 and 71.70; 72.20; 75.30; S8.15)
(b) Vol. **135**, No. 2, p. 605—614 (1986).

J. Fuentes and A. Rivero
*About the Ordering in Fe–Rh Alloys
(75.30; S1.1)
(b) Vol. **135**, No. 2, p. K125—K128 (1986).

D. Fraitová
An Analytical Theory of FMR in Bulk Metals (Parallel Configuration). IV. Exchange Dominated Surface Modes
(76.50; 75.30; S1.1)
(b) Vol. **136**, No. 1, p. 91—96 (1986).

S. L. Kakani and U. N. Upadhyaya
*Frequency Dependent Susceptibility of Rare Earth Ternary Superconductors
(74.10 and 75.30; S1.4)
(b) Vol. **136**, No. 1, p. 115—121 (1986).

P. V. Geld, V. V. Singer, I. Z. Radovskii, and S. V. Konovalov
Variable Valence of Ce and Pr in Liquid Alloys with Fe, Co, and Ni
(71.25; 61.25; 72.15; 75.30; S1.1; S1.2; S1.4)
(b) Vol. **136**, No. 1, p. 153—158 (1986).

D. Schmeltzer
*Coexistence of Superconductivity and Sinusoidal (Helical) Magnetic Order
(74.10 and 75.30)
(b) Vol. **136**, No. 1, p. 349—364 (1986).

A. V. Apostolov, M. T. Mikhov, T. L. Mydlarts, S. A. Nikitin, N. A. Scheludko, O. P. Tchistyakov, and O. P. Terekhova
*Magnetic Properties of Single Crystals of Highly Diluted Dysprosium–Yttrium Alloys
(75.30; S1.4)
(a) Vol. **96**, No. 1, p. 265—269 (1986).

C. Z. Yang and Z. Y. Li
*Surface Magnetic Properties of the Ising Ferromagnet with a Site-Bond Diluted Anisotropic Free Surface
(75.10 and 75.30)
(b) Vol. **136**, No. 2, p. 575—583 (1986).

Th. Grieb, K. Wacker, E. Naumann, W. Treutmann, G. Heger, and R. Geick
Magnetic Properties of $Rb_2Mn_xCr_{1-x}Cl_4$ Mixed Crystals
(75.25 and 75.60; 75.30; S9.16)
(b) Vol. **136**, No. 2, p. 585—591 (1986).

J. Abellán and M. Ortuño
*The Verwey Transition in Polycrystalline Cobalt–Iron Ferrites
(75.30; 75.50; S11.2)
(a) Vol. **96**, No. 2, p. 581—586 (1986).

A. Yazdani and H. Gamari-Seale
*The Effect of Disorder in the Magnetic Properties of the Pseudobinary Compound $Gd_{2-x}La_xAl$
(75.30; 75.50; S1.4)
(a) Vol. **96**, No. 2, p. 587—595 (1986).

T. Iwashita and N. Uryû
Ising Ferromagnet with Biquadratic Exchange Interaction and Uniaxial Anisotropy
(75.10; 75.30)
(b) Vol. **137**, No. 1, p. 65—72 (1986).

T. K. Kopeć and J. Klamut
*Effect of the Temperature and Concentration of Magnetic Atoms on Phase Diagrams and Order Parameters in Ferromagnetic Superconductors
(74.10 and 75.30; S1.4)
(b) Vol. **137**, No. 1, p. 73—79 (1986).

J. Barnás
*Non-Reciprocal Character of Wave Propagation in Magnetic Systems with Spontaneous Electric Polarization
(71.36 and 75.30; 78.20; S9.16)
(b) Vol. **137**, No. 1, p. K43—K47 (1986).

R. J. RADWAŃSKI
*The Intersublattice Molecular Fields in the Rare Earth–Cobalt Intermetallics
(75.30; S1.2; S1.4)
(b) Vol. **137**, No. 2, p. 487—493 (1986).

G. GANGADHAR REDDY and A. RAMAKANTH
*Collective Magnetic Order and Intermediate Valence in the Anderson Model
(71.25 and 75.30)
(b) Vol. **138**, No. 1, p. 171—179 (1986).

M. FÄHNLE and J. SOULETIE
*The Generalized Curie-Weiss Law
(75.30; 75.40; 75.50)
(b) Vol. **138**, No. 1, p. 181—188 (1986).

W. A. KACZMAREK and J. PIETRZAK
*Anomalous Temperature Dependence of Magnetocrystalline Anisotropy Constants of $BaIn_{1.5}Fe_{10.5}O_{19}$ Single Crystals
(75.30; 76.50; S11.2)
(a) Vol. **98**, No. 1, p. K43—K46 (1986).

A. V. ANDREEV, M. I. BARTASHEVICH, A. V. DERYAGIN, L. HAVELA, and V. SECHOVSKÝ
*Influence of Hydrogen on the Magnetic Properties of the U–Co System
(75.30; S1.2; S1.5)
(a) Vol. **98**, No. 1, p. K47—K51 (1986).

A. DUDA
Spin-Wave Spectrum in Uniaxial Ferromagnetic Thin Films with Nearest-Neighbour and Next-Nearest-Neighbour Exchange Couplings
(75.70; 75.30)
(b) Vol. **138**, No. 2, p. 509—519 (1986).

E. N. BULGAKOV and A. F. SADREEV
*Spin-Wave Damping in Uniaxial Ferromagnets at the Orientational Phase Transition
(75.30)
(b) Vol. **138**, No. 2, p. 521—530 (1986).

75.40. Critical-point effects, specific heats, short-range order (see also 65.40. Heat capacities)

Original Papers and Short Notes

W. Reich and H. Cofta
*Renormalization Group Free Energy for an Ising Ferromagnet on a Simple Cubic Lattice
(75.40)
(b) Vol. **134**, No. 1, p. K31—K34 (1986).

A. Czopnik, H. Mädge, and B. Staliński
*Specific Heat of $HoIn_3$ and $ErIn_3$ Compounds
(75.40; S1.4)
(a) Vol. **94**, No. 1, p. K13—K16 (1986).

A. A. El-Sharkawy, A. M. Abousehly, and E. S. M. Higgy
*Thermal Properties of Spinel Nickel–Zinc Ferrite in the Temperature Range 400 to 1000 K
(75.40; S11.2)
(a) Vol. **95**, No. 2, p. 605—611 (1986).

O. A. Nabutovskaya, E. V. Kholopov, and I. E. Paukov
*Nonpower-Law Behaviour in the Heat Capacity of CoO in the Vicinity of Strong Fluctuations
(75.40; S10.1)
(a) Vol. **96**, No. 1, p. K57—K60 (1986).

S. Kumar, P. K. Ahluwalia, and K. C. Sharma
Metallic Systems Containing Crystal-Field-Split Impurities. Thermodynamic Properties in the Presence of Magnetic Field
(65.50; 71.70; 75.40)
(b) Vol. **138**, No. 1, p. 117—121 (1986).

M. Fähnle and J. Souletie
The Generalized Curie-Weiss Law
(75.30; 75.40; 75.50)
(b) Vol. **138**, No. 1, p. 181—188 (1986).

W. Reichert and K. D. Usadel
*Renormalisation Group Approach for the Calculation of Thermodynamic Properties of Disordered Ising Systems
(75.40)
(b) Vol. **138**, No. 1, p. K25—K28 (1986).

Zhen-ya Li, Chuan-zhang Yang, and Jia-lin Zhong
*Surface Phase Diagram of a Layered Antiferromagnetic Ising System with a Site-Diluted Free Surface
(75.10 and 75.40)
(b) Vol. **138**, No. 2, p. K107—K112 (1986).

75.50. Studies of specific magnetic materials

Original Papers and Short Notes

G. C. Fletcher
The Band Structure of Co_3Ga and Co_2MnGa and the Electrical Conductivity of Related Alloys
(71.25 and 72.15; 75.50; S1.2)
(b) Vol. **133**, No. 1, p. 241—248 (1986).

J. Kadlecová, A. Handstein, T. Zemčík, and K. Závěta
*Electrical and Magnetic Properties of Annealed Amorphous $Fe_8Ni_{72}P_{10}B_{10}$ at Low Temperatures
(75.50; 72.15; 76.80; S1.1)
(a) Vol. **93**, No. 1, p. 213—220 (1986).

F. Rettenmeier and H. Kronmüller
*Reversible and Irreversible Structural Relaxations in Soft Ferromagnetic Amorphous Alloys. I. Magnetic After-Effects in $Co_{43}Fe_5Ni_{30}Si_{8.8}B_{13.2}$ Alloys
(75.50; S1.1)
(a) Vol. **93**, No. 1, p. 221—231 (1986).

B. Sidzhimov, N. Stanev, P. Tscholakov, and S. Neov
*Magnetic Properties of the Pseudobinary Intermetallic Compounds $Tb_xY_{1-x}Fe$ ($x = 0.25, 0.50, 0.75$)
(75.50; S1.1)
(a) Vol. **93**, No. 1, p. K37—K40 (1986).

G. Bottoni, D. Candolfo, A. Cecchetti, F. Masoli, E. Mello Ceresa, and A. Papoian
*Magnetic Parameters and Crystal Lattice Modifications of Doped CrO_2 Particles Subjected to Thermal Treatment
(61.70 and 75.50; 61.10; S10.1)
(a) Vol. **93**, No. 1, p. K41—K45 (1986).

L. Merten and R. Kotzott
Elliptically Polarized Magnon-Polaritons with Longitudinal and Transverse Components in Uniaxial Ferrites
(71.36; 75.50; 78.30)
(b) Vol. **133**, No. 2, p. 547—556 (1986).

Yu-Jin Chang and Xiang-Rong Qian
A High Resolution Microscopic Study on Nd–Fe–B Permanent Magnetic Alloys
(64.70 and 75.60; 61.16; 75.50; S1.1)
(a) Vol. **93**, No. 2, p. 573—583 (1986).

F. Rettenmeier, E. Kisdi-Koszo, and H. Kronmüller
*Reversible and Irreversible Structural Relaxations in Soft Ferromagnetic Amorphous Alloys. II. Magnetic After-Effects in $Fe_{100-x}B_x$
(75.50; S1.1)
(a) Vol. **93**, No. 2, p. 597—604 (1986).

K. Hisatake, T. Miyazaki, M. Takahashi, K. Maeda, T. Hattori, and I. Matsubara
*Anomalous Behavior of Permeability around the Curie Temperature in Amorphous Fe–Ni–Si–B Alloys
(75.50; S1.1)
(a) Vol. **93**, No. 2, p. 605—611 (1986).

A. Rojek, G. Sułkowski, A. Zygmunt, and G. Kozłowski
*Low-Temperature Properties of Antiferromagnetic $GdRh_xSn_y$ Compound
(75.50; 72.15; S1.4)
(b) Vol. **134**, No. 1, p. 125—129 (1986).

J. M. Wesselinowa
*The s–d Model Applied to $CdCr_2Se_4$ and $HgCr_2Se_4$
(75.10 and 75.50; S8.16)
(b) Vol. **134**, No. 1, p. K39—K42 (1986).

S. Manhas, A. Manoogian, G. Lamarche, and J. C. Woolley
*Lattice Parameters and Magnetic Properties of the $Cd_xZn_yMn_zSe$ Alloy System
(75.50; 61.55; 76.30; S1.63)
(a) Vol. **94**, No. 1, p. 213—221 (1986).

Ram Bilas and D. Bahadur
*Effect of Iron Substitution in Manganese Aluminosilicate Glass and Glass Ceramics
(75.50; 76.30; S10.15)
(a) Vol. **94**, No. 1, p. K41—K45 (1986).

K. Swiderczak, A. Zagórski, and J. Król
*Amorphous Ferromagnets with Two Magnetic Components. A BPW Approximation
(75.10 and 75.50)
(b) Vol. **134**, No. 2, p. 581—585 (1986).

T. Iwashita and N. Uryû
*The Curie Temperature of an Anisotropic Heisenberg Ferromagnet with Mixed Spins
(75.10 and 75.50)
(b) Vol. **134**, No. 2, p. 587—595 (1986).

A. Rojek and C. Sułkowski
Electric, Magnetic, and Superconducting Properties of Type-II Stannides $ErRh_{1.1}Sn_{3.6}$ and $LuRh_{1.2}Sn_{4.0}$
(72.15 and 74.70; 75.50; S1.2; S1.4)
(b) Vol. **134**, No. 2, p. K133—K137 (1986).

M. Hervieu, D. Groult, and B. Raveau
*HREM Study of Hexaferrite $BaFe_{12}O_{19}$ Samples Irradiated by 1.8 GeV Argon Ions
(61.55; 61.80; 75.50; S11.2)
(a) Vol. **94**, No. 2, p. 467—481 (1986).

W. MATZ and H. HERRMANN
*Non-Linear Dependence of Average Magnetic Moment on Boron Concentration in Amorphous Fe–B Alloys
(75.50; 76.80; S1.1)
(a) Vol. **94**, No. 2, p. 617—622 (1986).

R. Z. SADYKHOV, L. M. VALIEV, N. B. NABIEVA, and Z. M. NAMAZOV
*Magnetic Properties of $Fe_{1+x}Cr_{2-x}S_4$ ($0 < x \leqq 0.5$)
(75.50; S1.1; S1.2)
(a) Vol. **94**, No. 2, p. K103—K106 (1986).

M. MATLAK and K. SKROBIŚ
*Curie Temperature of a Doped Ferromagnetic Semiconductor
(75.10 and 75.50; S8; S10.1)
(b) Vol. **135**, No. 1, p. 155—161 (1986).

S. MOTOGI
*Polarization and Resonance Properties of Magnetoacoustic Waves in Ferromagnets with Dissipation
(75.50 and 75.80; 76.50; S1.1; S11.2)
(a) Vol. **95**, No. 1, p. 197—208 (1986).

I. ŠKORVÁNEK, A. ZENTKO, and T. ZEMČÍK
*Influence of the Surface Insulating Coating on the Magnetic Properties of $Fe_{82}B_{12}Si_6$ Amorphous Ribbons
(75.50; 76.80; S1.1)
(a) Vol. **95**, No. 1, p. 209—212 (1986).

K.-D. DURST, H. KRONMÜLLER, F. T. PARKER, and H. OESTERREICHER
*Temperature Dependence of Coercivity of Cellular Sm_2Co_{17}–$SmCo_5$ Permanent Magnets
(75.50; 75.60; S1.2; S4)
(a) Vol. **95**, No. 1, p. 213—219 (1986).

F. GOEDSCHE
*Electrical Resistivity and Thermoelectric Power of $Fe_{50}Co_{50}$
(72.15 and 75.50; S1.1; S1.2)
(b) Vol. **135**, No. 2, p. K153—K158 (1986).

A. AHARONI, J. B. DEPPE, and S. TYAGI
*Reciprocal Differential Susceptibility of a Ferromagnet
(75.50)
(a) Vol. **95**, No. 2, p. K127—K129 (1986).

Y. MIYAMOTO, T. YAMAMOTO, M. KOIZUMI, and M. SHIMADA
*Magnetic and Electrical Properties of Cubic Perovskite $Sr(Fe_{1-x}Ni_x)O_3$ ($0 \leqq x \leqq 0.4$) Synthesized under High Oxygen Pressure
(72.80 and 75.50; S10.15)
(a) Vol. **95**, No. 2, p. K187—K191 (1986).

M. Posselt, G. Diener, and M. Seidel
*Ideal Magnetization in Sintered Iron Materials
(75.50; 75.60; S1.1)
(b) Vol. **136**, No. 1, p. 123—130 (1986).

S. D. Stojanović, M. J. Škrinjar, and D. V. Kapor
*On Magnetic Solitons in Anharmonic Ferromagnets
(75.50)
(b) Vol. **136**, No. 1, p. K23—K26 (1986).

J. Kučera, B. Million, and J. Růžičková
Magnetic Anomalies of ^{59}Fe and ^{60}Co Diffusion in Iron and of ^{48}V Diffusion in Cobalt. Quasi-Chemical Model
(66.30; 75.50; S1.1; S1.2)
(a) Vol. **96**, No. 1, p. 177—184 (1986).

V. V. Lomonosov, V. D. Sedykh, and T. Zemčík
Thermally Activated Time Fluctuations of Nuclear Spin Orientation in Amorphous $Fe_{81}B_{19}$ Alloy
(75.25; 75.50; 76.80; S1.1)
(a) Vol. **96**, No. 1, p. 271—280 (1986).

B. S. Sidzhimov, N. V. Stanev, P. T. Tscholakov, and S. D. Neov
Structure and Magnetic Properties of the Pseudobinary Intermetallic Compounds $Tb(Fe_xCo_{1-x})_3$
(61.55; 75.50; S1)
(a) Vol. **96**, No. 1, p. K15—K19 (1986).

J. Abellán and M. Ortuño
The Verwey Transition in Polycrystalline Cobalt–Iron Ferrites
(75.30; 75.50; S11.2)
(a) Vol. **96**, No. 2, p. 581—586 (1986).

A. Yazdani and H. Gamari-Seale
The Effect of Disorder in the Magnetic Properties of the Pseudobinary Compound $Gd_{2-x}La_xAl$
(75.30; 75.50; S1.4)
(a) Vol. **96**, No. 2, p. 587—595 (1986).

J. D. Stephenson and B. Strocka
*White Synchrotron Radiation Topography of $Ba_{1.03}Fe_{11.97}O_{19}$ and $Ba_{1.06}Fe_{10.92}Mn_{1.02}O_{19}$ Single Crystals
(61.10 and 75.50; 61.70; S11.2)
(a) Vol. **96**, No. 2, p. 597—602 (1986).

K. Sumiyama, Y. Kawawake, and Y. Nakamura
*Amorphous Fe–Cu–Ag Alloys Produced by Vapor Quenching
(61.40 and 75.50; 76.80; S1.1; S1.2; S1.3)
(a) Vol. **96**, No. 2, p. K107—K110 (1986).

G. V. Loseva and L. I. Ryabinkina
*Electrical and Magnetic Properties of the Antiferromagnetic Semiconductor α-Mn_xS
(72.60 and 75.50; S8)
(a) Vol. **96**, No. 2, p. K195—K197 (1986).

G. L. WHITTLE, A. M. STEWART, and A. B. KAISER
*Magnetic and Electrical Properties of $(FeTM)_{85}B_{15}$ (TM = Cr, Ni) Metallic Glasses
(75.50; 72.15; 76.80; S1.1)
(a) Vol. **97**, No. 1, p. 199—211 (1986).

I. OKAMOTO, T. MIYAZAKI, and M. TAKAHASHI
*Spin-Glass in Rapidly Quenched $Cu_{100-x}Fe_x$ ($0 \leqq x \leqq 9.0$) Alloys
(75.50; 75.20; S1.1; S1.2)
(a) Vol. **97**, No. 1, p. 223—230 (1986).

M. SHIOJIRI, T. MAEDA, T. ISSHIKI, Y. HIROTA, S. SEKIMOTO, Y. IKEDA, and K. IWAUCHI
*Crystal Structures and Magnetic Properties of $BaTi_{2-x}Sn_xFe_4O_{11}$
(61.60 and 75.50; S11)
(a) Vol. **97**, No. 1, p. 231—241 (1986).

R. MOSKALEWICZ and W. ZYCH
*Application of the DTMG-DTA(M) Technique for the Curie Point and Crystallisation Temperature Determination of $Fe_{80-x}Mn_xB_{20}$ Amorphous Alloys
(75.50; S1.1)
(a) Vol. **97**, No. 1, p. K43—K48 (1986).

M. K. BOROMBAEV, R. Z. LEVITIN, A. S. MARKOSYAN, Z. SMETANA, V. V. SNEGIREV, and P. SVOBODA
*Magnetic and Crystallographic Properties of $Gd(Cu_{1-x}Ni_x)_2$ and $Gd(Cu_{1-x}Al_x)_2$ Intermetallic Compounds
(61.60 and 75.50; S1.2; S1.4)
(a) Vol. **97**, No. 2, p. 501—509 (1986).

A. COLLOMB, O. ABDELKADER, A. ISALGUÉ et J. P. MIGNOT
*Propriétés magnétiques des ferrites hexagonaux: $BaMg_2$–W et $BaCo_2$–W
(Magnetic Properties of Hexagonal Ferrites: $BaMg_2$–W and $BaCo_2$–W)
(75.50; S11.2)
(a) Vol. **97**, No. 2, p. 511—519 (1986).

G. A. GOVOR
*The Effect of Titanium Additions on Phase Transitions in Manganese Arsenide Single Crystals
(75.50; S1.62)
(a) Vol. **97**, No. 2, p. K145—K148 (1986).

T. KANOMATA, K. SHIRAKAWA, and T. KANEKO
*Effect of Hydrostatic Pressure on the Curie Temperature of the Hexagonal Compound PtMnGa
(75.50; S1.2)
(a) Vol. **97**, No. 2, p. K149—K152 (1986).

Z. KĄKOL and H. FIGIEL
*Co Contribution to Anisotropy of R–Co Type Pseudobinaries
(75.50; S1.2)
(b) Vol. **138**, No. 1, p. 151—160 (1986).

O. V. Gritsenko and G. M. Zhidomirov
A New Correlation Energy Interpolation Functional for Many-Electron Systems in the Homogeneous Electron Gas Model. The Ratio of Paramagnetic and Ferromagnetic State Energies versus Gas Density
(75.10; 75.20; 75.50; S2)
(b) Vol. **138**, No. 1, p. 161—169 (1986).

M. Fähnle and J. Souletie
The Generalized Curie-Weiss Law
(75.30; 75.40; 75.50)
(b) Vol. **138**, No. 1, p. 181—188 (1986).

M. K. Borombaev, R. Z. Levitin, A. S. Markosyan, Z. Smetana, P. Svoboda, and V. Šíma
*Magnetic and Crystallographic Properties of $Gd(Cu_{1-x}Co_x)_2$
(61.60 and 75.50; S1.2; S1.4)
(a) Vol. **98**, No. 1, p. 221—227 (1986).

K. Sumiyama, N. Ohshima, and Y. Nakamura
*Magnetic Properties of Metastable α-Mn-Type $Mn_{1-y}Fe_y$ Alloys Produced by Vapor Quenching
(75.50; 61.10; 76.80; S1.1)
(a) Vol. **98**, No. 1, p. 229—238 (1986).

Y. D. Yao, T. Y. Yen, and J. Y. Chen
*Electrical Resistivity and Magnetization Studies of Some Commercial Steels
(72.15 and 75.50; S1.1)
(a) Vol. **98**, No. 1, p. 285—289 (1986).

H. Gruber, E. Krautz, H. P. Fritzer, K. Gatterer, and A. Popitsch
Changes of Electrical Conductivity, Magnetic Susceptibility, and IR Spectra in the Ternary System $Mo_{n-x}W_xO_{3n-1}$
(72.80; 75.50; 78.30; S10.15)
(a) Vol. **98**, No. 1, p. 297—304 (1986).

S. Gharbage, J.-C. Bissey, and Y. Servant
Electron Paramagnetic Resonance (EPR) and Related Magneto-Microwave Birefringence (MMB) Powder Spectra of Copper Nitrate $Cu(NO_3)_2 \cdot 2.5\ H_2O$
(76.30; 75.50; 76.90; S11.1)
(b) Vol. **138**, No. 2, p. 655—659 (1986).

M. Tarabieh, M. Ahmad, and A. Zihlif
*Initial Magnetic Susceptibility and Electrical Resistivity of $Fe_{92}Zr_8$ Amorphous Alloy
(72.15 and 75.50; S1.1)
(a) Vol. **98**, No. 2, p. 545—550 (1986).

P. Simonnin, R. Tourbot, B. Boucher, M. Perrin, and J. Vanhaute
*Thermal, Electrical, and Magnetic Properties of the Amorphous Alloys RE_xSi_{1-x} (RE = Gd or Tb; $x = 0.87, 0.59, 0.18$)
(75.50; 61.40; 64.75; 72.60; S1.4; S1.61)
(a) Vol. **98**, No. 2, p. 551—566 (1986).

V. I. DANILOV, V. M. DROBIN, and V. N. TROFIMOV
*Ferromagnetic Behaviour of Non-Magnetic Samples
(75.50; S1.2; S4; S5; S12)
(a) Vol. **98**, No. 2, p. K137—K140 (1986).

75.60. Domain effects, magnetization curves, and hysteresis

Original Papers and Short Notes

R. A. Kosiński and J. Strzeszewski
*Magnetostatic Interactions of Moving Vertical Bloch Lines in a Domain Wall
(75.60)
(b) Vol. **133**, No. 2, p. 533—538 (1986).

A. N. Voloshinskii, N. V. Ryzhanova, L. Yu. Vishnekov, and A. G. Obukhov
Off-Diagonal Conductivity and Magnetization of PdFe, PdCo, and PdNi Alloys in Coherent Potential Approximation Calculated on the Basis of Realistic Models of Component Densities of States
(72.15; 71.20; 75.60; S1.1; S1.2)
(b) Vol. **133**, No. 2, p. 735—743 (1986).

Yu-Jin Chang and Xiang-Rong Qian
*A High Resolution Microscopic Study on Nd–Fe–B Permanent Magnetic Alloys
(64.70 and 75.60; 61.16; 75.50; S1.1)
(a) Vol. **93**, No. 2, p. 573—583 (1986).

D. K. Fowler, D. C. Creagh, and G. V. H. Wilson
NMR Free Induction Decay Measurements in Iron
(76.60; 75.60; S1.1)
(a) Vol. **94**, No. 2, p. K97—K101 (1986).

M. C. Desjardins-Deruelle, J. P. Lascaray, D. Coquillat, and R. Triboulet
Magneto-Optical Measurements in $Zn_{1-x}Mn_xTe$ with High Manganese Concentration
(71.35 and 71.70; 75.60; 78.20; S8.15)
(b) Vol. **135**, No. 1, p. 227—232 (1986).

K.-D. Durst, H. Kronmüller, F. T. Parker, and H. Oesterreicher
Temperature Dependence of Coercivity of Cellular Sm_2Co_{17}–$SmCo_5$ Permanent Magnets
(75.50; 75.60; S1.2; S4)
(a) Vol. **95**, No. 1, p. 213—219 (1986).

S. U. Jen
*Model Calculation on the Reduction of the Domain Wall Pinning Force by an AC Field
(75.60)
(a) Vol. **95**, No. 1, p. K21—K24 (1986).

M. Posselt, G. Diener, and M. Seidel
Ideal Magnetization in Sintered Iron Materials
(75.50; 75.60; S1.1)
(b) Vol. **136**, No. 1, p. 123—130 (1986).

N. K. Gobran, S. Saleh, B. F. Hanna, and R. Kamel
Kinetics of Crystallization for $Fe_{40}Ni_{40}B_{20}$ Metallic Glass
(64.70; 61.40; 62.20; 75.60; S1.1)
(a) Vol. **96**, No. 1, p. 33—37 (1986).

M. Tejedor and J. F. Fuertes
*Néel's Domain Structure on [011] Iron Whiskers
(75.60; S1.1)
(a) Vol. **96**, No. 1, p. K1—K4 (1986).

Th. Grieb, K. Wacker, E. Naumann, W. Treutmann, G. Heger, and R. Geick
*Magnetic Properties of $Rb_2Mn_xCr_{1-x}Cl_4$ Mixed Crystals
(75.25 and 75.60; 75.30; S9.16)
(b) Vol. **136**, No. 2, p. 585—591 (1986).

R. W. Chantrell, M. Fearon, and E. P. Wohlfarth
*The Time-Dependent Magnetic Behaviour of Fine Particle Systems
(75.60)
(a) Vol. **97**, No. 1, p. 213—221 (1986).

S. Ichinose
*Bound Domain Wall Pairs in Wolffram's Red Salt
(75.10 and 75.60; S12.1)
(b) Vol. **137**, No. 2, p. 479—485 (1986).

N. Amin and S. Arajs
*Magnetic Behaviour of Nearly Spherical Fe_3O_4 Particles in Low Magnetic Fields at Elevated Temperatures
(75.60; S10.1)
(a) Vol. **97**, No. 2, p. 479—484 (1986).

A. Roman
*Determination of the Losses Induced by Irreversible Barkhausen Jumps of the 180° Domain Wall Moving with a Sinusoidal Motion in the Multidomain Wall Model
(75.60)
(a) Vol. **97**, No. 2, p. 485—493 (1986).

L. F. Magana, M. A. Escobar, and R. Valenzuela
*Effect of the Grain Size Distribution on the Ferromagnetic Hysteresis Loop
(75.60; S11.2)
(a) Vol. **97**, No. 2, p. 495—500 (1986).

K. Elk and V. Christoph
*On the Influence of the Single Particle Coercivities on the Virgin Curves of Polycrystalline Magnets
(75.60)
(a) Vol. **98**, No. 1, p. K53—K56 (1986).

A. H. Wafik and I. A. Ibrahim
*Effect of Plastic Deformation on the Irreversible Magnetization in Nickel–Chromium Wire
(75.60; S1.2)
(a) Vol. **98**, No. 2, p. K141—K145 (1986).

H. J. Blythe, E. Klugmann, and F. Dworschak
*Magnetic Relaxation in Cobalt Following Low-Temperature Electron Irradiation
(61.80 and 75.60; S1.2)
(a) Vol. **98**, No. 2, p. K147—K151 (1986).

Th. Moch, B. Lippold, and A. Handstein
*Spin Glass Like Behaviour of a Sputtered $Cu_{94.6}Tb_{5.4}$ Alloy
(75.60; S1.2)
(a) Vol. **98**, No. 2, p. K161—K164 (1986).

75.70. Magnetic films and plates

Original Papers and Short Notes

V. I. Malyutin, V. E. Osukhovskii, A. A. Ivanov, L. A. Chebotkevich, I. V. Lobov, and Yu. D. Vorobiev
*The Effect of Volume and Surface Inhomogeneities on the Coercivity of Thin Magnetic Films
(75.70; S1.1)
(a) Vol. **93**, No. 2, p. 585—595 (1986).

T. Stobiecki and M. Przybylski
*Spontaneous Hall Effect and Magnetization of Fe–(B, Zr) and Co–Mo Amorphous Films
(75.70; 73.60; S1.1; S1.2)
(b) Vol. **134**, No. 1, p. 131—139 (1986).

A. Urbaniak-Kucharczyk
*The Magnetic Contribution to Electrical Resistivity in Thin Gadolinium Films
(73.60 and 75.70; S1.4)
(b) Vol. **134**, No. 2, p. 751—756 (1986).

W. Kaczmarek, I. Gościańska, J. Pietrzak, H. Ratajczak, and J. Baszyński
Ferromagnetic Resonance in Amorphous Gd–Fe Films
(76.50; 75.70; S1.1; S1.4)
(a) Vol. **95**, No. 1, p. 221—227 (1986).

K. P. Ghatak and M. Mondal
*Theoretical Analysis of the Magnetic Susceptibility of the Electrons in Ultrathin Films of Bi under Magnetic Quantization
(75.70; 75.20; S4)
(b) Vol. **135**, No. 2, p. K119—K124 (1986).

P. Görnert, Z. Šimša, J. Šimšová, I. Tomáš, R. Hergt, and J. Kub
*Growth and Properties of $Y_{3-u}Pb_uFe_{5-x-y-z}Co_xTi_yGa_zO_{12}$ LPE Films
(75.70; 68.55; S11.2)
(a) Vol. **95**, No. 2, p. 613—619 (1986).

N. Lenge and H. Kronmüller
*Low Temperature Magnetization of Sputtered Amorphous Fe–Ni–B Films
(75.70; 61.40; S1.1)
(a) Vol. **95**, No. 2, p. 621—633 (1986).

J. Kub
A Study of the Lattice Relaxation around Crack in LPE Garnet Films
(61.70 and 68.60; 75.70; S11.2)
(a) Vol. **96**, No. 1, p. 249—253 (1986).

T. Wielinga and J. C. Lodder
*Reversal Mechanism in Perpendicular Media with Columnar Microstructure
(75.70; S1.2)
(a) Vol. **96**, No. 1, p. 255—264 (1986).

T. Luciński and B. Szymański
*Hall Effect and Magnetic Anisotropy in Amorphous $Fe_{1-x}Si_x$ Films
(72.15 and 75.70; S1.1)
(a) Vol. **96**, No. 1, p. K53—K56 (1986).

H. Ratajczak and I. Gościańska
*Effect of Annealing on Coercive Field in Amorphous $Fe_{40}Ni_{40}B_{20}$ Thin Films
(75.70; 75.80; S1.1)
(a) Vol. **96**, No. 2, p. 563—571 (1986).

T. Tyliszczak, A. Budkowski, and J. A. Sawicki
*Study of Magnetism and Interface Processes between Thin Iron Film and Nickel Substrate by in-Situ Conversion Electron Mössbauer Spectroscopy
(75.70; 64.70; 76.80; S1.1; S1.2)
(a) Vol. **96**, No. 2, p. 573—579 (1986).

R. Krishnan and M. Tessier
*Enhanced Magnetization in Multilayer Ni–Ag Films
(75.70; S1.2; S1.3)
(a) Vol. **97**, No. 1, p. K39—K41 (1986).

A. Urbaniak-Kucharczyk
The Magnetic Contribution to the Electrical Resistivity in Thin Films of Diluted Ferromagnetic Alloys
(73.60; 75.70)
(a) Vol. **97**, No. 2, p. K197—K200 (1986).

G. Wiatrowski, T. Balcerzak, L. Wojtczak, and J. Mielnicki
*The Application of Differential Operator Method to the Calculation of Magnetization in Thin Ferromagnetic Films
(75.10 and 75.70)
(b) Vol. **138**, No. 1, p. 189—195 (1986).

R. Hergt, H. Pfeiffer, and L. Fritzsch
Formation of Texture in Co/Cr Films for Perpendicular Magnetic Recording
(68.55; 75.70; S1.2)
(a) Vol. **98**, No. 1, p. 69—80 (1986).

A. Duda
*Spin-Wave Spectrum in Uniaxial Ferromagnetic Thin Films with Nearest-Neighbour and Next-Nearest-Neighbour Exchange Couplings
(75.70; 75.30)
(b) Vol. **138**, No. 2, p. 509—519 (1986).

75.80. Magnetomechanical and magnetoelectric effects, magnetostriction For galvanomagnetic effects, see 72.15 an 72.20 For magneto-optical effects, see 78.20

Original Papers and Short Notes

M. VÁZQUEZ, A. HERNANDO, and H. KRONMÜLLER
*Critical Exponents of the Magnetostriction in Amorphous Alloys
(75.80; S1.1)
(b) Vol. **133**, No. 1, p. 167—170 (1986).

J. GONZÁLEZ, M. VÁZQUES, and J. M. BARANDIARAN
*On the Dependence of the Magnetization Curve on Applied Tensile Stress in Amorphous Alloys with Positive Magnetostriction
(75.80; S1.1)
(a) Vol. **93**, No. 2, p. K165—K170 (1986).

A. RAOUFI and S. ARAJS
*Effects of Magnetic Field and Tensile Stress on the Electrical Resistivity of Some Amorphous Ferromagnets
(72.15 and 75.80; S1.1)
(a) Vol. **95**, No. 1, p. 193—196 (1986).

S. MOTOGI
*Polarization and Resonance Properties of Magnetoacoustic Waves in Ferromagnets with Dissipation
(75.50 and 75.80; 76.50; S1.1; S11.2)
(a) Vol. **95**, No. 1, p. 197—208 (1986).

A. WĘGRZYN, T. JAGIELIŃSKI, S. OHNUMA, and T. MASUMOTO
*Saturation Magnetostriction and Forced Volume Magnetostriction in Amorphous $(Fe_{1-x}Co_x)_{90}Zr_{10}$ and $(Fe_{1-x}Ni_x)_{90}Zr_{10}$ Alloys. Experimental Results
(75.80; S1.1; S1.2)
(a) Vol. **95**, No. 2, p. 599—603 (1986).

J. ULNER
On the Theory of the Jahn-Teller Effect Induced by an External Magnetic Field in Orthorhombic $J = 1$ Paramagnets
(71.70; 75.80)
(b) Vol. **136**, No. 1, p. 299—308 (1986).

H. RATAJCZAK and I. GOŚCIAŃSKA
Effect of Annealing on Coercive Field in Amorphous $Fe_{40}Ni_{40}B_{20}$ Thin Films
(75.70; 75.80; S1.1)
(a) Vol. **96**, No. 2, p. 563—571 (1986).

75.90. Other topics in magnetic properties and materials

Original Papers and Short Notes

P. L. Rossiter and R. H. Mair
The Electrical Properties of Rapidly Quenched Cu–Ni Alloys
(72.15; 75.90; S1.2)
(b) Vol. **134**, No. 2, p. 335—342 (1986).

M. Cinal and S. Olszewski
*Magnetic Atoms in Metallic Compounds Approached on the Basis of the Quantum-Statistical ab Initio Theory
(75.90; S1.1; S1.2)
(b) Vol. **138**, No. 1, p. K17—K20 (1986).

76. Magnetic resonances and relaxation in condensed matter; Mössbauer effect

76.20. General theory of resonances and relaxation

Original Papers

L. L. Buishvili, N. P. Fokina, and L. T. Chelidze
*Cross-Relaxation of Nuclear Spins at Low Temperatures
(76.20)
(b) Vol. **133**, No. 1, p. 127—134 (1986).

R. R. Nigmatullin
*Fractional Moments. New Source of Information in Radiospectroscopy
(76.20 and 77.40)
(b) Vol. **133**, No. 2, p. 713—720 (1986).

M. E. Foglio and P. Schlottmann
Intermediate Valence Model of the Hyperfine Field and Relaxation Rate of Cerium in Gadolinium Metal
(71.70; 75.10; 76.20; S1.4)
(b) Vol. **136**, No. 1, p. 273—280 (1986).

I. Ursu, F. Balibanu, D. E. Demco, and M. Bogdan
*Spin Relaxation in the Presence of Coherent Averaging Pulse Sequences
(76.20)
(b) Vol. **136**, No. 1, p. 309—317 (1986).

76.30. Electron paramagnetic resonance and relaxation

Review Article

J. KLIAVA
*EPR of Impurity Ions in Disordered Solids. Distributions of the Spin Hamiltonian Parameters
(76.30; 61.40)
(b) Vol. **134**, No. 2, p. 411—455 (1986).
Erratum
(b) Vol. **137**, No. 2, p. 709 (1986).

Original Papers and Short Notes

V. P. SETH, A. YADAV, and R. S. BANSAL
*Solution of the Spin-Hamiltonian with Orthorhombic HF and *g*-Tensors ($S = \frac{1}{2}$). II. Application to VO^{2+} Centres in Oxide Glasses
(76.30; 71.70; S11.1)
(b) Vol. **133**, No. 1, p. 297—300 (1986).

S. K. HOFFMANN, J. GOSLAR, and L. S. SZCZEPANIAK
*Low-Symmetry Doped Cu(II) Complexes in $(gly)_2CaCl_2 \cdot 4\,H_2O$ Single Crystal: EPR Studies
(76.30; 71.70; S12)
(b) Vol. **133**, No. 1, p. 321—328 (1986).

M. KRUPSKI, A. A. MIRZAKHANYAN, W. HILCZER, and E. G. SHAROYAN
*Effect of Pressure and Temperature on the EPR Spectrum of Cr^{3+} Ions in α-$LiIO_3$ Single Crystals
(76.30; 71.70; S11.1)
(b) Vol. **133**, No. 1, p. 355—361 (1986).

T. KAWAKUBO
Color Centers in Irradiated Para-Dichlorobenzene Crystals
(61.70; 76.30; 78.50; S12)
(a) Vol. **93**, No. 1, p. 177—184 (1986).

V. A. RADTSIG and A. A. BOBYSHEV
Twofold Coordinated Si and Ge Atoms and "Hydrogen" Paramagnetic Centers in Amorphous SiO_2 and GeO_2
(71.55; 61.70; 76.30; S10.1)
(b) Vol. **133**, No. 2, p. 621—627 (1986).

E. TRIF and AL. NICULA
*EPR of Nd^{3+} Ions in Y-Type Zeolites
(76.30; S11)
(b) Vol. **133**, No. 2, p. 683—686 (1986).

J. KURIATA, N. GUSKOS, and P. PASTUSIAK
*Dynamic Effects in the Temperature Dependence of the "Intrinsic" $\bar{b}_2$ Parameter for $MeF_2{:}Gd^{3+}$
(76.30; S9.12)
(b) Vol. **133**, No. 2, p. K133—K136 (1986).

S. M. RYABCHENKO and YU. G. SEMENOV
EPR and Spin-Flip Raman Scattering in the n-Type Magnetically Mixed Semiconductors
(71.70; 76.30; 78.30)
(b) Vol. **134**, No. 1, p. 281—288 (1986).

Z. WILAMOWSKI, B. OCZKIEWICZ, P. KACMAN, and J. BLINOWSKI
*Asymmetry of the EPR Absorption Line in CdF_2 with Y
(76.30; S9)
(b) Vol. **134**, No. 1, p. 303—306 (1986).

S. MANHAS, A. MANOOGIAN, G. LAMARCHE, and J. C. WOOLLEY
Lattice Parameters and Magnetic Properties of the $Cd_xZn_yMn_zSe$ Alloy System
(75.50; 61.55; 76.30; S1.63)
(a) Vol. **94**, No. 1, p. 213—221 (1986).

F. CALLENS, R. VERBEECK, L. MARTENS, P. MATTHYS, and E. BOESMAN
*Electron Spin Resonance Absorption of Centres with an Orthorhombic *g*-Tensor in Polycrystalline Substances
(76.30)
(a) Vol. **94**, No. 1, p. 267—274 (1986).

RAM BILAS and D. BAHADUR
Effect of Iron Substitution in Manganese Aluminosilicate Glass and Glass Ceramics
(75.50; 76.30; S10.15)
(a) Vol. **94**, No. 1, p. K41—K45 (1986).

G. VÖLKEL and W. BRUNNER
*Electron Spin–Lattice Relaxation of Trapped EPR Active Centres in O_2-Contaminated Polyacetylene
(76.30; S12.1)
(a) Vol. **94**, No. 2, p. 673—678 (1986).

H. R. ASATRYAN, A. S. KUZANYAN, A. G. PETROSYAN, A. K. PETROSYAN, and E. G. SHAROYAN
Single Crystal Growth and Investigation of Optical and ESR Absorption Spectra of Zirconium-Doped YAG
(71.70; 76.30; 78.50; S11.2)
(b) Vol. **135**, No. 1, p. 343—352 (1986).

R. ALCALÁ, R. CASES, and P. J. ALONSO
*EPR of Nickel Centers in $SrCl_2$:Ni
(76.30; S9.12)
(b) Vol. **135**, No. 1, p. K63—K67 (1986).

M. GODLEWSKI and D. HOMMEL
Eu^{2+} Photocharge Transfer Processes in ZnS Crystals Determined by Photo-ESR Measurements
(71.55; 76.30; 78.55; S8.11)
(a) Vol. **95**, No. 1, p. 261—268 (1986).

P. SUBRAMANIAN and N. HARIHARAN
*EPR Spectra of Single Crystal Bis(Glycine) Calcium(II) Dichloride Tetrahydrate with Copper(II) Ion
(76.30; S12)
(b) Vol. **135**, No. 2, p. 731—740 (1986).

A. Hamaïdia, A. Hachimi, J. Margerie, and J. F. Hémidy
EPR of Colour Centres in Alkaline Metal Doped Calcium and Strontium Fluorides
(61.70; 76.30; 78.50; S9.12)
(b) Vol. **137**, No. 1, p. 47—56 (1986).

S. Hedewy and S. K. Hoffmann
*Electron Paramagnetic Resonance of Ferric Acetyl Acetonate
(76.30; S12)
(a) Vol. **97**, No. 1, p. 129—133 (1986).

J. Kreissl
*A Quantitative EPR Analysis of ZnS:Mn Powders of Different Crystal Structure
(76.30; 61.70; S8.11)
(a) Vol. **97**, No. 1, p. 191—198 (1986).

J. Walczak, H. Dębiński, B. Muraszko, J. Kuriata, and L. Sadłowski
*EPR Study of Copper(I) Sulphide
(64.60 and 76.30; S1.63)
(a) Vol. **97**, No. 1, p. 291—295 (1986).

V. N. Vasyukov
*Dependence of the Low Temperature EPR Spectrum of Cu^{2+} Ion on the Microwave Field Frequency and Temperature
(76.30; 71.70; S11)
(b) Vol. **137**, No. 2, p. 623—631 (1986).

E. V. Kholopov, A. M. Panich, N. K. Moroz, and Yu. G. Kriger
One Dimensional Mechanism of Antiferroelectric Ordering in a Non-Polar Crystal
(64.70 and 77.80; 65.40; 76.30; S9.16)
(b) Vol. **137**, No. 2, p. K121—K125 (1986).

A. A. Bugai, V. A. Girii, V. M. Maksimenko, Yu. V. Pomozov, and V. I. Shakhovtsov
Investigation of the Production of Quenching Thermodefects in Silicon
(71.55 and 72.80; 76.30; S5.11)
(a) Vol. **97**, No. 2, p. 549—555 (1986).

S. Benakki, A. Goltzene, C. Schwab, Wang Guangyu, and Zou Yuanxi
*Electron Paramagnetic Resonance Analysis of Strain-Induced Defects in Semi-Insulating GaAs
(61.70 and 76.30; S7.12)
(b) Vol. **138**, No. 1, p. 143—149 (1986).

M. Souiri, A. Goltzene, and C. Schwab
Correlations between Electron Paramagnetic Resonance and Bonding Parameters. Evidence of a Lattice Relaxation
(71.45 and 71.70; 76.30; S7; S8; S9; S10)
(b) Vol. **138**, No. 1, p. 321—330 (1986).

76.40. Diamagnetic and cyclotron resonances

Short Notes

V. I. Gatalskaya
*Cyclotron Resonance Anisotropy of Charge Carriers in Irradiated Germanium
(76.40; 71.25; S5.12)
(b) Vol. **133**, No. 1, p. K31—K33 (1986).

V. I. Gatalskaya
On the Mechanism of Charge Carrier Scattering in Germanium with Point Defects under Cyclotron Resonance
(72.20; 76.40; S5.12)
(b) Vol. **133**, No. 2, p. K137—K141 (1986).

76.50. Ferromagnetic, antiferromagnetic, and ferrimagnetic resonances; spin-wave resonance (see also 75.30. Spin waves)

Original Papers and Short Notes

K. M. SULTANOV, SH. A. KULIEV, and N. G. ABDULLAEV
Coupled Waves in Conducting Spin Glasses
(71.45 and 75.30; 76.50)
(b) Vol. **135**, No. 1, p. 163—171 (1986).

E. WOLSKA and J. BASZYŃSKI
Prereactional Transformations in the Topotactic Conversion γ-FeOOH → γ-Fe_2O_3
(64.70; 61.10; 76.50; 78.30; S11.2)
(a) Vol. **95**, No. 1, p. 87—92 (1986).

S. MOTOGI
Polarization and Resonance Properties of Magnetoacoustic Waves in Ferromagnets with Dissipation
(75.50 and 75.80; 76.50; S1.1; S11.2)
(a) Vol. **95**, No. 1, p. 197—208 (1986).

W. KACZMAREK, I. GOŚCIAŃSKA, J. PIETRZAK, H. RATAJCZAK, and J. BASZYŃSKI
*Ferromagnetic Resonance in Amorphous Gd–Fe Films
(76.50; 75.70; S1.1; S1.4)
(a) Vol. **95**, No. 1, p. 221—227 (1986).

D. FRAITOVÁ
*An Analytical Theory of FMR in Bulk Metals (Parallel Configuration). IV. Exchange Dominated Surface Modes
(76.50; 75.30; S1.1)
(b) Vol. **136**, No. 1, p. 91—96 (1986).

W. MACIEJEWSKI
Spin Wave Resonance Approach to Interdiffusion in Coupled Magnetic Layers. Theoretical Study
(66.30 and 68.48; 76.50)
(b) Vol. **136**, No. 2, p. 537—549 (1986).

W. A. KACZMAREK and J. PIETRZAK
Anomalous Temperature Dependence of Magnetocrystalline Anisotropy Constants of $BaIn_{1.5}Fe_{10.5}O_{19}$ Single Crystals
(75.30; 76.50; S11.2)
(a) Vol. **98**, No. 1, p. K43—K46 (1986).

PRABHAT K. SINGH and P. K. SINGH
*Effect of Dimensional Resonance on Outside Spin-Wave Manifold ΔH_{eff}-Values in Ferrites
(76.50; S11.2)
(a) Vol. **98**, No. 2, p. 541—544 (1986).

76.60. Nuclear magnetic resonance and relaxation

Review Article

K. V. S. RAMA RAO and S. RAMAPRABHU
*Nuclear Quadrupole Resonance Studies on Inorganic Solids
(76.60; S8; S9; S10; S11)
(a) Vol. **93**, No. 1, p. 17—27 (1986).

Original Papers and Short Notes

A. S. LITOVCHENKO, V. A. CHERNENKO, I. V. MATYASH, and O. D. ISHUTINA
Influence of Pressure on Dehydroxylation Effect in Muscovite
(64.75; 76.60; 77.20; S11)
(a) Vol. **93**, No. 1, p. K9—K13 (1986).

A. E. MEFED and V. A. ATSARKIN
*The "Cascade" Narrowing of the NMR Line in Solids by Radio-Frequency Fields and Slow Molecular Motions
(76.60; 63.20; S12)
(a) Vol. **93**, No. 1, p. K21—K24 (1986).

R. JAKUBAS, Z. CZAPLA, Z. GALEWSKI, L. SOBCZYK, O. J. ŻOGAL, and T. LIS
Structure and Phase Transition in $(CH_3NH_3)_3Sb_2Cl_9$
(61.50 and 64.70; 76.60; 77.20; S11)
(a) Vol. **93**, No. 2, p. 449—455 (1986).

F. ENGELKE, D. MICHEL, W. WINDSCH, and J. PETERSSON
*^{15}N-NMR Investigations of Dynamical Processes near the Ferroelectric Phase Transition of TSCC
(76.60 and 77.80; S12)
(b) Vol. **134**, No. 1, p. 29—37 (1986).

D. K. FOWLER, D. C. CREAGH, and G. V. H. WILSON
*NMR Free Induction Decay Measurements in Iron
(76.60; 75.60; S1.1)
(a) Vol. **94**, No. 2, p. K97—K101 (1986).

S. G. KOZLOVA, YU. G. KRIGER, P. P. SAMOILOV, and S. P. GABUDA
*Spin-Lattice Relaxation and Anisotropic Reorientation of PF_6^- in Antiferroelectric $NH_4PF_6 \cdot NH_4F$
(76.60; S11)
(b) Vol. **135**, No. 1, p. K53—K56 (1986).

S. BEDENK, J. LUŽNIK, J. PIRNAT, Z. TRONTELJ, and W. WINDSCH
*Bromine Nuclear Quadrupole Resonance in Tri-Sarcosine Calcium Bromide
(76.60; S12)
(a) Vol. **95**, No. 1, p. K33—K35 (1986).

76.70. Magnetic double resonances and cross effects

Original Papers

A. Bartl and A. Stasko
*ENDOR Studies on Conjugated Polymers
(76.30 and 76.70; S12.1)
(a) Vol. **95**, No. 2, p. 659—663 (1986).

P. G. Baranov, V. A. Vetrov, N. G. Romanov, and V. Topa
*ODMR in Triplet Excited States of ns^2 Ions in Alkali Halide Crystals
(71.55 and 76.70; 78.55; S9.11)
(b) Vol. **136**, No. 2, p. 699—707 (1986).

R. B. Bossoli and L. E. Halliburton
*^{27}Al Hyperfine and Quadrupole Interactions for the $[AlO_4]^\circ$ Center in Quartz
(71.70 and 76.70; 61.70; S10.1)
(b) Vol. **136**, No. 2, p. 709—714 (1986).

P. Studzinski and J.-M. Spaeth
*ENDOR Investigation of Cr^{3+} Centers in the Tetragonal Phase of $RbCdF_3$
(76.70; 64.70; 71.70; S9.16)
(b) Vol. **136**, No. 2, p. 735—742 (1986).

76.80. Mössbauer effect; other γ-ray spectroscopy

Original Papers and Short Notes

M. A. Haas
*Hot Nuclear Fluorescence in the LiI Crystal. Possibilities of Detection
(76.80; S9.11)
(b) Vol. **134**, No. 1, p. 63—69 (1986).

J. Roset, A. Fernández, X. Obradors, and J. Tejada
*Mössbauer Emission Studies of ^{57}Co-Doped ZnO, Cu_2O, and $TiCoO_3$
(76.80; S10; S10.15)
(b) Vol. **134**, No. 1, p. 297—302 (1986).

G. V. Smirnov, Yu. V. Shvydko, U. van Bürck, and R. L. Mössbauer
*Experimental Study of the Influence of Lattice Vibrations on the Suppression Effect at Room Temperature
(61.10 and 76.80; 63.20; S10.15)
(b) Vol. **134**, No. 2, p. 465—475 (1986).

A. Ya. Dzyublik
*Effect of Forced Vibrations on the Scattering of X-Rays and Mössbauer Radiation by a Crystal. II. Dynamical Effects
(61.10 and 76.80; 63.20; 78.30)
(b) Vol. **134**, No. 2, p. 503—513 (1986).

W. Matz and H. Herrmann
Non-Linear Dependence of Average Magnetic Moment on Boron Concentration in Amorphous Fe–B Alloys
(75.50; 76.80; S1.1)
(a) Vol. **94**, No. 2, p. 617—622 (1986).

K. K. Kadyrzhanov, A. K. Zhetbaev, M. Sh. Aimanov, and A. N. Ozernoi
Mössbauer Investigations of Phase Formation in Nitrogen-Implanted Iron
(61.70; 76.80; S1.1)
(a) Vol. **94**, No. 2, p. 823—825 (1986).

J. Fontcuberta, J. A. Cusidó, and J. Tejada
*On the Electric Field Effect in the Mössbauer Emission Studies
(76.80; 71.55; S10.1)
(b) Vol. **135**, No. 1, p. K27—K31 (1986).

I. Škorvánek, A. Zentko, and T. Zemčík
Influence of the Surface Insulating Coating on the Magnetic Properties of $Fe_{82}B_{12}Si_6$ Amorphous Ribbons
(75.50; 76.80; S1.1)
(a) Vol. **95**, No. 1, p. 209—212 (1986).

O. P. Singh and V. P. Gupta
Electronic Properties of Tin Monochalcogenides (SnS, SnSe, SnTe)
(71.25; 76.80; S8)
(b) Vol. **136**, No. 1, p. K41—K44 (1986).

M. Krause, C. Michalk, B. Lippold, M. Benedix, and J. Suwalski
Slow Paramagnetic Relaxation in Oxalate (2,2'-Bipyridine) Iron
(75.20; 61.40; 76.80; S1.1)
(a) Vol. **97**, No. 2, p. 521—526 (1986).

D. Baltrūnas, S. Motiejūnas, and E. I. Rogacheva
*Effect of the Deviation from Stoichiometry on the Mössbauer Parameters of SnTe
(64.75 and 76.80; S8)
(a) Vol. **97**, No. 2, p. K131—K133 (1986).

J. Golczewski and W. Frank
Rate-Equation Treatment of PAC-Monitored Defect Reactions in Irradiated Metals
(61.70; 76.80; S1)
(b) Vol. **138**, No. 1, p. 123—133 (1986).

P. Rochegude and J. Foct
Interstitial Atom Ordering in Binary Fe–N Solid Solutions Studied by Mössbauer Spectrometry
(64.75; 76.80; S1.1)
(a) Vol. **98**, No. 1, p. 51—62 (1986).

K. Sumiyama, N. Ohshima, and Y. Nakamura
Magnetic Properties of Metastable α-Mn-Type $Mn_{1-y}Fe_y$ Alloys Produced by Vapor Quenching
(75.50; 61.10; 76.80; S1.1)
(a) Vol. **98**, No. 1, p. 229—238 (1986).

A. I. Chumakov and G. V. Smirnov
On the Possibility of Multiple Increase of the Resolution of Layer-by-Layer Analysis of Materials
(61.16; 76.80; 79.20)
(a) Vol. **98**, No. 2, p. 339—343 (1986).

76.90. Other topics in magnetic resonances and relaxation

Original Paper

77. Dielectric properties and materials (for conductivity phenomena, see 72.20 and 72.80)

77.20. Permittivity

Review Article

J. Shanker, G. G. Agrawal, and N. Dutt
*Electronic Polarizabilities and Photoelastic Behaviour of Ionic Crystals
(77.20 and 77.30; 78.20; S8; S9)
(b) Vol. **138**, No. 1, p. 9—30 (1986).

Original Papers and Short Notes

T. Nattermann
*Dielectric Tails in K_2ZnCl_4: No Evidence for an Intrinsic Choatic State
(77.20; 71.45; S9.16)
(b) Vol. **133**, No. 1, p. 65—70 (1986).

G. Sorge, H. Maack, and L. A. Shuvalov
Low-Frequency Dielectric Properties of Rb_2ZnCl_4
(77.80; 77.20; S9.16)
(a) Vol. **93**, No. 1, p. 315—320 (1986).

A. S. Litovchenko, V. A. Chernenko, I. V. Matyash, and O. D. Ishutina
Influence of Pressure on Dehydroxylation Effect in Muscovite
(64.75; 76.60; 77.20; S11)
(a) Vol. **93**, No. 1, p. K9—K13 (1986).

R. Jakubas, Z. Czapla, Z. Galewski, L. Sobczyk, O. J. Żogal, and T. Lis
Structure and Phase Transition in $(CH_3NH_3)_3Sb_2Cl_9$
(61.50 and 64.70; 76.60; 77.20; S11)
(a) Vol. **93**, No. 2, p. 449—455 (1986).

G. W. Bąk
*Low Frequency Dielectric Properties of Some Organic Materials
(77.20; S12)
(a) Vol. **93**, No. 2, p. K195—K199 (1986).

M. Arnold and P. Bussemer
*A Modified Bruggeman Theory of Optical Properties for Metal–Isolator Transitions
(71.30 and 77.20)
(b) Vol. **134**, No. 1, p. K43—K45 (1986).

A. Shiromany, G. M. S. Srivastava, and J. Shanker
*Generalized Huggins-Mayer Potentials and Properties of Alkali Halides under the Effect of High Pressure
(64.30 and 77.20; 65.50; S9.11)
(b) Vol. **135**, No. 1, p. 119—126 (1986).

77.30. Polarization and depolarization effects

Review Article

J. Shanker, G. G. Agrawal, and N. Dutt
*Electronic Polarizabilities and Photoelastic Behaviour of Ionic Crystals
(77.20 and 77.30; 78.20; S8; S9)
(b) Vol. **138**, No. 1, p. 9—30 (1986).

Original Papers and Short Notes

V. V. Krasnogolovets and P. M. Tomchuk
*Erratum to: The Coulomb Correlations and the Electric Field Local Heterogeneities on the Proton Conductivity of Hydrogen Bonded Chains
(71.38 and 77.30; S12.1)
(b) Vol. **133**, No. 1, p. 431 (1986).

R. Coelho, L. Levy, and D. Sarrail
On the Natural Decay of Corona Charged Insulating Sheets
(73.25 and 77.50; 77.30; S12)
(a) Vol. **94**, No. 1, p. 289—298 (1986).

J. Prakash
*Thermally Stimulated Depolarization Measurements in AsO_2^--Doped Potassium Halides at Different Heating Rates
(77.30 and 77.40; 78.40; S9.11)
(b) Vol. **135**, No. 2, p. 741—748 (1986).

G. D. Guseinov, S. N. Mustafaeva, R. G. Guseinova, and E. G. Abdullaev
The Nature of Electric Memory in Lithium-Intercalated Chain-Structure $TlInSe_2$ Single Crystals
(72.80; 77.30; S8.16)
(a) Vol. **95**, No. 2, p. K159—K164 (1986).

L. N. Kantorovich
*Multipole Theory of the Polarization of Solids by Point Defects. III. Transition Energy Calculations, the Approximation of Non-Point Charges
(71.55 and 77.30; S9.11)
(b) Vol. **137**, No. 1, p. 229—240 (1986).

L. Wendler and R. Pechstedt
Dynamical Screening, Collective Excitations, and Electron–Phonon Interaction in Heterostructures and Semiconductor Quantum Wells. General Theory
(71.38 and 71.45; 63.20; 73.40; 77.30; 78.20)
(b) Vol. **138**, No. 1, p. 197—217 (1986).

J. Prakash
*Role of the Background Current in Analysing the ITC Spectrum
(77.30 and 77.40)
(a) Vol. **98**, No. 1, p. 247—252 (1986).

A. Kessler and U. Klein
*Conductivity and Polarization of NH_4Cl between 150 and 500 K
(66.90 and 77.30; 61.70; S9)
(a) Vol. **98**, No. 2, p. 617—627 (1986).

77.40. Dielectric loss and relaxation

Original Papers and Short Notes

N. CH. MITRA and A. L. BHATTACHARYYA
Specification of Dipole Complexes in Dy_2O_3–CaO Systems from Dielectric Relaxation Studies (61.70; 77.40; S10.15)
(a) Vol. **98**, No. 2, p. 477—486 (1986).

77.50. Dielectric breakdown and space-charge effects

Original Papers and Short Notes

77.55. Dielectric thin films

Original Papers and Short Notes

77.60. Piezoelectricity and electrostriction (for piezo-optical effects, see 78.20)

Original Papers and Short Notes

J. Bracinik and J. Štelina
*Sound Generation in Photoresistive Semiconductors
(62.65 and 77.60; S7.12)
(a) Vol. **95**, No. 2, p. K179—K182 (1986).

P. Bury, J. Ďurček, and K. Sakalauskas
*On the Piezoactivity of Si MOS Structures
(73.40 and 77.60; S5.11)
(a) Vol. **95**, No. 2, p. K207—K209 (1986).

S. Ghosh and R. B. Saxena
*Parametric Instabilities of Laser Beams in Material with Strain-Dependent Dielectric Constant
(62.65 and 77.60; S11.1)
(a) Vol. **96**, No. 1, p. 111—119 (1986).

G. Martin and K. Franke
An Examination of the Validity of a Common Model Used for the Analysis of Surface Acoustic Wave Interdigital Transducers
(68.25; 62.30; 77.60)
(a) Vol. **96**, No. 2, p. 489—499 (1986).

A. B. Ilyaev, B. S. Umarov, L. A. Shabanova, and M. F. Dubovik
*Temperature Dependence of Electromechanical Properties of LGS Crystals
(77.60; S10)
(a) Vol. **98**, No. 2, p. K109—K114 (1986).

77.70. Pyroelectric and electrocaloric effects

Original Papers and Short Notes

R. Poprawski and W. Salejda
The Phenomenological Model of Phase Transitions in NH_4HSeO_4 Crystals
(77.80; 65.50; 77.70; S11.1)
(a) Vol. **95**, No. 2, p. 459—465 (1986).

A. Scharmann and G. Schmirler
*Influence of Pyroelectric Properties on the Thermally Stimulated Exoelectron Emission (TSEE) of BeO
(77.70 and 79.75; S10.1)
(a) Vol. **97**, No. 1, p. 265—274 (1986).

G. Sorge and H. Hempel
*Optical Investigations of $LiKSO_4$ Single Crystals in the Temperature Range between 150 up to 730 K
(77.70 and 78.20; 77.80; S11.1)
(a) Vol. **97**, No. 2, p. 431—440 (1986).

77.80. Ferroelectricity and antiferroelectricity

Original Papers and Short Notes

S. N. Drozhdin, L. N. Kamysheva, and Z. A. Liberman
*Investigation of the Internal Field of Doped TGS Crystals
(77.80; S12)
(a) Vol. **94**, No. 1, p. K69—K73 (1986).

R. Styrkowiec
*The Influence of the Electrode on the Threshold Field of Domain Wall Motion in Rochelle Salt
(77.80; S11)
(a) Vol. **94**, No. 1, p. K79—K84 (1986).

A. M. Balagurov, N. C. Popa, and B. N. Savenko
*Symmetry Analysis of the Low-Temperature Domain Structure in Ferroelastic $LiKSO_4$
(77.80; S11.1)
(b) Vol. **134**, No. 2, p. 457—464 (1986).

B. Wyncke, F. Brehat, and H. Arbouz
Far-Infrared Reflectivity Spectra of Ammonium Dihydrogen Phosphate in the Para- and Antiferroelectric Phases
(78.30; 77.80; S11.1)
(b) Vol. **134**, No. 2, p. 523—531 (1986).

E. H. Birks
*The Electrocaloric Effect in $Pb(Sc_{0.5}Nb_{0.5})O_3$ Ceramic
(77.80; S10.15)
(a) Vol. **94**, No. 2, p. 523—527 (1986).

V. I. Torgashev, Yu. I. Yuzyuk, F. Smutný, and M. Połomska
Raman Spectra of a $LiN(H_xD_{1-x})_4SO_4$ Mixed Crystal
(78.30; 77.80; S11.1)
(b) Vol. **135**, No. 1, p. 93—104 (1986).

A. M. El-Dib and H. F. Hassan
*Energy Gap, Effective Mass, and Critical Phenomena in $NaNO_2$ Crystal
(71.25 and 77.80; 65.70; S11.1)
(b) Vol. **135**, No. 1, p. 253—259 (1986).

T. Hammad, F. El-Kabbany, and Y. Badr
*Ferroelectric and Dielectric Properties of Thin $NaNO_2$ Layers
(77.80; S11.1)
(a) Vol. **95**, No. 1, p. 121—125 (1986).

S. S. Sastry, C. R. K. Murty, and A. Subrahmanyam
*Dielectric Properties of Tetra-Methylammonium Tri- and Tetra-Halo Mercurates
(77.20 and 77.80; S11)
(a) Vol. **95**, No. 1, p. K57—K61 (1986).

S. H. Tang, E. C. Looi, and S. Radhakrishna
Raman and Infrared Studies of Tetramethylammonium Tetrachlorozincate
(78.30; 77.80; S12)
(b) Vol. **135**, No. 2, p. 519—522 (1986).

L. N. Kamysheva, S. N. Drozhdin, and O. M. Serdyuk
*The Influence of Defects on the Processes of Polarization Relaxation in TGS Crystals
(61.70 and 77.80; S12)
(a) Vol. **97**, No. 1, p. K29—K34 (1986).

E. V. Kholopov, A. M. Panich, N. K. Moroz, and Yu. G. Kriger
*One-Dimensional Mechanism of Antiferroelectric Ordering in a Non-Polar Crystal
(64.70 and 77.80; 65.40; 76.30; S9.16)
(b) Vol. **137**, No. 2, p. K121—K125 (1986).

A. M. Biradar, S. S. Bawa, S. B. Samanta, and S. Chandra
*Measurements of a Free Value of Helical Pitch in a Thin Sample of Ferroelectric Liquid Crystal
(61.30 and 77.80; S12)
(a) Vol. **97**, No. 2, p. 427—430 (1986).

G. Sorge and H. Hempel
Optical Investigations of $LiKSO_4$ Single Crystals in the Temperature Range between 150 up to 730 K
(77.70 and 78.20; 77.80; S11.1)
(a) Vol. **97**, No. 2, p. 431—440 (1986).

H. Konwent
*On the Application of a New Type Double-Well Potential in the Theory of Ferroelectric Phase Transitions
(77.80)
(b) Vol. **138**, No. 1, p. K7—K11 (1986).

F. El-Kabbany, Y. Badr, G. Said, and S. Taha
*Ferroelectricity Associated with the Metastable Phase III of $AgNO_3$
(77.80; 64.70; S11.1)
(a) Vol. **98**, No. 1, p. 117—125 (1986).

A. Kessler and U. Klein
*Spontaneous and Field Induced Current Peaks at the Order–Disorder Phase Transition of NH_4Cl
(77.80; S9)
(a) Vol. **98**, No. 1, p. K93—K98 (1986).

E. V. Kholopov
*Thermodynamic Features of the System of Packed Antiferroelectric Chains
(63.75 and 77.80; S9.16)
(b) Vol. **138**, No. 2, p. 415—422 (1986).

R. Jakubas, E. Narewski, and L. Sobczyk
*High Frequency Dielectric Dispersion in $[(CH_3)_3NH]_3Sb_2Cl_9$
(77.80; S12)
(a) Vol. **98**, No. 2, p. K115—K118 (1986).

77.90. Other topics in dielectric properties and materials

Original Paper

78. Optical properties and condensed-matter spectroscopy and other interactions of matter with particles and radiation (for phonon spectra, see 63)

78.20. Optical properties and materials

Review Articles

V. V. ROSSIN and V. G. SIDOROV
Reabsorption of Recombination Radiation in Semiconductors with High Internal Quantum Efficiency
(73.40 and 78.60; 72.40; 78.20; S7.12)
(a) Vol. **95**, No. 1, p. 15—40 (1986).

F. HENNEBERGER
*Optical Bistability at the Absorption Edge of Semiconductors
(78.20; 71.35; S7.16; S8.16)
(b) Vol. **137**, No. 2, p. 371—432 (1986).
Erratum
(b) Vol. **138**, No. 2, p. 744 (1986).

J. SHANKER, G. G. AGRAWAL, and N. DUTT
Electronic Polarizabilities and Photoelastic Behaviour of Ionic Crystals
(77.20 and 77.30; 78.20; S8; S9)
(b) Vol. **138**, No. 1, p. 9—30 (1986).

Original Papers and Short Notes

O. GOEDE, W. HEIMBRODT, and F. SITTEL
*Photoacoustic Spectroscopy of Isovalent Impurities in II–VI Semiconductors
(78.20; 71.55; 78.50; S8.11)
(a) Vol. **93**, No. 1, p. 277—282 (1986).

J. SZCZYRBOWSKI, K. SCHMALZBAUER, and H. HOFFMANN
*Optical Properties and Dynamic Current Density for Polycrystalline Metallic Films
(73.60 and 78.20; 78.65; S1.3)
(a) Vol. **93**, No. 1, p. 329—339 (1986).

L. K. SAMANTA, D. K. GHOSH, and G. C. BHAR
*Some Potential Mixed Chalcopyrite Crystals for Nonlinear Laser Devices
(78.20; S8.16)
(a) Vol. **93**, No. 1, p. K51—K54 (1986).

Z. CIMPL and F. KOSEK
*Refractive Index of $As_{2-x}Sb_xS_3$ and $As_{2-x}Sb_xSe_3$ Systems
(78.20; S8)
(a) Vol. **93**, No. 1, p. K55—K58 (1986).

K. T. STOYCHEV
*Non-Linear Optical Susceptibilities of Molecular Crystals in the Region of Vibronic Excitations. III. Non-Totally Symmetric Vibrations
(78.20; 63.20; 71.35)
(b) Vol. **133**, No. 2, p. 557—562 (1986).

C. TRALLERO GINER and M. ANTÓN
*Quantum Theory of Free-Carrier Absorption in Quasi-Two-Dimensional Semiconducting Structures
(78.20; 71.38; 78.65; S5.12; S7.12)
(b) Vol. **133**, No. 2, p. 563—572 (1986).

LUO ZUN-DU
*One Possible Mechanism of Deexcitation for Localized Electronic States — Lattice Relaxation in Virtual Photon Process
(63.20 and 78.20)
(b) Vol. **133**, No. 2, p. K77—K80 (1986).

P. K. KATANA
*Multiphoton Interband Absorption of Light in Semiconductors in External Parallel Electric and Magnetic Fields
(78.20)
(b) Vol. **134**, No. 1, p. 195—201 (1986).

S. I. GUBAREV
Valence Band Splittings in Wurtzite-Type Semimagnetic Semiconductors
(71.25 and 71.70; 78.20; S8.16)
(b) Vol. **134**, No. 1, p. 211—222 (1986).

A. M. BRODSKII
Nonlinear Optics of Metal Surface
(71.45 and 73.20; 78.20)
(b) Vol. **134**, No. 1, p. 251—256 (1986).

SH. M. SHARIPOV, K. M. MUKIMOV, and L. A. ERNAZAROVA
*The Heavy Rare-Earth Contribution to the Electronic Structure of the RFe_2 and R_2Fe_{17} Intermetallic Compounds
(78.20; 71.25; S1.1)
(b) Vol. **134**, No. 1, p. K59—K62 (1986).

F. LUKEŠ
*Ellipsometry of Silicon with Natural Surface Film at 632.8 nm
(78.20; 78.65; S5.11)
(a) Vol. **94**, No. 1, p. 223—230 (1986).

M. KULIK and J. ŻUK
*Ellipsometric Studies of Radiation Damage in Ion Bombarded GaP
(61.80 and 78.20; S7.11)
(a) Vol. **94**, No. 1, p. 427—431 (1986).

I. I. Sasovskaya and V. P. Korabel
*Optical Properties of α- and β-CuZn Brasses in the Region of Quantum Absorption
(78.20; 64.70; 71.25; 72.30; S1.2)
(b) Vol. **134**, No. 2, p. 621—630 (1986).

N. Kukhtarev, B. Pavlik, and T. Semenets
*Selfdiffrication and Phase Conjugation of Laser Beams in Electrooptic Crystals
(78.20 and 78.45)
(a) Vol. **94**, No. 2, p. 623—633 (1986).

W. Wesch and G. Götz
Rapid Annealing of Ion-Implanted GaAs
(61.70; 61.80; 73.60; 78.20; S7.12)
(a) Vol. **94**, No. 2, p. 745—766 (1986).

M. C. Desjardins-Deruelle, J. P. Lascaray, D. Coquillat, and R. Triboulet
Magneto-Optical Measurements in $Zn_{1-x}Mn_xTe$ with High Manganese Concentration
(71.35 and 71.70; 75.60; 78.20; S8.15)
(b) Vol. **135**, No. 1, p. 227—232 (1986).

G. Valiukonis, D. A. Guseinova, G. Krivaitė, and A. Šileika
*Optical Spectra and Energy Band Structure of Layer-Type $A^{IV}B^{VI}$ Compounds
(71.25 and 78.20; S8)
(b) Vol. **135**, No. 1, p. 299—307 (1986).

L. A. Yuguryan and A. G. Makhanek
Coherent Four-Photon Spectroscopy of Two-Photon Resonances of Trivalent Rare-Earth Ions
(71.70; 78.20)
(b) Vol. **135**, No. 1, p. 353—367 (1986).

S. Marcinkevičius, G. Babonas, B. V. Korzun, and I. G. Tregub
*Optical Activity of Semiconductors with Chalcopyrite Structure
(78. 20; S7.16; S8.16)
(b) Vol. **135**, No. 2, p. 615—621 (1986).

V. E. Mashchenko and A. I. Ziborov
Recombination Radiation and Electroreflection of Excitons in $Zn_{1-x}Cd_xSe$ Solid Solutions
(71.35; 78.20; 78.55; S8.15)
(b) Vol. **135**, No. 2, p. 623—631 (1986).

A. M. Elkorashy
*Indirect Forbidden Fundamental Absorption Edge in Germanium Selenide Single Crystals
(78.20; S8)
(b) Vol. **135**, No. 2, p. 707—713 (1986).

L. Wendler
S-Polarized Nonlinear Surface Polaritons. Effects of a Transition Layer
(71.36 and 73.20; 68.30; 78.20)
(b) Vol. **135**, No. 2, p. 759—774 (1986).

A. A. ALY, M. GAAFAR, N. S. ALY, and M. M. ABDEL-GAWAD
*Variances in Molecular Refractivities and Anisotropies in Monoclinic Crystals
(78.20; S12)
(a) Vol. **96**, No. 2, p. K181—K186 (1986).

G. JUNGK and A. RÖSELER
*Optical Properties of Metals: Au
(72.30 and 78.20; 78.65; S1.3)
(b) Vol. **137**, No. 1, p. 117—123 (1986).

B. ESSER and F. MARLOW
On the Influence of Noise on the Density of States and Absorption in Disordered Molecular Systems
(71.20 and 71.35; 78.20)
(b) Vol. **137**, No. 1, p. 169—181 (1986).

J. BARNAŚ
Non-Reciprocal Character of Wave Propagation in Magnetic Systems with Spontaneous Electric Polarization
(71.36 and 75.30; 78.20; S9.16)
(b) Vol. **137**, No. 1, p. K43—K47 (1986).

H. PUFF and H. W. STREITWOLF
*Optical Selection Rules of the Soliton Lattice
(78.20)
(b) Vol. **137**, No. 1, p. K59—K62 (1986).

B. A. VAID and K. C. SHARMA
Structural Disorder Model for Amorphous Semiconductors: a-Se
(61.40; 65.50; 78.20; S5)
(b) Vol. **137**, No. 2, p. 433—440 (1986).

F. LUKEŠ, E. SCHMIDT, J. HUMLÍČEK, P. DUB, and F. KOSEK
*Absorption Edge of SnSe
(78.20; 71.25; S8)
(b) Vol. **137**, No. 2, p. 569—579 (1986).

J. DONECKER and J. KLUGE
*Excitonic Absorption in Silicon in the Presence of Low Magnetic Fields
(78.20; 71.35; S5.11)
(b) Vol. **137**, No. 2, p. 591—595 (1986).

G. SORGE and H. HEMPEL
*Optical Investigations of $LiKSO_4$ Single Crystals in the Temperature Range between 150 up to 730 K
(77.70 and 78.20; 77.80; S11.1)
(a) Vol. **97**, No. 2, p. 431—440 (1986).

A. K. S. THAKUR
*On the Validity of Thin Film Models
(78.20 and 78.65)
(a) Vol. **97**, No. 2, p. K123—K125 (1986).

J. Y. Bigot, M. Frindi, F. Fidorra, R. Levy, and B. Hönerlage
*Nonlinear Propagation of Nanosecond Laser Pulses through Polycrystalline CuCl Layers
(78.20; 71.35; S9)
(b) Vol. **138**, No. 2, p. 621—631 (1986).

L. Artus, Y. Bertrand, C. Ance, and A. Lopez-Soler
*Low-Temperature Reflectivity Spectra of $AgGaS_2$ and $AgGaSe_2$ Compounds
(78.20; S8.16)
(b) Vol. **138**, No. 2, p. 633—636 (1986).

J. Gutowski, J. Schott, and I. Broser
Resonant Brillouin Scattering at Bound-Exciton States in CdS
(71.35; 78.20; S8.11)
(b) Vol. **138**, No. 2, p. 673—683 (1986).

78.30. Infrared and Raman spectra and scattering

Original Papers and Short Notes

J. WESZKA, M. RENUCCI, and A. ZWICK
*Some Aspects of Raman Scattering in Cd_3As_2 Single Crystals
(78.30; S7)
(b) Vol. **133**, No. 1, p. 57—64 (1986).

M. GOIRAN, A. ZWICK, R. CARLES, J. LEOTIN, E. ARUSHANOV, and M. VON ORTENBERG
*Cadmium Antimonide Lattice Excitations
(78.30; S7)
(b) Vol. **133**, No. 1, p. K1—K4 (1986).

D. WRUCK and F. SPIEGELBERG
Metastable Species of Oxygen-Related Donors in Silicon
(71.55; 78.30; S5.11)
(b) Vol. **133**, No. 1, p. K39—K43 (1986).

N. N. BELEVICH, A. I. LUKOMSKII, L. A. MAKOVETSKAYA, and A. P. CHERNYAKOVA
*Absorption Edge of Stoichiometric and Nonstoichiometric $CuInSe_2$ Crystals
(78.30; 71.55; S8.16)
(b) Vol. **133**, No. 1, p. K49—K52 (1986).

NGUYEN VAN DEN, A. HEINRICH, K. KLOSTERMANN, and G. SOBE
*IR Absorption in CrSiO Thin Films
(78.30; 63.50; 78.65; S10.15)
(a) Vol. **93**, No. 1, p. 163—170 (1986).

P. J. S. FOOT and B. A. NEVETT
Electronic Conduction in p- and n-Type $NiPS_3$
(72.20; 78.30; S8)
(a) Vol. **93**, No. 1, p. 283—292 (1986).

M. G. KAPLUNOV, E. B. YAGUBSKII, L. P. ROSENBERG, and YU. G. BORODKO
*Erratum to: Optical Properties of the Two Crystal Modifications of the Organic Conductor $(BEDT\text{-}TTF)_2I_3$
(78.30; S12)
(a) Vol. **93**, No. 1, p. 399 (1986).

L. MERTEN and R. KOTZOTT
Elliptically Polarized Magnon-Polaritons with Longitudinal and Transverse Components in Uniaxial Ferrites
(71.36; 75.50; 78.30)
(b) Vol. **133**, No. 2, p. 547—556 (1986).

T. N. MAMONTOVA, KH. K. MUKANOV, G. Z. VINOGRADOVA, N. G. MAISASHVILI, and I. Z. BABIEVSKAYA
*A Study of the Optical Properties and Photoluminescence of Glasses in the Ge–P–S System
(78.30 and 78.55; S11)
(a) Vol. **93**, No. 2, p. 635—644 (1986).

V. Riede, H. Neumann, H. Sobotta, R. Schwabe, W. Seifert, and S. Schwetlick
*The Localized Vibrational Mode of Nitrogen in GaAs
(78.30; S7.12)
(a) Vol. **93**, No. 2, p. K151—K156 (1986).

J. Kalus, H. Gerlach, J. Schleifer, F. Wörlen, G. Voss, M. Godlewska, I. Natkaniec, N. Karl, and M. Prager
Inelastic Incoherent Neutron Scattering and Raman Scattering of 2,3-Dimethylnaphthalene
(63.50; 78.30; S12)
(b) Vol. **134**, No. 1, p. 53—62 (1986).

Y. Shibata, T. Suemoto, and M. Ishigame
*Raman Scattering Studies of Mobile Ions in Superionic Conductor Hollandites
(78.30; S10)
(b) Vol. **134**, No. 1, p. 71—79 (1986).

A. Parisini, P. P. Lottici, C. Razzetti, and R. Bacewicz
*Resonance Effects in the Raman Spectrum of $CdGa_2Se_4$
(78.30; 63.20; S8.16)
(b) Vol. **134**, No. 1, p. 81—86 (1986).

S. M. Ryabchenko and Yu. G. Semenov
EPR and Spin-Flip Raman Scattering in the n-Type Magnetically Mixed Semiconductors
(71.70; 76.30; 78.30)
(b) Vol. **134**, No. 1, p. 281—288 (1986).

F. El-Kabbany, Y. Badr, M. Tosson, S. Taha, and S. Mahrous
A Detailed IR Study of the Order–Disorder Phase Transition of $AgNO_3$
(64.60; 78.30; S11.1)
(a) Vol. **94**, No. 1, p. 35—43 (1986).

I. Kosacki, K. Hibner, A. P. Litvinchuk, and M. Ya. Valakh
*Lattice Dynamics and Ionic Conductivity of PbF_2:6% LiF Superionic Crystals
(66.30 and 78.30; 61.70; S9)
(b) Vol. **134**, No. 2, p. 495—501 (1986).

A. Ya. Dzyublik
Effect of Forced Vibrations on the Scattering of X-Rays and Mössbauer Radiation by a Crystal. II. Dynamical Effects
(61.10 and 76.80; 63.20; 78.30)
(b) Vol. **134**, No. 2, p. 503—513 (1986).

A. N. Botvich, V. F. Shabanov, A. N. Vtyurin, and T. A. Pozdnyakova
Photon–Phonon Interaction Influence on the External Vibration Spectra of Molecular Crystals
(63.20 and 71.36; 78.30; S12)
(b) Vol. **134**, No. 2, p. 515—522 (1986).

R. Pechstedt and L. Leine
Effects of a Thin Transition Layer on the Dispersion Relation of Surface Polaritons on a Randomly Rough Surface
(71.36 and 73.20; 68.30; 78.30; S4)
(b) Vol. **135**, No. 2, p. 749—758 (1986).

T. Juhász, J. S. Bakos, and Cs. Kuti
Direct Lifetime Measurement of the A_{1g} Phonon Mode in Calcite
(63.20; 78.30; S11.1)
(b) Vol. **135**, No. 2, p. K99—K102 (1986).

N. M. Gasanly, N. N. Melnik, V. I. Tagirov, and A. A. Yushin
*Special Features of First- and Second-Order Raman and Infrared Spectra of $GaS_{1-x}Se_x$ Mixed Crystals
(78.30; 63.20; S8)
(b) Vol. **135**, No. 2, p. K107—K112 (1986).

B. G. Svensson and J. L. Lindström
Annealing Studies of the 862 cm^{-1} Infrared Band in Silicon
(61.70; 78.30; S5.11)
(a) Vol. **95**, No. 2, p. 537—542 (1986).

A. G. Abdullaev and T. D. Ibragimov
*One-Phonon Absorption in GaS Thin Films
(63.20 and 78.30; S8)
(b) Vol. **136**, No. 1, p. K13—K15 (1986).

G. Irmer, J. Monecke, B. Kh. Bairamov, and V. V. Toporov
*Phonon Shifts in GaP Due to Temperature and Pressure Rise Induced by a Laser Beam
(63.20 and 78.30; S7.11)
(b) Vol. **136**, No. 2, p. 481—488 (1986).

M. Sieskind, M. Ayadi, and G. Zachmann
*Infrared Lattice Vibration, Dielectric Dispersion, and Lattice Dynamics of BaFCl
(63.20 and 78.30; S9.15)
(b) Vol. **136**, No. 2, p. 489—495 (1986).

P. Kleinert and M. Giehler
Theory of Free-Carrier Infrared Absorption in GaAs
(72.10; 63.20; 71.45; 78.30; S7.12)
(b) Vol. **136**, No. 2, p. 763—777 (1986).

Hui Xie, Zhi-cheng Wang, and Jun-Xin Fang
*Study of Material Dispersion in Amorphous Silica Optical Fibers
(78.30; 78.20; S10.1)
(a) Vol. **96**, No. 2, p. 483—487 (1986).

S. Sinha and A. P. B. Sinha
Effect of Substrate Temperature on Bond Angle Distortion in a-Si:H Films. A Raman Scattering Study
(68.55; 78.30; S5.11)
(a) Vol. **96**, No. 2, p. K117—K120 (1986).

CHEN GUANGHUA, ZHEN CHENZHI, ZHANG FANGQING, CHENG JINLONG, and CHEN WEI
*ESR and IR Studies on B-Doped Hydrogenated Amorphous Silicon–Nitrogen Film
(76.30 and 78.30; 78.65; S7)
(a) Vol. **96**, No. 2, p. K187—K189 (1986).

SH. NUROV, V. M. BURLAKOV, E. A. VINOGRADOV, N. M. GASANLY, and B. M. DZHAVADOV
*Vibrational Spectra of $TlInS_2$, $TlIn_{0.95}Ga_{0.05}S_2$, and $TlIn(S_{0.8}Se_{0.2})_2$ Crystals in the Vicinity of Phase Transitions
(64.60 and 78.30; S8.16)
(b) Vol. **137**, No. 1, p. 21—32 (1986).

P. KLEINERT
Ionized-Impurity-Mediated Free-Carrier Absorption in n-GaAs Slabs
(63.20 and 71.45; 78.30; S7.12)
(b) Vol. **137**, No. 1, p. 125—135 (1986).

A. MEMON and D. B. TANNER
*Long Wavelength Optical Phonons in Mixed Alkali Halide Powder Crystals
(63.20 and 78.30; S9.11)
(b) Vol. **137**, No. 1, p. K9—K13 (1986).

A. A. KAMINSKII, E. L. BELOKONEVA, B. V. MILL, S. E. SARKISOV, and K. KURBANOV
*Crystal Structure, Absorption, Luminescence Properties, and Stimulated Emission of Ga Gehlenite $(Ca_{2-x}Nd_xGa_{2+x}Si_{1-x}O_7)$
(78.30 and 78.55; 61.60; 78.45; S10)
(a) Vol. **97**, No. 1, p. 279—290 (1986).

M. ZAŁUŻNY and A. ŁUKASIK
*Theory of the Infrared Spectrum of Size-Quantized Bi Films
(78.30; 71.25; 78.65; S4)
(b) Vol. **137**, No. 2, p. 607—617 (1986).

A. P. KOVCHAVTSEV, G. L. KURISHEV, and K. O. POSTNIKOV
Inelastic Electron Tunneling Spectroscopy of InAs–Superthin Insulator–Au Structures
(63.20 and 73.40; 78.30; S7.12)
(a) Vol. **97**, No. 2, p. 421—425 (1986).

M. S. HUSSAIN, E. E. KHAWAJA, and G. D. KHATTAK
*Optical and Infrared Studies of Sodium-Ditellurate Glasses Containing Chromium Oxide
(78.30 and 78.50; S11.1)
(a) Vol. **97**, No. 2, p. 451—454 (1986).

M. STÖLZER, M. STORDEUR, H. SOBOTTA, and V. RIEDE
IR Transmission Investigations of $(Bi_{1-x}Sb_x)_2Te_3$ Single Crystals
(78.20; 71.25; 78.30; S8)
(b) Vol. **138**, No. 1, p. 259—266 (1986).

CHANG YOUMING, LÜ SHENGQING, and QIAN RENYUAN
Polymorphism and Electronic Properties of 3-Ethyl-5-[2-(3-Ethyl-2-Benzothiazolinylidene)-Ethylidene]-Rhodanine
(61.65; 72.40; 78.30; S12)
(a) Vol. **98**, No. 1, p. 37—42 (1986).

P. DANESH, I. SAVATINOVA, E. ANACHKOVA, and ST. GEORGIEV
*Raman Scattering in Glow Discharge Si:H:Cl Films
(78.30; 61.40; 68.55; S5.11)
(a) Vol. **98**, No. 1, p. 127—132 (1986).

H. GRUBER, E. KRAUTZ, H. P. FRITZER, K. GATTERER, and A. POPITSCH
Changes of Electrical Conductivity, Magnetic Susceptibility, and IR Spectra in the Ternary System $Mo_{n-x}W_xO_{3n-1}$
(72.80; 75.50; 78.30; S10.15)
(a) Vol. **98**, No. 1, p. 297—304 (1986).

E. SCHULTHEISS, A. SCHARMANN, and D. SCHWABE
Lattice Vibrations in $BiLiF_4$ and $YLiF_4$
(63.20; 78.30; S9.16)
(b) Vol. **138**, No. 2, p. 465—475 (1986).

E. J. K. B. BANDA
*Optical Absorption of $FePSe_3$ in the near Infrared, Visible, and near Ultraviolet Regions
(78.30 and 78.40; S8.16)
(b) Vol. **138**, No. 2, p. K125—K129 (1986).

M. COUZI, S. KHAÏROUN, and A. TRESSAUD
Structural Phase Transitions in $Rb_2KM^{III}F_6$ Elpasolites. II. Raman Scattering and Group-Theoretical Studies of Rb_2KFeF_6 and Rb_2KYF_6
(64.70; 63.20; 78.30; S9.16)
(a) Vol. **98**, No. 2, p. 423—434 (1986).

78.40. Visible and ultraviolet spectra

Original Papers and Short Notes

S. G. GUSEINOV, G. D. GUSEINOV, N. Z. GASANOV, and S. B. KYAZIMOV
*Special Features of Exciton Absorption Spectra of $A^{III}B^{III}X_2^{VI}$-Type Layer-Semiconductor Crystals
(71.35 and 78.40; S8.16)
(b) Vol. **133**, No. 1, p. K25—K30 (1986).

R. CAPELLETTI, R. FAVA, M. MANFREDI, and M. SOLZI
Decay Scheme of 439 and 478 nm Emission Bands in the KCl:Eu^{2+} System
(78.55; 71.55; 78.40; S9.11)
(a) Vol. **93**, No. 1, p. 263—269 (1986).

D. WOLF and S. KLEMMER-SACK
The Luminescence of Nd^{3+} in the Polymorphic Perovskites $Ba_6B_2W_3O_{18}$ (B = Gd, Lu, Y). A Case with a Low-Lying Charge Transfer Transition
(78.55; 71.55; 78.40; S11)
(a) Vol. **93**, No. 1, p. 271—275 (1986).

V. A. KISELEV, B. V. NOVIKOV, A. E. CHEREDNICHENKO, and E. A. UBUSHIEV
*Effect of Thin Space-Charge Layers on Exciton Reflectance
(71.35 and 78.40; S8.11)
(b) Vol. **133**, No. 2, p. 573—581 (1986).

K. A. DMITRENKO, S. G. SHEVEL, L. V. TARANENKO, and A. V. MARINTCHENKO
*The Temperature Dependence (4.2 to 293 K) of the Resonance Energies of Excitonic Transitions in II–VI Compounds
(71.35 and 78.40; S8.11; S8.12; S8.13)
(b) Vol. **134**, No. 2, p. 605—613 (1986).

M. LENGLET, P. FOULATIER, J. DÜRR and J. ARSÈNE
Characterization of Cu–O Bonds in $CuMM'O_4$ (M = Fe, Cr; M' = Al, Ga, Mn) Mixed Oxydes. Correlation with the Jahn-Teller Effect
(61.50; 78.40; 78.70; S11.2)
(a) Vol. **94**, No. 2, p. 461—466 (1986).

N. N. ABLYAZOV, A. G. ARESHKIN, V. G. MELEKHIN, L. G. SUSLINA, and D. L. FEDOROV
*Fluctuation-Induced Broadening of Exciton Reflection Spectra in $A^{II}B^{VI}$ Solid Solutions
(71.35 and 78.40; 71.25; S8.15)
(b) Vol. **135**, No. 1, p. 217—225 (1986).

B. HÖNERLAGE, F. TOMASINI, J. Y. BIGOT, M. FRINDI, and J. B. GRUN
*Nonlinear Optical Effects Studied by Time-Resolved Reflection Measurements
(71.35 and 78.40; S9)
(b) Vol. **135**, No. 1, p. 271—282 (1986).

V. A. Kiselev, B. V. Novikov, A. S. Batyrev, E. A. Ubushiev, and A. E. Cherednichenko
Localization of Excitons in Space-Charge Layers
(71.35; 73.30; 78.40; S8.11; S8.12)
(b) Vol. **135**, No. 2, p. 597—604 (1986).

A. V. Sherman
*Exciton Absorption Spectrum (II)
(71.35 and 78.40)
(b) Vol. **135**, No. 2, p. 697—705 (1986).

J. Prakash
Thermally Stimulated Depolarization Measurements in AsO_2^--Doped Potassium Halides at Different Heating Rates
(77.30 and 77.40; 78.40; S9.11)
(b) Vol. **135**, No. 2, p. 741—748 (1986).

K. Kambas, A. Anagnostopoulos, B. Ploss, and J. Spyridelis
*On the Optical Absorption Edge in the $CdInGaS_4$ Layered Compound at Low Temperatures
(78.40; S8.16)
(a) Vol. **95**, No. 2, p. K139—K144 (1986).

E. Radzhabov and P. Figura
*Optical Properties of Oxygen-Vacancy Centers in Fluorite
(78.40 and 78.50; S9.12)
(b) Vol. **136**, No. 1, p. K55—K59 (1986).

R. Madelon, J. Hurel, and E. Paumier
Exciton Absorption Spectrum in a $ZnSiP_2$ Crystal
(71.35; 78.40; S7.16)
(b) Vol. **136**, No. 2, p. 679—684 (1986).

W. Giriat
*Optical Absorption in $Mn_xCd_{1-x}S$
(78.40; S8.15)
(b) Vol. **136**, No. 2, p. K129—K131 (1986).

O. B. Tagiev and E. N. Ibrahimova*
*Optical Properties of $EuGa_2S_4$ near the Fundamental Absorption Edge
(78.40; S8.16)
(a) Vol. **97**, No. 1, p. K49—K52 (1986).

V. S. Dneprovskii, V. D. Egorov, D. S. Khechinashvili, and H. X. Nguyen
*Room Temperature Self-Screening and Screening of Excitons in GaSe
(71.35 and 78.40; S8)
(b) Vol. **138**, No. 1, p. K39—K42 (1986).

E. J. K. B. Banda
*Optical Absorption of $FePSe_3$ in the near Infrared, Visible, and near Ultraviolet Regions
(78.30 and 78.40; S8.16)
(b) Vol. **138**, No. 2, p. K125—K129 (1986).

I. V. Bodnar and A. I. Lukomskii
*The Concentration Dependence of the Band Gap for $CuGa_xIn_{1-x}S_2$ and $AgGa_xIn_{1-x}S_2$ Solid Solutions (71.25 and 78.40; S8.16)
(a) Vol. **98**, No. 2, p. K165—K169 (1986).

78.45. Stimulated emission

Original Papers and Short Notes

A. P. Dudka, A. A. Kaminskii, and V. I. Simoňov
Refinement of $NaGdGeO_4$, $NaYGeO_4$, and $NaLuGeO_4$ Single-Crystal Structures
(61.55; 71.70; 78.45; 78.55; S11.1)
(a) Vol. **93**, No. 2, p. 495—502 (1986).

H. Bachert, W. Pittroff, I. Rechenberg, S. Stoeff, and K. Vogel
The Influence of Mg Doping on the Optical and Electrical Properties of GaAs/(AlGa)As Injection Lasers
(66.30 and 68.55; 73.40; 78.45; S7.12; S7.15)
(a) Vol. **94**, No. 1, p. 321—327 (1986).

N. Kukhtarev, B. Pavlik, and T. Semenets
*Selfdiffraction and Phase Conjugation of Laser Beams in Electrooptic Crystals
(78.20 and 78.45)
(a) Vol. **94**, No. 2, p. 623—633 (1986).

V. Schulz and L. Merten
Euler's Formulae and Sturm's Sequence and the Stimulating Regions for Stimulated Raman Effect in Simple Piezoelectric Cubic Crystals
(71.10 and 78.30; 63.20; 71.36; 78.45)
(b) Vol. **135**, No. 1, p. 105—118 (1986).

Yu. V. Naboikin, S. N. Andrianov, P. V. Zinoviev, Yu. V. Malyukin, E. N. Rudenko, V. V. Samartsev, N. B. Silaeva, and Yu. E. Sheibut
*Optical Superradiance in Pyrene-Doped Biphenyl Crystals and Effect of Phonons on Its Formation
(63.20 and 78.45; S12)
(b) Vol. **135**, No. 2, p. 503—512 (1986).

L. L. Kulyuk, S. I. Radautsan, E. V. Russu, A. V. Siminel, V. G. Smirnov, and E. E. Strumban
*Photoluminescence and Laser Emission in $In_{0.53}Ga_{0.47}As$/InP Layers
(78.45 and 78.65; S7.11; S7.15)
(a) Vol. **96**, No. 1, p. 289—293 (1986).

A. A. Kaminskii
*Stimulated-Emission Spectroscopy of Er^{3+} Ions in Cubic $(Y, Ln)_3Al_5O_{12}$ and Monoclinic $K(Y, Ln)W_2O_8$ Single Crystals
(78.45; 78.55; S10)
(a) Vol. **96**, No. 2, p. K175—K179 (1986).

A. A. Kaminskii, E. L. Belokoneva, B. V. Mill, S. E. Sarkisov, and K. Kurbanov
Crystal Structure, Absorption, Luminescence Properties, and Stimulated Emission of Ga Gehlenite $(Ca_{2-x}Nd_xGa_{2+x}Si_{1-x}O_7)$
(78.30 and 78.55; 61.60; 78.45; S10)
(a) Vol. **97**, No. 1, p. 279—290 (1986).

A. A. KAMINSKII
*Stimulated Emission Spectroscopy of Ln^{3+} Ions in Tetragonal $LiLuF_4$ Fluoride
(78.45; 71.70; 78.55; S9.16)
(a) Vol. **97**, No. 1, p. K53—K58 (1986).

J. PASTRŇÁK, F. KAREL, J. OSWALD, and O. PETŘÍČEK
The Electric Field Dependence of Photoconductivity Spectra in AlGaAs/GaAs DH Laser Diodes in Relation with Laser Parameters
(72.40 and 73.40; 78.45; S7.12; S7.15)
(a) Vol. **97**, No. 2, p. 657—665 (1986).

A. A. KAMINSKII and S. E. SARKISOV
Stimulated-Emission Spectroscopy of Pr^{3+} Ions in Monoclinic BaY_2F_8 Fluoride
(71.55; 78.45; 78.55; S9.16)
(a) Vol. **97**, No. 2, p. K163—K168 (1986).

A. A. KAMINSKII, K. KURBANOV, and A. G. PETROSYAN
*Spectral Composition and Kinetics of 2 μm Stimulated Emission of Ho^{3+} Ions in Sensitized $Y_3Al_5O_{12}$ and $Lu_3Al_5O_{12}$ Single Crystals
(78.45; 78.55; S11)
(a) Vol. **98**, No. 1, p. K57—K62 (1986).

78.50. Impurity and defect absorption in solids

Original Papers and Short Notes

S. Asano, Y. Nakao, N. Yamashita and I. Matsuyama
Luminescence and Non-Radiative $^3T_{1u} \rightarrow {}^3A_{1u}$ Transition in the BaS:Bi^{3+} Phosphor
(78.55; 71.55; 78.50; S8)
(b) Vol. **133**, No. 1, p. 333—344 (1986).

T. Kawakubo
Color Centers in Irradiated Para-Dichlorbenzene Crystals
(61.70; 76.30; 78.50; S12)
(a) Vol. **93**, No. 1, p. 177—184 (1986).

E. Wendler, W. Wesch, and G. Götz
*Near-Edge Optical Absorption Behaviour in Weakly Damaged Ion-Implanted GaAs
(61.70 and 78.50; S7.12)
(a) Vol. **93**, No. 1, p. 207—212 (1986).

O. Goede, W. Heimbrodt, and F. Sittel
Photoacoustic Spectroscopy of Isovalent Impurities in II–VI Semiconductors
(78.20; 71.55; 78.50; S8.11)
(a) Vol. **93**, No. 1, p. 277—282 (1986).

K. D. Nierzewski, B. Macalik, and H. Opyrchał
Effect of Low Temperature γ-Irradiation on I–V Dipoles in KCl:Eu^{2+} Crystals
(61.70 and 61.80; 78.50; S9.11)
(b) Vol. **133**, No. 2, p. K91—K93 (1986).

C. Zaldo and F. Agulló-López
Optical Monitoring of the Precipitation-Induced Hardening in NaCl:Sr and NaCl:Ca
(61.70; 78.50; 78.55; S9.11)
(a) Vol. **93**, No. 2, p. 535—542 (1986).

Ya. I. Latushko, L. F. Makarenko, V. P. Markevich, and L. I. Murin
Electrical and Optical Characterization of Thermal Donors in Silicon
(71.55; 78.50; S5.11)
(a) Vol. **93**, No. 2, p. K181—K184 (1986).

L. Bosi and M. Nimis
Pertubative Effects on the F Centre Lifetime in NaF Crystals Doped with Magnesium
(61.70; 78.50; S9.11)
(b) Vol. **134**, No. 1, p. K17—K21 (1986).

P. R. Collins and W. J. Fredericks
*Note on the Absorption Spectrum of KBr:Cd^{2+}
(78.50; S9.11)
(b) Vol. **134**, No. 1, p. K67—K70 (1986).

R. C. Bhuniya and S. E. Rao
Colour Centres and Luminescence in KCl:Ni Crystals
(61.70 and 78.60; 78.50; S9.11)
(b) Vol. **134**, No. 2, p. 561—568 (1986).

W. Ulrici
*Optical Absorption of EL2° and $EL2^+$ in Semi-Insulating GaAs:Cr
(78.50; S7.12)
(b) Vol. **134**, No. 2, p. 699—706 (1986).

A. A. Kaminskii, A. A. Kornienko, and M. I. Chertanov
Parametrization of Electric-Dipole Intensities in f^N Systems Due to Electron-Correlation Effects
(71.70; 78.50; S9; S11)
(b) Vol. **134**, No. 2, p. 717—729 (1986).

G. Prasad and K. V. Rao
*F-Band Absorption and Thermoluminescence of X-Ray Irradiated KCl Single Crystals Excited with Laser Light
(78.50 and 78.60; S9.11)
(a) Vol. **94**, No. 2, p. 659—663 (1986).

H. Opyrchał, K. D. Nierzewski, B. Macalik, and M. Mladenova
γ-Ray Induced Coloration of KCl:Eu^{2+} Crystals
(61.70; 78.50; 78.60; S9.11)
(b) Vol. **135**, No. 1, p. 141—148 (1986).

E. Rzepka, S. Lefrant, and L. Taurel
Optical Study of the Defect Stability in Tl-Doped Potassium Iodide
(61.70; 78.30; 78.50; 78.60; S9.11)
(b) Vol. **135**, No. 1, p. 149—154 (1986).

H. R. Asatryan, A. S. Kuzanyan, A. G. Petrosyan, A. K. Petrosyan, and E. G. Sharoyan
Single Crystal Growth and Investigation of Optical and ESR Absorption Spectra of Zirconium-Doped YAG
(71.70; 76.30; 78.50; S11.2)
(b) Vol. **135**, No. 1, p. 343—352 (1986).

K. Schmitt
*Absorption and Uniaxial Stress Experiments in the B, C, and D' Bands of Pb^{2+} Centres in KBr, RbBr, KI, and RbI
(78.50; 71.70; S9.11)
(b) Vol. **135**, No. 1, p. 389—398 (1986).

P. A. Nagaraja, A. S. Jacob, and B. J. Reddy
Electronic Spectra of Cu^{2+} and VO_4^{3-} in Vanadinite
(71.70; 78.50; S11)
(b) Vol. **135**, No. 1, p. K39—K42 (1986).

E. J. K. B. BANDA
*Optical Absorption of $CoPS_3$ in the Energy Range 0.73 to 5.40 eV
(71.70 and 78.50; S11)
(b) Vol. **135**, No. 1, p. K43—K47 (1986).

A. A. ALYBAKOV, V. A. GUBANOVA, and M. M. KIDIBAEV
*Anomalous Change of Absorption Spectra of X-Irradiated LiF:U and LiF:U, Na Crystals when Heating
(78.50; 61.80; S9.11)
(b) Vol. **135**, No. 1, p. K49—K52 (1986).

S. KISHIDA, K. MATSUURA, H. NAGASE, H. MORI, F. TAKEDA, and I. TSURUMI
*The Photosensitive Optical Absorption Bands in Zn-Treated and Neutron-Irradiated ZnSe Single Crystals
(78.50; 61.70; S8.12)
(a) Vol. **95**, No. 1, p. 155—164 (1986).

R. SATI, O. P. THAKUR, and J. MEDEIROS E SILVA
*Optical Absorption Line Shape of R′ Centres in Alkali Halides
(78.50; S9.11)
(b) Vol. **135**, No. 2, p. 723—730 (1986).

K. CHETOUANE, J. HUGEL, and J. C. PARLEBAS
*Optical Absorption in Binary Semiconductors Containing Substitutional Impurities
(78.50; 71.55)
(b) Vol. **135**, No. 2, p. 805—812 (1986).

S. E. SARKISOV and A. A. KAMINSKII
Luminescence of Nd^{3+} Ions in Semiconductor $Bi_{12}GeO_{20}$ Crystals
(71.55 and 78.55; 61.60; 78.20; 78.50; S11.1)
(a) Vol. **95**, No. 2, p. 641—649 (1986).

E. RADZHABOV and P. FIGURA
*Optical Properties of Oxygen-Vacancy Centers in Fluorite
(78.40 and 78.50; S9.12)
(b) Vol. **136**, No. 1, p. K55—K59 (1986).

B. C. GRABMAIER and R. OBERSCHMID
Properties of Pure and Doped $Bi_{12}GeO_{20}$ and $Bi_{12}SiO_{20}$ Crystals
(61.70 and 71.55; 72.40; 78.50; 78.55; S10)
(a) Vol. **96**, No. 1, p. 199—210 (1986).

L. E. REZNIK
Electron–Phonon Interaction and Structure of Molecular Luminescence Center in Crystals
(63.20 and 71.70; 71.38; 78.50; 78.55; S11.1)
(b) Vol. **136**, No. 2, p. 417—424 (1986).

V. BETHKENHAGEN and K. HÜBNER
Cluster-Bethe-Lattice Approach to the Electronic Structure of Intrinsic Defects in v-SiO_2
(71.55; 71.20; 78.50; 78.55; S10.1)
(b) Vol. **136**, No. 2, p. 721—734 (1986).

K. Inabe and N. Takeuchi
On the X-Ray Production of F-Centers in BaFCl Doped with Mono- or Trivalent Impurities
(61.70; 78.50; S9.15)
(b) Vol. **136**, No. 2, p. K97—K99 (1986).

E. Radzhabov
Optical Transitions of Chalcogen-Vacancy Centers in Ionic Crystals
(71.55; 78.50; 78.55; S9.11)
(b) Vol. **136**, No. 2, p. K139—K143 (1986).

A. F. Rakov
Investigation of Colour Centres in γ-Irradiated $Y_3Al_5O_{12}$ Crystals at Room Temperature
(61.70; 78.50; S11.2)
(a) Vol. **96**, No. 2, p. K169—K173 (1986).

A. Hamaïdia, A. Hachimi, J. Margerie, and J. F. Hémidy
EPR of Colour Centres in Alkaline Metal Doped Calcium and Strontium Fluorides
(61.70; 76.30; 78.50; S9.12)
(b) Vol. **137**, No. 1, p. 47—56 (1986).

T. F. Veremeichik and I. N. Kalinkina
The Effect of Spin–Orbit Interaction on the Intensity of Transitions from $^4A_2$ and ^2E States in Strong Crystal Fields
(71.70; 78.50; S10.1)
(b) Vol. **137**, No. 1, p. 207—215 (1986).

M. M. Kidibaev, A. A. Alybakov, and V. A. Gubanova
*Absorption Bands of U^{4+} and U^{3+} Ions in X-Irradiated LiF: U, Me Crystals
(78.50; 61.80; S9.11)
(b) Vol. **137**, No. 1, p. K73—K75 (1986).

V. E. Godwin and M. Tomak
Electron Energy Levels in $GaAs/Ga_{1-x}Al_xAs$ Heterojunctions. II. Optical Properties
(71. 55 and 73.40; 78.50; S7.12; S7.15)
(b) Vol. **137**, No. 2, p. 619—622 (1986).

M. S. Hussain, E. E. Khawaja, and G. D. Khattak
*Optical and Infrared Studies of Sodium-Ditellurate Glasses Containing Chromium Oxide
(78.30 and 78.50; S11.1)
(a) Vol. **97**, No. 2, p. 451—454 (1986).

78.55. Photoluminescence

Original Papers and Short Notes

S. ASANO, Y. NAKAO, N. YAMASHITA et I. MATSUYAMA
*Luminescence et transition non radiative $^3T_{1u} \rightarrow {}^3A_{1u}$ dans le luminophore BaS: Bi^{3+}
(Luminescence and Non-Radiative $^3T_{1u} \rightarrow {}^3A_{1u}$ Transition in the BaS:Bi^{3+} Phosphor)
(78.55; 71.55; 78.50; S8)
(b) Vol. **133**, No. 1, p. 333—344 (1986).

C. K. JAYASANKAR, E. ANTIC-FIDANCEV, M. LEMAITRE BLAISE, and P. PORCHER
Optical Properties of EuF_3 and GdF_3 and Paramagnetic Susceptibility of EuF_3
(71.70; 75.20; 78.55; S9)
(b) Vol. **133**, No. 1, p. 345—353 (1986).

M. COLOCCI, F. FERMI, R. QUERZOLI, and A. VINATTIERI
*Time-Resolved Luminescence of $CdGaInS_4$ under Pulsed Laser Excitation at 30 K
(78.55; S8.16)
(b) Vol. **133**, No. 1, p. 379—388 (1986).

G. AMBRAZEVIČIUS and G. BABONAS
*Temperature Dependence of Photoluminescence Excitation Spectra in $Cd_{1-x}Mn_xTe$
(78.55; S8.15)
(b) Vol. **133**, No. 1, p. K45—K48 (1986).

R. CAPELLETTI, R. FAVA, M. MANFREDI, and M. SOLZI
*Decay Scheme of 439 and 478 nm Emission Bands in the KCl:Eu^{2+} System
(78.55; 71.55; 78.40; S9.11)
(a) Vol. **93**, No. 1, p. 263—269 (1986).

D. WOLF and S. KEMMLER-SACK
*The Luminescence of Nd^{3+} in the Polymorphic Perovskites $Ba_6B_2W_3O_{18}$ (B = Gd, Lu, Y). A Case with a Low-Lying Charge Transfer Transition
(78.55; 71.55; 78.40; S11)
(a) Vol. **93**, No. 1, p. 271—275 (1986).

I. B. ERMOLOVICH and V. V. MILENIN
*Nature of Deep Luminescence Centres in CdSe and $CdSe_xTe_{1-x}$
(78.55; 71.70; S8.12; S8.15)
(b) Vol. **133**, No. 2, p. 611—620 (1986).

H. CONZELMANN, A. HANGLEITER, and J. WEBER
Thallium-Related Isoelectric Bound Excitons in Silicon. A Bistable Defect at Low Temperatures
(71.70; 71.35; 78.55; S5.11)
(b) Vol. **133**, No. 2, p. 655—668 (1986).

D. S. Domanevskii, V. V. Krasovskii, M. V. Prokopenya, V. A. Vilkotskii, and S. V. Zhokhovets
The Origin of the Minority Impurity States in Heavily Doped Semiconductors
(71.55; 78.55; S7.11; S7.12)
(b) Vol. **133**, No. 2, p. 693—700 (1986).

A. P. Dudka, A. A. Kaminskii, and V. I. Simonov
Refinement of $NaGdGeO_4$, $NaYGeO_4$, and $NaLuGeO_4$ Single-Crystal Structures
(61.55; 71.70; 78.45; 78.55; S11.1)
(a) Vol. **93**, No. 2, p. 495—502 (1986).

C. Zaldo and F. Agulló-López
Optical Monitoring of the Precipitation-Induced Hardening in NaCl:Sr and NaCl:Ca
(61.70; 78.50; 78.55; S9.11)
(a) Vol. **93**, No. 2, p. 535—542 (1986).

T. N. Mamontova, Kh. K. Mukanov, G. Z. Vinogradova, N. G. Maisashvili, and I. Z. Babievskaya
*A Study of the Optical Properties and Photoluminescence of Glasses in the Ge–P–S System
(78.30 and 78.55; S11)
(a) Vol. **93**, No. 2, p. 635—644 (1986).

F.-G. Kirscht, V. I. Nikitenko, H. Richter, E.-A. Steinman, and E. B. Yakimov
*Photoluminescence of Preannealed Plastically Deformed Silicon Crystals
(61.70 and 78.55; S5.11)
(a) Vol. **93**, No. 2, p. K143—K146 (1986).

G. Massé
Cation Vacancy Acceptor in $AgGaSe_2$ and Some I–III–VI_2 Compounds
(71.55; 78.55; S8.16)
(a) Vol. **93**, No. 2, p. K177—K179 (1986).

A. N. Trukhin
*Luminescence of Fused Silica Containing Sodium
(78.55; S10.1)
(a) Vol. **93**, No. 2, p. K185—K190 (1986).

A. J. Pal and T. N. Misra
*Delayed Luminescence of Organic Mixed Crystals. Guest Concentration Effect in the p-Terphenyl Host–Pyrene Guest System
(78.55; 71.35; S12)
(b) Vol. **134**, No. 1, p. 243—249 (1986).

A. N. Georgobiani, V. B. Gutan, I. M. Tiginyanu, V. V. Ursaki, and T. F. Filina
*Exitonic Luminescence in $CdGa_2S_4$ Single Crystals
(78.55 and 78.60; 71.35; S8.16)
(b) Vol. **134**, No. 1, p. K47—K49 (1986).

K. D. Glinchuk, V. I. Guroshev, and A. V. Prokhorovich
*Photoluminescence in Chromium-Doped Semi-Insulating GaAs after Irradiation by 2.2 MeV Electrons and Annealing
(78.55; 61.80; S7.12)
(a) Vol. **94**, No. 1, p. 203—206 (1986).

K. Mochizuki and M. Takakusaki
*Growth of In-Doped ZnS_xSe_{1-x} Single Crystals and Their Photoluminescence
(78.55; 61.70; 72.80; S8.15)
(a) Vol. **94**, No. 1, p. 243—250 (1986).

P. Roura, A. Lousa, A. Cornet, J. L. Morenza, and J. R. Morante
*Photoluminescence Measurements in Small Grain Size Polycrystalline Zn_3P_2
(78.55; S7)
(a) Vol. **94**, No. 1, p. K47—K50 (1986).

S. O. Cholakh, V. A. Pustovarov, and B. V. Shulgin
The Effects of Electron Excitation Energy Transfer to the Impurity Centre in LiH Crystals
(71.35 and 71.55; 78.55; S2)
(b) Vol. **134**, No. 2, p. 741—744 (1986).

F. J. Bryant, W. E. Hagston, and A. Krier
*The Electroluminescence of Rare-Earth Implanted ZnS and ZnSe Diodes
(78.55 and 78.60; 73.40; S8.11; S8.12)
(a) Vol. **94**, No. 2, p. 701—712 (1986).

P. A. Pundur, J. G. Shavalgin, and V. A. Gritsenko
On the Nature of Deep Centres Responsible for the Memory Effect and Luminescence of a-SiN_x with $x \leqq 4/3$
(71.55; 78.55; 78.60; S7)
(a) Vol. **94**, No. 2, p. K107—K112 (1986).

Sh. M. Efendiev, N. G. Darvishov, and F. K. Yusifov
*Photoluminescence and Excitation Spectra in $Sr_xCa_{1-x}MoO_4$ Solid Solutions
(78.55; S10.15)
(a) Vol. **94**, No. 2, p. K113—K117 (1986).

M. Garrigues and S. K. Krawczyk
*Characterization of the InP Surface by Photoluminescence Imaging
(78.55; S7.11)
(a) Vol. **94**, No. 2, p. K163—K166 (1986).

V. E. Mashchenko, V. F. Kharsik, and S. V. Brezhneva
Secondary Emission of Excitons in CuCl Polycrystals
(71.35; 78.55; S9)
(b) Vol. **135**, No. 1, p. 201—206 (1986).

H. Zimmermann and R. Boyn
*Excitation and Deexcitation of Tm^{3+} Centres in ZnS Involving "External" Excited States
(71.55 and 78.55; S8.11)
(b) Vol. **135**, No. 1, p. 379—387 (1986).

M. Godlewski and D. Hommel
Eu^{2+} Photocharge Transfer Processes in ZnS Crystals Determined by Photo-ESR Measurements
(71.55; 76.30; 78.55; S8.11)
(a) Vol. **95**, No. 1, p. 261—268 (1986).

V. S. Varichenko, A. M. Zaitsev, and V. F. Stelmakh
*Luminescence of Natural IIa Diamond Implanted with Nitrogen Ions
(78.55 and 78.60; 61.70; S5)
(a) Vol. **95**, No. 1, p. K25—K28 (1986).

V. V. Vasilev, I. P. Mikhailovskii, and K. K. Svitashev
*Effect of Hydrogen on Photoluminescence Spectra of Silicon Nitride Amorphous Films
(78.55; S7)
(a) Vol. **95**, No. 1, p. K37—K42 (1986).

D. Crandles and E. Fortin
The Trap Distribution and the Luminescence Band in $ZnIn_2S_4$
(71.55; 78.55; S8.16)
(a) Vol. **95**, No. 1, p. K47—K51 (1986).

V. E. Mashchenko and A. I. Ziborov
Recombination Radiation and Electroreflection of Excitons in $Zn_xCd_{1-x}Se$ Solid Solutions
(71.35; 78.20; 78.55; S8.15)
(b) Vol. **135**, No. 2, p. 623—631 (1986).

W. Heimbrodt and O. Goede
Energy Transfer Processes between Te_n Centres in ZnS:Te and CdS:Te
(71.35 and 71.55; 78.55; S8.11; S8.15)
(b) Vol. **135**, No. 2, p. 795—804 (1986).

H. Weinert, A.Glismann, and J. Voigt
*"Memory" Effect of Bound Exciton Luminescence in GaP:N
(78.55; 71.35; S7.11)
(b) Vol. **135**, No. 2, p. K133—K138 (1986).

D. Cassi, M. Manfredi, and M. Solzi
*On the 427 nm Emission Band in the KCl:Eu^{2+} System
(78.55; S9.11)
(b) Vol. **135**, No. 2, p. K143—K147 (1986).

S. E. Sarkisov and A. A. Kaminskii
*Luminescence of Nd^{3+} Ions in Semiconductor $Bi_{12}GeO_{20}$ Crystals
(71.55 and 78.55; 61.60; 78.20; 78.50; S11.1)
(a) Vol. **95**, No. 2, p. 641—649 (1986).

J. P. Aicardi et G. Aguero
*Luminescences excitoniques et du bord de bande du thiogallate d'argent ($AgGaS_2$)
(Excitonic and Edge Luminescence of Silver Thiogallate ($AgGaS_2$))
(78.55 and 78.60; 71.35; 71.55; S8.16)
(a) Vol. **95**, No. 2, p. 679—689 (1986).

J. W. Tomm, K. Herrmann, A. E. Yunovich, and I. V. Koleshnikov
*YAG-Laser Excited Photoluminescence in p-CdTe
(78.55; S8.13)
(a) Vol. **95**, No. 2, p. K131—K134 (1986).

O. GOEDE, W. HEIMBRODT, and V. WEINHOLD
*Luminescence and Excitation Spectroscopy of MnS Thin Films
(78.55; 78.65; S8.11)
(b) Vol. **136**, No. 1, p. K49—K54 (1986).

B. C. GRABMAIER and R. OBERSCHMID
Properties of Pure and Doped $Bi_{12}GeO_{20}$ and $Bi_{12}SiO_{20}$ Crystals
(61.70 and 71.55; 72.40; 78.50; 78.55; S10)
(a) Vol. **96**, No. 1, p. 199—210 (1986).

P. N. METLINSKII, S. I. RADAUTSAN, I. M. TIGINYANU, V. G. TYRZIU, and V. V. URSAKI
Deep Levels in $HgGa_2Se_4$ Single Crystals
(71.55; 72.40; 78.55; S8.16)
(a) Vol. **96**, No. 1, p. 295—299 (1986).

L. E. REZNIK
Electron–Phonon Interaction and Structure of Molecular Luminescence Center in Crystals
(63.20 and 71.70; 71.38; 78.50; 78.55; S11.1)
(b) Vol. **136**, No. 2, p. 417—424 (1986).

P. G. BARANOV, V. A. VETROV, N. G. ROMANOV, and V. TOPA
ODMR in Triplet Excited States of ns^2 Ions in Alkali Halide Crystals
(71.55 and 76.70; 78.55; S9.11)
(b) Vol. **136**, No. 2, p. 699—707 (1986).

V. BETHKENHAGEN and K. HÜBNER
Cluster-Bethe-Lattice Approach to the Electronic Structure of Intrinsic Defects in v-SiO_2
(71.55; 71.20; 78.50; 78.55; S10.1)
(b) Vol. **136**, No. 2, p. 721—734 (1986).

U. W. POHL
*Manganese Correlated Luminescence in Zinc Sulfide Detected by Site Selected Spectroscopy
(78.55; S8.11)
(b) Vol. **136**, No. 2, p. K123—K128 (1986).

E. RADZHABOV
Optical Transitions of Chalcogen-Vacancy Centers in Ionic Crystals
(71.55; 78.50; 78.55; S9.11)
(b) Vol. **136**, No. 2, p. K139—K143 (1986).

R. HERZOG, G. SCHRAMM, and T. FLADE
*Impurity Analysis of Silicon Epitaxial Layers by the 4.2 K Photoluminescence Spectroscopy
(73.60 and 78.55; 78.65; S5.11)
(a) Vol. **96**, No. 2, p. 629—636 (1986).

E. GEHRKE, K. WANDEL, and H. WEINERT
*Experimental Investigations on the Influence of Surface and Interface Recombination on the Minority Carrier Lifetime in LPE-GaP:N, Si, VPE-GaP:N, Te, and VPE-$GaAs_{0.12}P_{0.88}$:N, Te
(73.40 and 78.55; S7.11; S7.15)
(a) Vol. **96**, No. 2, p. 683—689 (1986).

R. B. Pode
Formation of the Z_1 Colour Centre by Mechanical Bleaching in γ-Irradiated KCl:Ca
(61.70; 78.55; S9.11)
(a) Vol. **96**, No. 2, p. K147—K149 (1986).

A. A. Kaminskii
Stimulated-Emission Spectroscopy of Er^{3+} Ions in Cubic (Y, Ln)$_3$Al$_5$O$_{12}$ and Monoclinic K(Y, Ln)W$_2$O$_8$ Single Crystals
(78.45; 78.55; S10)
(a) Vol. **96**, No. 2, p. K175—K179 (1986).

D. V. Korbutyak, Yu. V. Kryuchenko, L. A. Demchina, V. G. Litovchenko, and V. I. Sugakov
*Influence of Macroscopic Inclusions on Excitonic Luminescence of InSe Crystals
(78.55; 71.35; 71.36; S8)
(b) Vol. **137**, No. 1, p. 217—228 (1986).

V. Z. Polinger and S. I. Boldyrev
The Multimode Jahn-Teller Effect in the Luminescence Spectrum of the Tetrahedral $^2T_2 \otimes (e + t_2 + t_2)$ Impurity System ZnS:Cu^{2+}
(63.20 and 71.70; 78.55; S8.11)
(b) Vol. **137**, No. 1, p. 241—253 (1986).

E. Müller and W. Gebhardt
*Position and Lifetime of Photoluminescence in Cd$_{1-x}$Mn$_x$Te and Zn$_{1-x}$Mn$_x$Te. Exchange Dependent Effects
(78.55; S8.15)
(b) Vol. **137**, No. 1, p. 259—267 (1986).

K. Thonke, U. Schall, and R. Sauer
*Photoluminescence Study of the Gallium Defect Spectrum at ≈ 0.928 eV (Ga3) in Irradiated Silicon
(71.55 and 78.55; 71.35; S5.11)
(b) Vol. **137**, No. 1, p. 291—303 (1986).

U. Schall, K. Thonke, and R. Sauer
*Photoluminescence Study of the Gallium Defect Spectrum at ≈ 1.049 eV (Ga2) in Irradiated Silicon
(71.55 and 78.55; 71.35; S5.11)
(b) Vol. **137**, No. 1, p. 305—317 (1986).

P. Enders
*Photon Recycling in Double Heterostructures. I. The Case of Perfect Optical Confinement
(78.55; 72.20; 72.90)
(b) Vol. **137**, No. 1, p. 353—360 (1986).

G. Blasse, E. W. J. L. Oomen, and J. Liebertz
*On the Luminescence of the Bismuth Borates BiB$_3$O$_6$ and Bi$_3$B$_5$O$_{12}$
(78.55; 78.60; S10)
(b) Vol. **137**, No. 1, p. K77—K81 (1986).

A. A. KAMINSKII, E. L. BELOKONEVA, B. V. MILL, S. E. SARKISOV, and K. KURBANOV
*Crystal Structure, Absorption, Luminescence Properties, and Stimulated Emission of Ga Gehlenite ($Ca_{2-x}Nd_xGa_{2+x}Si_{1-x}O_7$)
(78.30 and 78.55; 61.60; 78.45; S10)
(a) Vol. 97, No. 1, p. 279—290 (1986).

P. GALL, A. KALLEL, N. LAURET, J. P. PEYRADE, B. BROUSSEAU, and J. MAZZASCHI
An Experimental Evidence of the Destruction of Complex Defects by the Introduction of Dislocations in InP
(61.70; 78.55; S7.11)
(a) Vol. 97, No. 1, p. K25—K27 (1986).

A. A. KAMINSKII
Stimulated Emission Spectroscopy of Ln^{3+} Ions in Tetragonal $LiLuF_4$
(78.45; 71.70; 78.55; S9.16)
(a) Vol. 97, No. 1, p. K53—K58 (1986).

P. A. APANASEVICH, V. I. KRUGLOV, and I. E. TRALLE
Interaction of Crystal Impurity Centres with a Coherent Electromagnetic Field
(71.38 and 71.55; 78.55)
(b) Vol. 137, No. 2, p. 641—653 (1986).

P. ENDERS
*Photon Recycling in Double Heterostructures. II. The Case of Non-Perfect Optical Confinement
(78.55; 72.20; 72.90)
(b) Vol. 137, No. 2, p. 701—708 (1986).

YA. D. LEBEDEV and M. D. SHAGALOV
*Electrophysical Properties of Gallium Nitride in the Model of a Nonhomogeneous Semiconductor
(72.40 and 78.55; 71.55; S7.1)
(a) Vol. 97, No. 2, p. 539—548 (1986).

A. A. KAMINSKII and S. E. SARKISOV
Stimulated-Emission Spectroscopy of Pr^{3+} Ions in Monoclinic BaY_2F_8 Fluoride
(71.55; 78.45; 78.55; S9.16)
(a) Vol. 97, No. 2, p. K163—K168 (1986).

H. WEINERT, B. DIEGNER, and J. KUGLER
*Photoluminescence Study on the Influence of Substrate Doping on the Properties of Nominally Undoped VPE-GaAs
(61.70 and 78.55; S7.12)
(a) Vol. 97, No. 2, p. K177—K182 (1986).

G. PRASAD and K. V. RAO
*Spectral Characteristics of Luminescence Exhibited by X-Ray Irradiated KCl and NaCl Single Crystals Subsequently Excited with Laser Light
(78.55; 61.80; S9.11)
(a) Vol. 97, No. 2, p. K183—K186 (1986).

A. Martin, F. J. López, E. Diéguez, F. Agulló-López, and H. Arend
*Luminescence Properties of Rb_2ZnCl_4 and K_2ZnCl_4 Single Crystals
(78.55 and 78.60; 71.35; S9.16)
(b) Vol. **138**, No. 1, p. 245—252 (1986).

A. A. Kaminskii, K. Kurbanov, and A. G. Petrosyan
Spectral Composition and Kinetics of 2 μm Stimulated Emission of Ho^{3+} Ions in Sensitized $Y_3Al_5O_{12}$ and $Lu_3Al_5O_{12}$ Single Crystals
(78.45; 78.55; S11)
(a) Vol. **98**, No. 1, p. K57—K62 (1986).

K. Kurbanov, E. S. Dosmagambetov, R. Uecker, D. Schultze, and A. A. Kaminskii
*Crystal Growth and Spectroscopic Properties of Nd^{3+} Ions in a New Nonstoichiometric Bismuth Phosphate $Bi_{5.8}PO_{11.2}$
(78.55; S11)
(a) Vol. **98**, No. 1, p. K79—K82 (1986).

D. Wolf and S. Kemmler-Sack
*Cathodo- and Photoluminescence in Rare Earth Activated Garnets of Type $Ca_3M_2Ge_{3-x}Si_xO_{12}$ (M = Al, Ga, Sc, Y)
(78.55 and 78.60; S11.2)
(a) Vol. **98**, No. 2, p. 567—575 (1986).

K. Kurbanov, E. S. Dosmagambetov, R. Ehlert, D. Schultze, and A. A. Kaminskii
*Crystal Growth and Spectroscopic Properties of Nd^{3+} Ions in Potassium Bismuth Niobate
(78.55; S10.15; S11.1)
(a) Vol. **98**, No. 2, p. K175—K177 (1986).

78.60. Other luminescence and radiative recombination For photoconduction and photovoltaic effects, see 72.40

Review Article

V. V. Rossin and V. G. Sidorov
*Reabsorption of Recombination Radiation in Semiconductors with High Internal Quantum Efficiency
(73.40 and 78.60; 72.40; 78.20; S7.12)
(a) Vol. **95**, No. 1, p. 15—40 (1986).

Original Papers and Short Notes

R. K. Gartia, R. Singha, and T. Shanker Rao
*Thermoluminescence of Topaz X-Irradiated at 80 K
(78.60; S11.1)
(a) Vol. **93**, No. 1, p. K59—K63 (1986).

A. Opanowicz
*On Thermoluminescence in Conductive Semiconductors
(78.60; S8.12)
(a) Vol. **93**, No. 1, p. K65—K67 (1986).

B. T. Deshmukh, B. H. Ghiya, and R. B. Pode
*Study of the Particle Size Effect on the Thermoluminescence of KCl
(78.60; 61.70; S9.11)
(b) Vol. **133**, No. 2, p. K125—K128 (1986).

R. Pandey and P. K. Ghosh
*AC Electroluminescence in Polycrystalline CaS:Ce Phosphors
(78.60; S8)
(a) Vol. **93**, No. 2, p. K173—K175 (1986).

S. M. Pillai and C. P. G. Vallabhan
*A Study on the Electroluminescence Spectrum of CaS:Sm and CaS:Dy
(78.60; 71.70; S8)
(b) Vol. **134**, No. 1, p. 383—390 (1986).

R. B. Pode and B. H. Ghiya
*On the Deformation Effect on Z_1 Centres in KCl:Sr
(61.70 and 78.60; S9.11)
(b) Vol. **134**, No. 1, p. K23—K26 (1986).

A. N. Georgobiani, V. B. Gutan, I. M. Tiginyanu, V. V. Ursaki, and T. F. Filina
*Excitonic Luminescence in $CdGa_2S_4$ Single Crystals
(78.55 and 78.60; 71.35; S8.16)
(b) Vol. **134**, No. 1, p. K47—K49 (1986).

C. Christodoulides
*Parameter Determination for General-Order Thermoluminescence Peaks Measured with Hyperbolic Heating
(78.60; 71.55)
(a) Vol. **94**, No. 1, p. 251—258 (1986).

M. K. Kerimov and B. A. Süleimanov
*Effect of Electric Field Strength on the Intensity Distribution of Electrostimulated Thermoluminescence of Polyethylene
(78.60; S12)
(a) Vol. **94**, No. 1, p. K51—K54 (1986).

J. Pułtorak
*Current Induced Shift of the Maximum Radiation Plane in LE GaAs:Si Diodes Obtained by LPE Technique
(73.40 and 78.60; S7.12)
(a) Vol. **94**, No. 1, p. K93—K96 (1986).

R. C. Bhuniya and S. E. Rao
*Colour Centres and Luminescence in KCl:Ni Crystals
(61.70 and 78.60; 78.50; S9.11)
(b) Vol. **134**, No. 2, p. 561—568 (1986).

A. B. Ahmed and R. K. Gartia
*Trapping Levels in $KAlSi_3O_8$
(78.60; 71.55; S11)
(a) Vol. **94**, No. 2, p. 645—651 (1986).

G. Prasad and K. V. Rao
*F-Band Absorption and Thermoluminescence of X-Ray Irradiated KCl Single Crystals Excited with Laser Light
(78.50 and 78.60; S9.11)
(a) Vol. **94**, No. 2, p. 659—663 (1986).

F. J. Bryant, W. E. Hagston, and A. Krier
*The Electroluminescence of Rare-Earth Implanted ZnS and ZnSe Diodes
(78.55 and 78.60; 73.40; S8.11; S8.12)
(a) Vol. **94**, No. 2, p. 701—712 (1986).

P. A. Pundur, J. G. Shavalgin, and V. A. Gritsenko
On the Nature of Deep Centres Responsible for the Memory Effect and Luminescence of a-SiN_x with $x \leqq 4/3$
(71.55; 78.55; 78.60; S7)
(a) Vol. **94**, No. 2, p. K107—K112 (1986).

A. M. Zaitsev, A. A. Melnikov, V. B. Shipilo, and E. M. Shishonok
*Cathodolumniescence of Electron Irradiated Cubic Boron Nitride
(61.80 and 78.60; S7.1)
(a) Vol. **94**, No. 2, p. K125—K127 (1986).

H. Opyrchał, K. D. Nierzewski, B. Macalik, and M. Mladenova
γ-Ray Induced Coloration of KCl:Eu^{2+} Crystals
(61.70; 78.50; 78.60; S9.11)
(b) Vol. **135**, No. 1, p. 141—148 (1986).

E. Rzepka, S. Lefrant, and L. Taurel
Optical Study of the Defect Stability in Tl-Doped Potassium Iodide
(61.70; 78.30; 78.50; 78.60; S9.11)
(b) Vol. **135**, No. 1, p. 149—154 (1986).

V. S. Varichenko, A. M. Zaitsev, and V. F. Stelmakh
*Luminescence of Natural IIa Diamond Implanted with Nitrogen Ions
(78.55 and 78.60; 61.70; S5)
(a) Vol. **95**, No. 1, p. K25—K28 (1986).

A. M. Zaitsev, V. B. Shipilo, E. M. Shishonok, and A. A. Melnikov
*Cathodoluminescence of Annealed Cubic Boron Nitride
(78.60; 61.70; S7)
(a) Vol. **95**, No. 1, p. K29—K32 (1986).

M. V. Chukhichev, V. I. Petrov, and R. Stegmann
*Observation of the $N_{\Gamma-x}$-Line in $GaAs_{0.42}P_{0.58}$:N
(78.60; 71.35; 71.55; S7.15)
(b) Vol. **135**, No. 2, p. K139—K142 (1986).

B. Akamatsu, A. M. Jean-Louis, A. C. Papadopoulo and P. Henoc
Electrical Activity of Dislocations Introduced in Optically Degraded InP
(61.70; 61.16; 78.60; S7.11)
(a) Vol. **95**, No. 2, p. 557—564 (1986).

J. P. Aicardi et. G. Aguero
*Luminescence excitoniques et du bord de bande du thiogallate d'argent ($AgGaS_2$)
(Excitonic and Edge Luminescence of Silver Thiogallate ($AgGaS_2$)
(78.55 and 78.60; 71.35; 71.55; S8.16)
(a) Vol. **95**, No. 2, p. 679—689 (1986).

B. T. Deshmukh, S. V. Bodadae, and S. V. Moharil
*Thermoluminescence in K_2ZnCl_4
(78.60; S9.16)
(a) Vol. **95**, No. 2, p. K145—K148 (1986).

B. P. Chandra, N. G. Deshmukh, and K. K. Shrivastava
*A Theoretical Approach to the Impulsive Excitation of Mechanoluminescence in Crystals
(78.60; S12)
(a) Vol. **96**, No. 1, p. 167—175 (1986).

L. N. Skuja and W. Entzian
*Cathodoluminescence of Intrinsic Defects in Glassy SiO_2, Thermal SiO_2 Films, and α-Quartz
(78.60; S10.1)
(a) Vol. **96**, No. 1, p. 191—198 (1986).

B. T. Deshmukh, S. K. Omanwar, and S. V. Moharil
*Thermoluminescence of Colloidal Centres in NaCl
(78.60; S9.11)
(a) Vol. **96**, No. 1, p. 211—215 (1986).

M. A. Ochando, I. Fernández, and J. Llopis
*Effect of Heat Treatments on the Cathodoluminescence of MgO:Al Crystals
(78.60; S10.1)
(a) Vol. **96**, No. 2, p. 547—554 (1986).

N. S. Bogdanyuk and S. S. Ostapenko
*Structure Peculiarities and Annealing Mechanism of the Red Flash-Like Luminescent Centre in CdS Single Crystals Irradiated with 1.2 MeV Electron Beam
(78.60; 61.80; 71.55; S8.11)
(a) Vol. **96**, No. 2, p. 621—628 (1986).

R. Stegmann, G. Neubert, E. Dobruinina, and M. V. Chukhichev
The Influence of the Energy Spectrum of the Nitrogen Bound Exciton on the Efficiency in $GaAs_{1-x}P_x$:N
(71.35; 78.60; S7.15)
(a) Vol. **96**, No. 2, p. 643—650 (1986).

G. Blasse, E. W. J. L. Oomen, and J. Liebertz
On the Luminescence of the Bismuth Borates BiB_3O_6 and $Bi_3B_5O_{12}$
(78.55; 78.60; S10)
(b) Vol. **137**, No. 1, p. K77—K81 (1986).

R. K. Gartia and A. B. Ahmed
*Low-Temperature Thermally Stimulated Luminescence in $KAlSi_3O_8$
(78.60, 71.55; S11)
(a) Vol. **97**, No. 1, p. 257—264 (1986).

M. Matyáš, Jr.
Deep Levels and Processes in the Course of Degradation of GaP:N LED's
(71.55; 73.40; 78.60; S7.11)
(a) Vol. **97**, No. 1, p. 297—305 (1986).

B. P. Chandra, P. K. Dubey, and S. C. Datt
*Mechano- and Electroluminescence of Heavily Doped ZnS:Mn Phosphors
(78.60; S8.11)
(a) Vol. **97**, No. 1, p. K59—K61 (1986).

Uma Subramanian and M. L. Mukherjee
*Thermoluminescence Study of CaF_2:Nd Single Crystals from LNT to 400 °C
(78.60; 71.55; S9.12)
(b) Vol. **137**, No. 2, p. 675—681 (1986).

N. A. Atari and R. Ramani
*Piezoluminescence and Thermoluminescence Spectral Shifts in γ-Irradiated KBr and KCl Crystals
(78.60; S9.11)
(a) Vol. **97**, No. 2, p. 461—468 (1986).

J. LLOPIS, J. PIQUERAS, J. A. GARCÍA, and A. REMÓN
*Red Cathodoluminescence from Deformed MgO Crystals
(78.60; S10.1)
(a) Vol. **97**, No. 2, p. K159—K162 (1986).

Y. KUTOMI, A. TOMITA, and N. TAKEUCHI
*Thermally Stimulated Exoelectron Emission and Thermoluminescence in $Li_2B_4O_7$ Crystallized Glasses
(78.60 and 79.75; S10)
(a) Vol. **97**, No. 2, p. K169—K172 (1986).

A. MARTIN, F. J. LÓPEZ, E. DIÉGUEZ, F. AGULLÓ-LÓPEZ, and H. AREND
*Luminescence Properties of Rb_2ZnCl_4 and K_2ZnCl_4 Single Crystals
(78.55 and 78.60; 71.35; S9.16)
(b) Vol. **138**, No. 1, p. 245—252 (1986).

B. T. DESHMUKH, S. V. BODADE, and S. V. MOHARIL
*Thermoluminescence of $K_2Cd_2(SO_4)_3$
(78.60; 64.70; S11.1)
(a) Vol. **98**, No. 1, p. 239—246 (1986).

A. B. AHMED and R. K. GARTIA
*Thermally Stimulated Luminescence of $NaAlSi_3O_8$ and Its Analysis
(78.60; S10)
(a) Vol. **98**, No. 1, p. K63—K67 (1986).

K. INABE, Y. UTSUI, and M. TAKEUCHI
*Thermoluminescence of BaFCl:Dy and BaFCl:Cu X-Irradiated at Room Temperature
(78.60; 61.70; S9.16)
(b) Vol. **138**, No. 2, p. 493—500 (1986).

D. WOLF and S. KEMMLER-SACK
*Cathodo- and Photoluminescence in Rare Earth Activated Garnets of Type $Ca_3M_2Ge_{3-x}Si_xO_{12}$ (M = Al, Ga, Sc, Y)
(78.55 and 78.60; 71.55; S11.2)
(a) Vol. **98**, No. 2, p. 567—575 (1986).

A. POPOV and R. ATANASOVA
*The Origin of Nonradiative Centers in Degraded LEDs from $GaAs_{0.6}P_{0.4}$
(73.40 and 78.60; 71.55; S7.15)
(a) Vol. **98**, No. 2, p. 629—632 (1986).

78.65. Optical properties of thin films

Original Papers and Short Notes

NGUYEN VAN DEN, A. HEINRICH, K. KLOSTERMANN, and G. SOBE
IR Absorption in CrSiO Thin Films
(78.30; 63.50; 78.65; S10.15)
(a) Vol. **93**, No. 1, p. 163—170 (1986).

J. SZCZYRBOWSKI, K. SCHMALZBAUER, and H. HOFFMANN
Optical Properties and Dynamic Current Density for Polycrystalline Metallic Films
(73.60 and 78.20; 78.65; S1.3)
(a) Vol. **93**, No. 1, p. 329—339 (1986).

N. SAITO and N. TANAKA
*Preparation and Properties of Magnetron Sputtered Fluorinated and Hydrogenated Amorphous Si–C Films
(73.60 and 78.65; 68.55; S6)
(a) Vol. **93**, No. 1, p. 341—345 (1986).

S. CHAUDHURI, S. K. BISWAS, and A. CHOUDHURY
*The Optical Energy Gap of Amorphous $In_{30}Se_{70}$ Films
(78.65; S8)
(a) Vol. **93**, No. 1, p. K47—K49 (1986).

C. TRALLERO GINER and M. ANTÓN
Quantum Theory of Free-Carrier Absorption in Quasi-Two-Dimensional Semiconducting Structures
(78.20; 71.38; 78.65; S5.12; S7.12)
(b) Vol. **133**, No. 2, p. 563—572 (1986).

M. E. ABU-ZAID, A. E. RAKHSHANI, A. A. AL-JASSAR, and Y. A. YOUSSEF
*Determination of the Thickness and Refractive Index of Cu_2O Thin Film Using Thermal and Optical Interferometry
(78.65; S10.1)
(a) Vol. **93**, No. 2, p. 613—620 (1986).

E. E. KHAWAJA and G. D. KHATTAK
*The Optical Absorption Edge in Vanadate Glasses
(78.65; S10.15)
(a) Vol. **93**, No. 2, p. 621—625 (1986).

W. DUŁAK and H. MĘCZYŃSKA
*Exciton Spectra of Thin CdS Films Prepared by Spray Pyrolysis
(78.65; 71.35; S8.11)
(a) Vol. **93**, No. 2, p. K171—K172 (1986).

F. LUKEŠ
Ellipsometry of Silicon with Natural Surface Film at 632.8 nm
(78.20; 78.65; S5.11)
(a) Vol. **94**, No. 1, p. 223—230 (1986).

O. GOEDE, W. HEIMBRODT, T. LAU, G. MATZKEIT, and B. SELLE
*Optical and RBS Studies of ZnS:Te Thin Films
(78.65; 61.70; S8.11)
(a) Vol. **94**, No. 1, p. 259—265 (1986).

V. S. VAVILOV, I. P. AKIMCHENKO, and V. V. KRASNOPEVTSEV
Ion Implantation as a Tool to Control Properties of Amorphous Hydrogenated Silicon
(61.70 and 73.60; 78.65; S5.11)
(a) Vol. **94**, No. 1, p. 343—350 (1986).

A. F. KHOKHLOV, A. I. MASHIN, A. V. ERSHOV, and YU. A. MORDVINOVA
*The Properties of Amorphous Silicon Doped with the Isovalent Germanium Impurity
(73.60 and 78.65; 61.70; S5.11)
(a) Vol. **94**, No. 1, p. 379—384 (1986).

T. BARBASZEWSKI, M. DRWIEGA, E. LIPINSKA, and F. STARZYK
Iodine Implantation in Poly-N-Vinylcarbazole — an Organic Semiconductor
(61.70; 61.80; 78.65; S12)
(a) Vol. **94**, No. 1, p. 419—422 (1986).

J. DRYZEK and A. CZAPLA
*Absorption Oscillations Due to the Size Effect in Thin Metallic Films
(78.65)
(a) Vol. **94**, No. 2, p. 665—671 (1986).

J. BHATTACHARYYA, S. CHAUDHURI, and A. K. PAL
*Studies on the Optical Properties and the Burstein-Moss Shift in Indium Tin Oxide Films
(78.65; 68.55; S10.15)
(a) Vol. **95**, No. 1, p. 239—248 (1986).

K. YASUDA, H. MASUDA, M. TAKEDA, and A. YOSHIDA
*Annealing Effects in Amorphous Hydrogenated Carbon Films Prepared by Plasma Deposition from Butane Gas
(78.65; 68.55; S5)
(a) Vol. **95**, No. 1, p. 249—254 (1986).

B. YOUS et A. DONNADIEU
*Tandems de composés de tungstène. Absorbeurs sélectifs de l'énergie solaire
(Bilayers of Tungsten Compounds. Selective Absorbers of Solar Energy)
(78.65; S1)
(a) Vol. **95**, No. 2, p. 651—657 (1986).

O. GOEDE, W. HEIMBRODT, and V. WEINHOLD
Luminescence and Excitation Spectroscopy of MnS Thin Films
(78.55; 78.65; S8.11)
(b) Vol. **136**, No. 1, p. K49—K54 (1986).

S. A. EL-SAHHAR, E. S. M. HIGGY, M. A. SOBEE, and H. M. TALAAT
*Some Optical Constants of Thin Crystalline Lead Films in the Visible Region
(78.65; S4)
(a) Vol. **96**, No. 1, p. 285—288 (1986).

L. L. Kulyuk, S. I. Radautsan, E. V. Russu, A. V. Siminel, V. G. Smirnov, and E. E. Strumban
*Photoluminescence and Laser Emission in $In_{0.53}Ga_{0.47}As/InP$ Layers
(78.45 and 78.65; S7.11; S7.15)
(a) Vol. **96**, No. 1, p. 289—293 (1986).

V. Hönig, V. Fedorov, G. Liebmann, and P. Süptitz
*Ellipsometric Investigation of Ag-Doping Profiles in Amorphous Chalcogenide Thin Films
(78.65; 68.55; S8)
(a) Vol. **96**, No. 2, p. 611—619 (1986).

R. Herzog, G. Schramm, and T. Flade
Impurity Analysis of Silicon Epitaxial Layers by the 4.2 K Photoluminescence Spectroscopy
(73.60 and 78.55; 78.65; S5.11)
(a) Vol. **96**, No. 2, p. 629—636 (1986).

Chen Guanghua, Zhen Chenzhi, Zhang Fangqing, Cheng Jinlong, and Chen Wei
ESR and IR Studies on B-Doped Hydrogenated Amorphous Silicon–Nitrogen Film
(76.30 and 78.30; 78.65; S7)
(a) Vol. **96**, No. 2, p. K187—K189 (1986).

S. Raghunath Reddy, A. K. Mallik, and S. R. Jawalekar
*Electrical and Optical Properties of Chemically Vapour Deposited Fluorine Doped SnO_2 Films
(73.60 and 78.65; S10.1)
(a) Vol. **96**, No. 2, p. K191—K194 (1986).

G. Jungk and A. Röseler
Optical Properties of Metals: Au
(72.30 and 78.20; 78.65; S1.3)
(b) Vol. **137**, No. 1, p. 117—123 (1986).

A. K. Abass, Z. A. Ahmed, and R. E. Tahir
*Interband Transitions of Chemically Deposited Pyrite FeS_2 Films in the Fundamental Absorption Region between 1 and 3.8 eV
(71.25 and 78.65; S8)
(a) Vol. **97**, No. 1, p. 243—247 (1986).

H. Hemme, H. Dötsch, and J. Middelberg
*Phase Matching of Optical Modes in Orthorhombic Garnet Films
(78.65; S11.2)
(a) Vol. **97**, No. 1, p. 249—256 (1986).

M. Załużny and A. Łukasik
Theory of the Infrared Spectrum of Size-Quantized Bi Films
(78.30; 71.25; 78.65; S4)
(b) Vol. **137**, No. 2, p. 607—617 (1986).

C. A. HOGARTH and Z. T. AL-DHHAN
*Optical Absorption in Thin Films of Cerium Dioxide and Cerium Dioxide Containing Silicon Monoxide
(78.65; S10.15)
(b) Vol. **137**, No. 2, p. K157—K160 (1986).

L. KOMITOV, G. HAUCK, and H. D. KOSWIG
Initially Splayed Nematic Liquid Crystal Layer under an Electric Field. Induction of π-Twist Metastable State
(61.30; 77.90; 78.65; S12)
(a) Vol. **97**, No. 2, p. 645—655 (1986).

A. K. S. THAKUR
*On the Validity of Thin Film Models
(78.20 and 78.65)
(a) Vol. **97**, No. 2, p. K123—K125 (1986).

H. P. HINOV
On the Theory of Partial and Total Internal Reflection of a Plane TM Wave from Symmetrically Realigned Nematic Liquid Crystal Layers
(61.30; 78.65)
(b) Vol. **138**, No. 1, p. 31—46 (1986).

H. HEMME, H. DÖTSCH, and H. DAMMANN
*Stress Induced Optical Anisotropy in Magnetic Garnet Films
(78.65; S11.2)
(a) Vol. **98**, No. 2, p. K119—K124 (1986).

B. P. RAI
*The Optical Absorption Edge of Thin GeO_2 Films Doped with Copper
(78.65; S10.1)
(a) Vol. **98**, No. 2, p. K171—K174 (1986).

78.70. Other interactions of matter with particles and radiation

Original Papers and Short Notes

L. A. Bugaev, I. I. Gegusin, V. N. Datsyuk, A. A. Novakovich, and R. V. Vedrinskii
Multiple Scattering Approach to the XANES Theory of Alkali Halide Crystals. I. Crystalline Potential in the X-Ray Absorption Spectra Problem
(71.10; 78.70; S9.11)
(b) Vol. **133**, No. 1, p. 195—202 (1986).

V. N. Baier, V. M. Katkov, and V. M. Strakhovenko
*Nondipole Radiation of Electrons at Planar Channeling
(61.80 and 78.70; S5; S5.11)
(b) Vol. **133**, No. 1, p. 211—222 (1986).

D. K. Papademetriou, D. Katsanos, and A. G. Doukas
*Plasmon Bands in X-Ray Spectra of Polycrystalline Beryllium
(71.45 and 78.70; S3)
(b) Vol. **133**, No. 1, p. 223—228 (1986).

B. Poumellec, F. Lagnel, J. F. Marucco, and B. Touzelin
*X-Ray Absorption K-Edge Spectra of Titanium in (Ti, Nb)O_2 Rutile Solid Solutions
(78.70; 71.70; S10.15)
(b) Vol. **133**, No. 1, p. 371—377 (1986).

O. P. Singh and V. P. Gupta
Modified PVV Theory and X-Ray K Absorption Edge Studies in Zinc Chalcogenides (Ionicity Determination and d Electron Effect on It)
(71.25; 78.70; S8.11; S8.12; S8.13)
(b) Vol. **133**, No. 1, p. K17—K19 (1986).

V. N. Brudnyi, V. A. Charchenko, N. G. Kolin, V. A. Novikov, A. D. Pogrebnyak, and Sh. M. Ruzimov
Electrical Properties and Positron Annihilation in Neutron-Irradiated n-InP
(61.80; 72.80; 78.70; S7.11)
(a) Vol. **93**, No. 1, p. 195—200 (1986).

J. Filipecki, E. Jakubczyk, B. Całusiński, and D. Gaudyn
Electron Phenomena Connected with Hydriding of the Metallic Glass $Fe_{81}B_{13.5}Si_{3.5}C_2$
(61.70; 72.15; 78.70; S1.1)
(a) Vol. **93**, No. 1, p. K25—K28 (1986).

Yu. M. Annenkov, S. B. Nurmagambetov, and A. B. Chormonov
*Nature of the Centres of Positron Annihilation in Intensively Irradiated Alkali Halides
(61.70 and 78.70; S9.11)
(b) Vol. **133**, No. 2, p. 505—509 (1986).

V. N. Baier, V. M. Katkov, and V. M. Strakhovenko
*Radiation Yield of High-Energy Electrons in Thick Crystals
(61.80 and 78.70; S1.1; S1.2; S5; S5.11; S5.12)
(b) Vol. **133**, No. 2, p. 583—592 (1986).

Cs. Szeles, A. Vértes, and G. Brauer
The Effect of Deposition Temperature on the Defect Structure of Ni Electrodeposits Studied by Positron Annihilation
(61.70; 78.70; S1.2)
(a) Vol. **93**, No. 2, p. 559—564 (1986).

J. Q. Shen, C. W. Lung, and K. L. Wang
Dislocation Core Models and Their Positron Annihilation Effects
(61.70; 78.70)
(b) Vol. **134**, No. 1, p. 97—102 (1986).

Anand P. Pathak and Prasanna K. J. Balagari
*The Effects of Defects on Channeling Radiation
(61.80 and 78.70)
(b) Vol. **134**, No. 1, p. 115—118 (1986).

V. N. Datsyuk, I. I. Gegusin, and R. V. Vedrinskii
Multiple Scattering Approach to the XANES Theory of Alkali Halide Crystals. II: Band Structure Approach and Cluster Approximations
(71.10; 71.25; 78.70; S9.11)
(b) Vol. **134**, No. 1, p. 175—184 (1986).

L. T. Tsankov, N. I. Ognjanov, and R. P. Chakurova
*Time Dependence of Mössbauer Radiation Transmitted through a RF Vibrating Resonant Medium
(78.70; 62.80)
(a) Vol. **94**, No. 1, p. 131—138 (1986).

A. A. Manenkov, G. N. Mikhailova, S. Yu. Sokolov, and S. G. Tikhodeev
Phonon Wind Generation in Ge by CO_2 Laser Induced Heating of Electron–Hole Drops
(71.35; 71.38; 72.30; 78.70; S5.12)
(b) Vol. **134**, No. 2, p. 631—639 (1986).

I. I. Gegusin, V. N. Datsyuk, A. A. Novakovich, L. A. Bugaev, and R. V. Vedrinskii
Multiple Scattering Approach to the XANES Theory of Alkali Halide Crystals. III. Comparison with the Experimental X-Ray Spectra
(71.10; 78.70; S9.11)
(b) Vol. **134**, No. 2, p. 641—650 (1986).

R. Markendorf, E. Zschech, W. Blau, and K. Kleinstück
*Chemical Shifts of EXAFS Backscattering Phases within the Framework of Neutral Pseudoatoms
(71.25 and 78.70; S1.1; S1.2)
(b) Vol. **134**, No. 2, p. K121—K124 (1986).

M. Lenglet, P. Foulatier, J. Dürr and J. Arsène
Characterization of Cu–O Bonds in $CuMM'O_4$ (M = Fe, Cr; M' = Al, Ga, Mn) Mixed Oxydes. Correlation with the Jahn-Teller Effect
(61.50; 78.40; 78.70; S11.2)
(a) Vol. **94**, No. 2, p. 461—466 (1986).

G. DAS, H. C. PADHI, and O. AIKALA
*Directional Compton Profiles of CaF_2
(78.70; S9.12)
(b) Vol. **136**, No. 2, p. 623—627 (1986).

S. K. BOSE, U. B. DUBEY, and V. N. TEWARI
*The Spectral Behaviour of a General Damped Anharmonic Oscillator
(78.70)
(b) Vol. **136**, No. 2, p. 693—698 (1986).

F. EGGERT and W. SCHOLZ
*A Rapid Deconvolution Method Based on the Bayes Theorem Applied to Energy Dispersive X-Ray Emission Analysis — Improvements and Error Influences
(78.70)
(a) Vol. **97**, No. 1, p. K9—K13 (1986).

F. CORNI and G. M. ANTONINI
EXAFS Investigation of Interdiffusive Kinetics in Cu–Au Thin Films
(66.30; 61.55; 78.70; S1.2; S1.3)
(a) Vol. **97**, No. 1, p. K21—K24 (1986).

M. DĘBOWSKA, A. BARANOWSKI, K. JERIE and W. ŚWIĄTKOWSKI
The Effect of Impregnation with Gold on Positron Lifetime Spectra in NaY Zeolite
(61.70; 68.20; 78.70; S2)
(a) Vol. **97**, No. 1, p. K35—K38 (1986).

B. POUMELLEC, J. F. MARUCCO, and B. TOUZELIN
*X K Edge Absorption of Titanium and L Edge of Tantalum in (Ti, Ta)O_2 Rutile Solid Solutions
(78.70; 71.35; 71.70; S10.15)
(b) Vol. **137**, No. 2, p. 519—531 (1986).

N. I. PAPANICOLAOU, N. C. BACALIS, and D. A. PAPACONSTANTOPOULOS
Calculation of Compton Profiles of Tantalum and Tungsten
(71.25; 78.70; S1)
(b) Vol. **137**, No. 2, p. 597—606 (1986).

K. HEMACHANDRAN and A. R. CHETAL
*Correlation between X-Ray Chemical Shift and Partial Charge in Some Minerals
(78.70; S1.62; S7)
(b) Vol. **138**, No. 1, p. 229—233 (1986).

A. A. LAVRENTEV, A. N. GUSATINSKII, A. V. SOLDATOV, and V. YU. SLIVKA
X-Ray Spectroscopic Investigation of the Electronic Energy Structure and Chemical Bond of the Tl_3AS_4 (A = P, As, Ta) Type Semiconductors
(71.25; 78.70; S8.16)
(b) Vol. **138**, No. 1, p. K49—K54 (1986).

78.90. Other topics in optical properties of condensed matter and other interactions of matter with particles and radiation

Original Papers and Short Notes

J. Golczewski, K. Maier, and W. Frank
Defect Reactions in Electron-Irradiated α-$Fe^{111}In$ Studied by Means of PAC Measurements
(61.70; 78.90; S1.1)
(b) Vol. **134**, No. 2, p. 551—559 (1986).

H. Pohlack
Optically Enhanced Schottky Barrier Photodetectors for IR Image Sensor Application
(73.30; 78.90; S1.61)
(a) Vol. **97**, No. 2, p. K211—K214 (1986).

79. Electron and ion emission by liquids and solids; impact phenomena

79.20. Impact phenomena, including electron spectra and sputtering

Original Papers and Short Notes

SZ. B. TÖRÖK
Chemical Characterization of Contaminating Airborne Dust Particles by X-Ray Microprobes
(61.55; 79.20)
(a) Vol. **94**, No. 2, p. 849—854 (1986).

V. FRITZSCHE and P. RENNERT
*A Reduced Angular Momentum Expansion in the Multiple-Scattering Theory
(79.20 and 79.60; 61.10; 61.14; 68.20)
(b) Vol. **135**, No. 1, p. 49—60 (1986).

S. MRÓZ and A. MRÓZ
Quantitative AES Analysis of PbTe and SnTe Samples with and without Matrix Effect Correction
(68.20; 79.20; S8)
(a) Vol. **95**, No. 2, p. 407—412 (1986).

M. I. BÎRJEGA, M. DINESCU, I. N. MIHĂILESCU, L. NANU, C. A. CONSTANTIN, I. TH. FLORESCU, M. POPESCU-POGRION, and C. SÂRBU
*On CW CO_2 Laser Oxidation and Crystallization of Thin Amorphous Sputtered Cr Films
(68.55 and 79.20; S1.2)
(a) Vol. **95**, No. 2, p. 423—432 (1986).

DAO KHAC AN and F. PAVLYÁK
Use of SIMS for Determining the Segregation Coefficient of Boron at the Si–SiO_2 Interface in Steam Oxidation Following Ion Implantation
(64.75; 79.20; S5.11; S10.1)
(a) Vol. **95**, No. 2, p. K117—K121 (1986).

M. PROCOP and B. SANDOW
*Au–Ge Based Ohmic Contacts on GaAs
(73.40 and 79.20; S7.12)
(a) Vol. **95**, No. 2, p. K211—K215 (1986).

T. JUNG and M. SCHMIDT
Properties of AlN_xO_y/GaAs Structures Made by Plasma Nitridation of Evaporated Aluminium Films
(68.55 and 73.60; 79.20; S7.1; S7.12)
(a) Vol. **96**, No. 1, p. 75—81 (1986).

H. MENNIGER, M. BEER, and H. LANGE
Surface Imperfection Induced Electrical Leakage Paths in Silicide Schottky Diodes Investigated by SEM/EBIC and AES
(61.70 and 73.30; 61.14; 79.20; S1.61)
(a) Vol. **96**, No. 1, p. 353—362 (1986).

P. K. BHAT, J. J. DUBOWSKI, and D. F. WILLIAMS
*Emission of Neutral and Charged Species from InSb Induced by Pulsed Laser Evaporation
(68.55 and 79.20; S7.13)
(a) Vol. **96**, No. 1, p. K9—K14 (1986).

S. P. CHENAKIN, V. T. CHEREPIN, A. L. PIVOVAROV, and M. A. VASILEV
Secondary Ion Emission from Amorphous Metallic Alloys
(61.40 and 61.55; 79.20; S1.1)
(a) Vol. **96**, No. 1, p. K21—K26 (1986).

T. JUNG and W. TITEL
*On the Influence of Ar^+ Ion Bombardement on the Si_{LVV} Auger Line in Silicon Nitride Films
(79.20; 68.55; S6)
(a) Vol. **98**, No. 1, p. 63—67 (1986).

M. WAUTELET
*Cohesion of Solids under Very High Electronic Excitation Conditions
(64.60 and 79.20; 63.20; 71.35)
(b) Vol. **138**, No. 2, p. 447—456 (1986).

A. I. CHUMAKOV and G. V. SMIRNOV
On the Possibility of Multiple Increase of the Resolution of Layer-by-Layer Analysis of Materials
(61.16; 76.80; 79.20)
(a) Vol. **98**, No. 2, p. 339—343 (1986).

E. CHRZANOWSKI, W. DOLIŃSKI, S. MRÓZ, and E. WOŹNIAK
Investigation of $Pb_{1-x}Sn_xTe$ Crystal Surfaces by Quantitative AES, LEED, and Kikuchi Pattern Observation
(68.20; 61.14; 79.20; S8.15)
(a) Vol. **98**, No. 2, p. 359—366 (1986).

79.40. Thermionic emission

Original Paper

K. BIEDRZYCKI
*New Aspects of Electron Emission from Virgin TGS Single Crystals
(77.80 and 79.40; S12)
(a) Vol. **93**, No. 2, p. 503—508 (1986).

79.60. Photoemission and photoelectron spectra

Original Papers and Short Notes

M. V. KRUGLOV, I. K. SOLOMIN, and A. V. LUNEV
*Determination of the Photoemission Generation Depth with Use of Experiments on the Dynamic Scattering of X-Rays
(79.60; 61.10; S5.11; S5.12)
(b) Vol. **133**, No. 1, p. 47—55 (1986).

B. VISWANATHAN, S. MADHAVAN, and C. S. SWAMY
*Charge Transfer Satellites in X-Ray Photoelectron Spectra of La_2CuO_4
(79.60; 71.50; S10.15)
(b) Vol. **133**, No. 2, p. 629—632 (1986).

J. ZEMEK and J. KRÁL
Chemical State of Ion-Implanted Nitrogen in Iron
(61.70; 79.20; 79.60; S1.1)
(a) Vol. **94**, No. 2, p. 845—848 (1986).

V. FRITZSCHE and P. RENNERT
*A Reduced Angular Momentum Expansion in the Multiple-Scattering Theory
(79.20 and 79.60; 61.10; 61.14; 68.20)
(b) Vol. **135**, No. 1, p. 49—60 (1986).

JUH TZENG LUE and SEN-YEN SHAW
Laminant Distribution of Hydrogenated Bonds in Amorphous Silicon Deposited by RF Sputtering
(72.40; 79.60; S5.11)
(a) Vol. **95**, No. 1, p. K75—K80 (1986).

U. BERG and T. CHASSÉ
Angular-Resolved X-Ray Photoelectron Spectra and Electronic Structure of SnS_2
(71.25; 79.60; S8)
(b) Vol. **135**, No. 2, p. 633—637 (1986).

O. ENGSTRÖM, H. PETTERSSON, and B. SERNELIUS
*Photoelectric Yield Spectra of Metal–Semiconductor Structures
(79.60; 73.40; S1; S1.3; S1.61; S5.11)
(a) Vol. **95**, No. 2, p. 691—701 (1986).

N. WAGNER, O. BRÜMMER, R. HOFMANN, M. HORX, M. JACOBS, and M. SUCHY
*Photoelectron Spectroscopic Studies of Anodic Surface Oxidation on Pt–Ir Alloys
(79.60; S1.3)
(a) Vol. **95**, No. 2, p. K149—K152 (1986).

V. P. PARKHUTIK, V. A. LABUNOV, and G. G. CHIGIR
Kinetics and Mechanism of Transient Thermal Oxidation of Silicon
(68.55; 73.20; 73.40; 79.60; S5.11; S10.1)
(a) Vol. **96**, No. 1, p. 11—18 (1986).

79.70. Field emission and field ionization

Original Papers

C. J. Workowski
*The Second-Order Correction to Field-Electron-Emission Formulae for an Extended Temperature Range
(79.70)
(b) Vol. **133**, No. 1, p. 183—193 (1986).

H.-J. Fitting and A. von Czarnowski
Electron Breakdown in Polar Insulating and Semiconducting Layers
(73.60; 71.38; 79.70; S8.11; S10.1)
(a) Vol. **93**, No. 1, p. 385—396 (1986).

J. Kundrotas and A. Dargys
Electron Tunneling from an Ultrathin Quantum Well in Constant and Alternating Electric Fields
(73.40; 79.70)
(b) Vol. **134**, No. 1, p. 267—274 (1986).

79.75. Exoelectron emission

Original Papers and Short Notes

M. Kamada and K. Tsutsumi
*Energy Spectra of Thermally Stimulated Exoelectrons from NaCl and NaCl:Cu X-Irradiated at 77 K
(79.75; S9.11)
(a) Vol. **94**, No. 1, p. 275—282 (1986).

A. Scharmann and G. Schmirler
*Influence of Pyroelectric Properties on the Thermally Stimulated Exoelectron Emission (TSEE) of BeO
(77.70 and 79.75; S10.1)
(a) Vol. **97**, No. 1, p. 265—274 (1986).

A. R. Krasnaya, L. N. Oster, and V. Ya. Yaskolko
*Exoelectron Spectroscopy of Dielectrics under Photon and Low Energy Electron Excitation
(79.75; S11.1)
(a) Vol. **97**, No. 2, p. K153—K157 (1986).

Y. Kutomi, A. Tomita, and N. Takeuchi
*Thermally Stimulated Exoelectron Emission and Thermoluminescence in $Li_2B_4O_7$ Crystallized Glasses
(78.60 and 79.75; S10)
(a) Vol. **97**, No. 2, p. K169—K172 (1986).

S1. Transition metals and their alloys and compounds (for oxides, see S10)

Original Papers and Short Notes

M. G. E. Brand and A. Lodder
Electromigration in Niobium. A Finite-Cluster-Model Study
(66.30; S1)
(b) Vol. **133**, No. 1, p. 119—125 (1986).

Yu. L. Shitikov, N. A. Tulina, and M. G. Zemlyanov
Special Features of Lattice Dynamics and Electron–Phonon Interaction in Re
(63.20 and 71.38; S1)
(b) Vol. **133**, No. 2, p. 469—474 (1986).

A. Czopnik, R. Pott, H. Mädge, N. Iliev, and B. Stalinski
Specific Heat and Thermal Expansion of $LaIn_3$
(65.40 and 65.70; S1)
(a) Vol. **93**, No. 2, p. K137—K141 (1986).

A. Klamt and T. Teichler
Quantum-Mechanical Calculations for Hydrogen in Niobium and Tantalum. I. Model Potential and Properties of Self-Trapped States
(61.70; 66.30; S1)
(b) Vol. **134**, No. 1, p. 103—114 (1986).

H. Siethoff
Dynamical Recovery of Cadmium, Zinc, and Magnesium at Intermediate Temperatures
(62.20; S1; S1.2; S3)
(a) Vol. **94**, No. 1, p. 183—189 (1986).

A. Klamt and H. Teichler
Quantum-Mechanical Calculations for Hydrogen in Niobium and Tantalum. II. Diffusion Coefficients
(66.30; 61.70; S1)
(b) Vol. **134**, No. 2, p. 533—544 (1986).

A. M. Dolgin and V. Z. Bengus
Kinetics of High-Velocity Processes of Low Temperature Jump-Like Deformation of Niobium
(62.20; S1)
(a) Vol. **94**, No. 2, p. 529—535 (1986).

J. Petrů
Computed X-Ray Emission Spectra for Rhodium under High Pressure
(78. 70; 71.25; S1)
(b) Vol. **135**, No. 1, p. 233—237 (1986).

A. Jezierski
Chemical Short-Range Order in the Computer Simulated Transition Metal–Metalloid Alloys
(61.40; S1)
(b) Vol. **135**, No. 2, p. 467—473 (1986).

G. Singh and R. P. S. Rathore
Generalised Morse Potential for B.C.C. Complex Metals
(62.20; S1; S4)
(b) Vol. **135**, No. 2, p. 513—518 (1986).

B. Yous and A. Donnadieu
Bilayers of Tungsten Compounds. Selective Absorbers of Solar Energy
(78.65; S1)
(a) Vol. **95**, No. 2, p. 651—657 (1986).

O. Engström, H. Pettersson, and B. Sernelius
Photoelectric Yield Spectra of Metal–Semiconductor Structures
(79.60; 73.40; S1; S1.3; S1.61; S5.11)
(a) Vol. **95**, No. 2, p. 691—701 (1986).

G. Singh and R. P. S. Rathore
Two- and Three-Body Interactions and Phonons in Some F.C.C. Metals
(63.20; 62.20; S1; S3)
(b) Vol. **136**, No. 1, p. 57—61 (1986).

B. S. Sidzhimov, N. V. Stanev, P. T. Tscholakov, and S. D. Neov
Structure of Magnetic Properties of the Pseudobinary Intermetallic Compounds $Tb(Fe_xCo_{1-x})_3$
(61.55; 75.50; S1)
(a) Vol. **96**, No. 1, p. K15—K19 (1986).

Wang Rong-yao and Luo Qi-guang
Structure of A15 Compounds
(61.55; 74.70; S1; S4)
(a) Vol. **96**, No. 2, p. 397—406 (1986).

M. S. Rahman Khan
Electrical Size Effects in Polycrystalline Rhodium Films
(73.60; S1)
(a) Vol. **96**, No. 2, p. 651—655 (1986).

P. Seidel, H.-J. Köhler, F. Müller, U. Moldenhauer, and U.-J. Tietz
The Pseudo Inverse AC Josephson Effect
(74.50; S1)
(a) Vol. **96**, No. 2, p. K203—K205 (1986).

F. Schmidl, M. Siegel, P. Seidel, and H.-J. Köhler
Some Aspects of the AC Josephson Effect at Nb/Amorphous Si/Nb Structures
(74.50; S1; S5.11)
(a) Vol. **97**, No. 1, p. K87—K89 (1986).

S1.1. Fe

Review Article

B. X. Liu
Ion Mixing and Metallic Alloy Phase Formation
(61.80 and 64.75; 61.70; S1.1; S1.2; S1.3; S4)
(a) Vol. **94**, No. 1, p. 11—34 (1986).

Original Papers and Short Notes

M. Vázquez, A. Hernando, and H. Kronmüller
Critical Exponents of the Magnetostriction in Amorphous Alloys
(75.80; S1.1)
(b) Vol. **133**, No. 1, p. 167—170 (1986).

K. Sumiyama, H. Ezawa, and Y. Nakamura
Metastable $Fe_{1-x}Ti_x$ Alloys Produced by Vapor Quenching
(64.75; 61.40; S1.1)
(a) Vol. **93**, No. 1, p. 81—86 (1986).

H. Reuther
Conversion Electron Mössbauer Spectroscopy of Nitrogen and Phosphorus Implanted Iron
(61.70; 76.80; S1.1)
(a) Vol. **93**, No. 1, p. 201—205 (1986).

J. Kadlecová, A. Handstein, T. Zemčík, and K. Závěta
Electrical and Magnetic Properties of Annealed Amorphous $Fe_8Ni_{72}P_{10}B_{10}$ at Low Temperatures
(75.50; 72.15; 76.80; S1.1)
(a) Vol. **93**, No. 1, p. 213—220 (1986).

F. Rettenmeier and H. Kronmüller
Reversible and Irreversible Structural Relaxations in Soft Ferromagnetic Amorphous Alloys. I. Magnetic After-Effects in $Co_{43}Fe_5Ni_{30}Si_{8.8}B_{13.2}$ Alloys
(75.50; S1.1)
(a) Vol. **93**, No. 1, p. 221—231 (1986).

J. Filipecki, E. Jakubczyk, B. Całusiński, and D. Gaudyn
Electron Phenomena Connected with Hydriding of the Metallic Glass $Fe_{81}B_{13.5}Si_{3.5}C_2$
(61.70; 72.15; 78.70; S1.1)
(a) Vol. **93**, No. 1, p. K25—K28 (1986).

B. K. Das, K. Chowdhury, D. L. Bhattacharya, and A. K. Tewari
Study of the Mössbauer Effect on Deformed Stainless Steel Foils
(76.80; S1.1)
(a) Vol. **93**, No. 1, p. K29—K31 (1986).

B. Sidzhimov, N. Stanev, P. Tscholakov, and S. Neov
Magnetic Properties of the Pseudobinary Intermetallic Compounds $Tb_xY_{1-x}Fe$ ($x = 0.25$, 0.50, 0.75)
(75.50; S1.1)
(a) Vol. **93**, No. 1, p. K37—K40 (1986).

V. N. Baier, V. M. Katkov, and V. M. Strakhovenko
Radiation Yield of High-Energy Electrons in Thick Crystals
(61.80 and 78.70; S1.1; S1.2; S5; S5.11; S5.12)
(b) Vol. **133**, No. 2, p. 583—592 (1986).

A. N. Voloshinskii, N. V. Ryzhanova, L. Yu. Vishnekov, and A. G. Obukhov
Off-Diagonal Conductivity and Magnetization of PdFe, PdCo, and PdNi Alloys in Coherent Potential Approximation Calculated on the Basis of Realistic Models of Component Densities of States
(72.15; 71.20; 75.60; S1.1; S1.2)
(b) Vol. **133**, No. 2, p. 735—743 (1986).

Yu-Jin Chang and Xiang-Rong Qian
A High Resolution Microscopic Study on Nd–Fe–B Permanent Magnetic Alloys
(64.70 and 75.60; 61.16; 75.50; S1.1)
(a) Vol. **93**, No. 2, p. 573—583 (1986).

V. I. Malyutin, V. E. Osukhovskii, A. A. Ivanov, L. A. Chebotkevich, I. V. Lobov, and Yu. D. Vorobiev
The Effect of Volume and Surface Inhomogeneities on the Coercivity of Thin Magnetic Films
(75.70; S1.1)
(a) Vol. **93**, No. 2, p. 585—595 (1986).

F. Rettenmeier, E. Kisdi-Koszo, and H. Kronmüller
Reversible and Irreversible Structural Relaxations in Soft Ferromagnetic Amorphous Alloys. II. Magnetic After-Effects in $Fe_{100-x}B_x$
(75.50; S1.1)
(a) Vol. **93**, No. 2, p. 597—604 (1986).

K. Hisatake, T. Miyazaki, M. Takahashi, K. Maeda, T. Hattori, and I. Matsubara
Anomalous Behaviour of Permeability around the Curie Temperature in Amorphous Fe–Ni–Si–B Alloys
(75.50; S1.1)
(a) Vol. **93**, No. 2, p. 605—611 (1986).

M. Miglierini and J. Sitek
Mössbauer Lineshape Analysis by the DISPA Method
(76.80; S1.1)
(a) Vol. **93**, No. 2, p. 627—634 (1986).

G. Graiss, A. Rizk, and A. H. Wafik
Influence of Carbon Content and Quenching Rate on the Amount of Martensite Formed in Two Kinds of Steel
(64.75; S1.1)
(a) Vol. **93**, No. 2, p. K123—K124 (1986).

A. ZENTKO, I. ŠKORVÁNEK, V. V. JEREMENKO, and V. D. CHECHERSKII
Mössbauer Study of Neutron Irradiated $Fe_{60}Ni_{20}B_{20}$ Amorphous Alloys
(76.80; 61.80; S1.1)
(a) Vol. **93**, No. 2, p. K161—K164 (1986).

J. GONZÁLEZ, M. VÁZQUEZ, and J. M. BARANDIARAN
On the Dependence of the Magnetization Curve on Applied Tensile Stress in Amorphous Alloys with Positive Magnetostriction
(75.80; S1.1)
(a) Vol. **93**, No. 2, p. K165—K170 (1986).

T. STOBIECKI and M. PRZYBYLSKI
Spontaneous Hall Effect and Magnetization of Fe–(B, Zr) and Co–Mo Amorphous Films
(75.70; 73.60; S1.1; S1.2)
(b) Vol. **134**, No. 1, p. 131—139 (1986).

B. A. OLI and J. S. ONONIWU
Effect of Non-Locality in the Calculation of Liquid Metal Resistivity of Transition Metals
(72.15; 71.25; S1.1; S1.2; S1.3)
(b) Vol. **134**, No. 1, p. 325—330 (1986).

A. PAJA and T. STOBIECKI
Electrical Resistivity of Amorphous $Fe_{1-x}Zr_x$ Alloys
(73.60; S1.1)
(b) Vol. **134**, No. 1, p. 331—334 (1986).

SH. M. SHARIPOV, K. M. MUKIMOV, and L. A. ERNAZAROVA
The Heavy Rare-Earth Contribution to the Electronic Structure of the RFe_2 and R_2Fe_{17} Intermetallic Compounds
(78.20; 71.25; S1.1)
(b) Vol. **134**, No. 1, p. K59—K62 (1986).

V. V. KOLOMYTKIN and U. R. KEVORKYAN
A Computer Simulation of Self-Interstital Migration along the [100] (010) Edge Dislocation Core in α-Fe
(66.30; S1.1)
(a) Vol. **94**, No. 1, p. 153—160 (1986).

P. P. PAL-VAL and S. KADEČKOVÁ
Amplitude Dependent Internal Friction in High-Purity Iron Single Crystals at Low Temperatures
(62.20; 61.70; S1.1)
(a) Vol. **94**, No. 1, p. K29—K32 (1986).

K. MASUDA-JINDO
On the Grain Boundary Segregation of sp-Valence Impurities in B.C.C. Transition Metal
(64.75; 71.55; S1.1)
(b) Vol. **134**, No. 2, p. 545—550 (1986).

J. Golczewski, K. Maier, and W. Frank
Defect Reactions in Electron-Irradiated α-$Fe^{111}In$ Studied by Means of PAC Measurements
(61.70; 78.90; S1.1)
(b) Vol. **134**, No. 2, p. 551—559 (1986).

R. Markendorf, E. Zschech, W. Blau, and K. Kleinstück
Chemical Shifts of EXAFS Backscattering Phases within the Framework of Neutral Pseudoatoms
(71.25 and 78.70; S1.1; S1.2)
(b) Vol. **134**, No. 2, p. K121—K124 (1986).

Yu. S. Nechaev, S. A. Vladimirov, N. A. Olshevskii, and V. S. Khlomov
On the Effect of High-Rate Strain on Diffusion Mass Transfer in Metals
(66.30; S1.1; S1.2; S1.3; S4)
(a) Vol. **94**, No. 2, p. 553—560 (1986).

W. Matz and H. Herrmann
Non-Linear Dependence of Average Magnetic Moment on Boron Concentration in Amorphous Fe-B Alloys
(75.50; 76.80; S1.1)
(a) Vol. **94**, No. 2, p. 617—622 (1986).

A. Turos
Defect-Impurity Interactions in Ion-Implanted Metals
(61.70; S1.1; S1.2; S1.3; S4)
(a) Vol. **94**, No. 2, p. 809—822 (1986).

K. K. Kadyrzhanov, A. K. Zhetbaev, M. Sh. Aimanov, and A. N. Ozernoi
Mössbauer Investigations of Phase Formation in Nitrogen-Implanted Iron
(61.70; 76.80; S1.1)
(a) Vol. **94**, No. 2, p. 823—825 (1986).

T. Ungár and G. Szenes
Investigation of N-Implanted Fe Single Crystals by X-Ray Line Broadening
(61.70; 61.10; S1.1)
(a) Vol. **94**, No. 2, p. 839—844 (1986).

J. Zemek and J. Král
Chemical State of Ion-Implanted Nitrogen in Iron
(61.70; 79.20; 79.60; S1.1)
(a) Vol. **94**, No. 2, p. 845—848 (1986).

D. K. Fowler, D. C. Creagh, and G. V. H. Wilson
NMR Free Induction Decay Measurements in Iron
(76.60; 75.60; S1.1)
(a) Vol. **94**, No. 2, p. K97—K101 (1986).

R. Z. Sadykhov, L. M. Valiev, N. B. Nabieva, and Z. M. Namazov
Magnetic Properties of $Fe_{1+x}Cr_{2-x}S_4$ $(0 < x \leqq 0.5)$
(75.50; S1.1; S1.2)
(a) Vol. **94**, No. 2, p. K103—K106 (1986).

P. Muth
On the Anisotropy of the Magnetoresistance of Nickel- and Iron-Based Alloys
(72.15; S1.1; S1.2)
(b) Vol. **135**, No. 1, p. K75—K78 (1986).

M. Shahmiri, D. J. Vaughan, and S. Murphy
Observation of Twin-Related Order Domains in $CuFePt_2$
(61.50; S1.1; S1.3)
(a) Vol. **95**, No. 1, p. 63—72 (1986).

B. Panigrahi, A. K. Tyagi, and R. V. Nandedkar
Surface Damage of Ferritic Steels by Helium Ion Irradiation
(61.80 and 64.80; S1.1)
(a) Vol. **95**, No. 1, p. 99—103 (1986).

A. Raoufi and S. Arajs
Effects of Magnetic Field and Tensile Stress on the Electrical Resistivity of Some Amorphous Ferromagnets
(72.15 and 75.80; S1.1)
(a) Vol. **95**, No. 1, p. 193—196 (1986).

S. Motogi
Polarization and Resonance Properties of Magnetoacoustic Waves in Ferromagnets with Dissipation
(75.50 and 75.80; 76.50; S1.1; S11.2)
(a) Vol. **95**, No. 1, p. 197—208 (1986).

I. Škorvánek, A. Zentko, and T. Zemčík
Influence of the Surface Insulating Coating on the Magnetic Properties of $Fe_{82}B_{12}Si_6$ Amorphous Ribbons
(75.50; 76.80; S1.1)
(a) Vol. **95**, No. 1, p. 209—212 (1986).

W. Kaczmarek, I. Gościańska, J. Pietrzak, H. Ratajczak, and J. Baszyński
Ferromagnetic Resonance in Amorphous Gd–Fe Films
(76.50; 75.70; S1.1; S1.4)
(a) Vol. **95**, No. 1, p. 221—227 (1986).

A. V. Deryabin, A. V. Tkov, V. K. Kazantsev, Yu. A. Chirkov, and I. V. Zakharov
Concentration Magnetic Transitions in Alloys with Mixed Exchange Interaction
(75.30; 65.40; 72.15; S1.1)
(a) Vol. **95**, No. 1, p. 229—237 (1986).

Z. A. Matysina, M. I. Milyan, and S. Yu. Zaginaichenko
The Solubility of Interstitial Impurities in Ordering Phases
(64.75; 61.70; S1.1; S1.3)
(a) Vol. **95**, No. 1, p. K9—K14 (1986).

H. Röll and J. Pabst
Induced Magnetic Form Factors for 3d Transition Metals and 3d–3d Alloys. I. Results of Band Structure Calculations
(71.25 and 75.25; S1.1; S1.2)
(b) Vol. **135**, No. 2, p. 691—695 (1986).

J. Fuentes and A. Rivero
About the Ordering in Fe–Rh Alloys
(75.30; S1.1)
(b) Vol. **135**, No. 2, p. K125—K128 (1986).

F. Goedsche
Electrical Resistivity and Thermoelectric Power of $Fe_{50}Co_{50}$
(72.15 and 75.50; S1.1; S1.2)
(b) Vol. **135**, No. 2, p. K153—K158 (1986).

M. I. Guseva and G. V. Gordeeva
Phase Transitions in the Surface Layers of Materials under Ion Implantation
(61.80 and 68.20; 64.70; S1.1)
(a) Vol. **95**, No. 2, p. 385—390 (1986).

G. Gawlik, J. Jagielski, and A. Podgórski
Fatigue-Life Enhancement of Nitrogen-Implanted Chromium Steel
(62.20; 61.70; S1.1)
(a) Vol. **95**, No. 2, p. 453—458 (1986).

A. Węgrzyn, T. Jagieliński, S. Ohnuma, and T. Masumoto
Saturation Magnetostriction and Forced Volume Magnetostriction in Amorphous $(Fe_{1-x}Co_x)_{90}Zr_{10}$ and $(Fe_{1-x}Ni_x)_{90}Zr_{10}$ Alloys. Experimental Results
(75.80; S1.1; S1.2)
(a) Vol. **95**, No. 2, p. 599—603 (1986).

N. Lenge and H. Kronmüller
Low Temperature Magnetization of Sputtered Amorphous Fe–Ni–B Films
(75.70; 61.40; S1.1)
(a) Vol. **95**, No. 2, p. 621—633 (1986).

D. Fraitová
An Analytical Theory of FMR in Bulk Metals (Parallel Configuration). IV. Exchange Dominated Surface Modes
(76.50; 75.30; S1.1)
(b) Vol. **136**, No. 1, p. 91—96 (1986).

M. Posselt, G. Diener, and M. Seidel
Ideal Magnetization in Sintered Iron Materials
(75.50; 75.60; S1.1)
(b) Vol. **136**, No. 1, p. 123—130 (1986).

P. V. Geld, V. V. Singer, I. Z. Radovskii, and S. V. Konovalov
Variable Valence of Ce and Pr in Liquid Alloys with Fe, Co, and Ni
(71.25; 61.25; 72.15; 75.30; S1.1; S1.2; S1.4)
(b) Vol. **136**, No. 1, p. 153—158 (1986).

J. Pabst and H. Röll
Induced Magnetic Form Factors for 3d Transition Metals and 3d–3d Alloys. II. Induced Spin Form Factors
(71.25 and 75.25; S1.1; S1.2)
(b) Vol. **136**, No. 1, p. 225—231 (1986).

N. K. Gobran, S. Saleh, B. F. Hanna, and R. Kamel
Kinetics of Crystallization for $Fe_{40}Ni_{40}B_{20}$ Metallic Glass
(64.70; 61.40; 62.20; 75.60; S1.1)
(a) Vol. **96**, No. 1, p. 33—37 (1986).

J. Kučera, B. Million, and J. Růžičková
Magnetic Anomalies of ^{59}Fe and ^{60}Co Diffusion in Iron and of ^{48}V Diffusion in Cobalt. Quasi-Chemical Model
(66.30; 75.50; S1.1; S1.2)
(a) Vol. **96**, No. 1, p. 177—184 (1986).

I. Baker and D. J. Gaydosh
Displacement Fringes in FeAl
(61.70; S1.1)
(a) Vol. **96**, No. 1, p. 185—190 (1986).

V. V. Lomonosov, V. D. Sedykh, and T. Zemčík
Thermally Activated Time Fluctuations of Nuclear Spin Orientation in Amorphous $Fe_{81}B_{19}$ Alloy
(75.25; 75.50; 76.80; S1.1)
(a) Vol. **96**, No. 1, p. 271—280 (1986).

M. Tejedor and J. F. Fuertes
Néel's Domain Structure on [011] Iron Whiskers
(75.60; S1.1)
(a) Vol. **96**, No. 1, p. K1—K4 (1986).

S. P. Chenakin, V. T. Cherepin, A. L. Pivovarov, and M. A. Vasilev
Secondary Ion Emission from Amorphous Metallic Alloys
(61.40 and 61.55; 79.20; S1.1)
(a) Vol. **96**, No. 1, p. K21—K26 (1986).

S. U. Jen and Y. D. Yao
Thermal Expansion of FePd Alloys from 50 to 1000 °C
(65.70; S1.1)
(a) Vol. **96**, No. 1, p. K27—K30 (1986).

T. Luciński and B. Szymański
Hall Effect and Magnetic Anisotropy in Amorphous $Fe_{1-x}Si_x$ Films
(72.15 and 75.70; S1.1)
(a) Vol. **96**, No. 1, p. K53—K56 (1986).

S. Garg and H. C. Gupta
Phonon Dispersion of Disordered Fe–Al Alloy
(63.50; S1.1)
(b) Vol. **136**, No. 2, p. 453—456 (1986).

J. Pabst and H. Röll
Induced Magnetic Form Factors for 3d Transition Metals and 3d–3d Alloys. II. Diamagnetic Form Factors
(71.25 and 75.20; S1.1; S1.2)
(b) Vol. **136**, No. 2, p. 637—642 (1986).

M. A. Mostafa, J. Balogh, and E. Kuzmann
Mössbauer Study of $(Fe_{1-x}Co_x)_{77}B_{13}Si_{10}$ Amorphous Alloys
(61.40; 76.80; S1.1)
(a) Vol. **96**, No. 2, p. 445—449 (1986).

H. Ratajczak and I. Gościańska
Effect of Annealing on Coercive Field in Amorphous $Fe_{40}Ni_{40}B_{20}$ Thin Films
(75.70; 75.80; S1.1)
(a) Vol. **96**, No. 2, p. 563—571 (1986).

T. Tyliszczak, A. Budkowski, and J. A. Sawicki
Study of Magnetism and Interface Processes between Thin Iron Film and Nickel Substrate by in-Situ Conversion Electron Mössbauer Spectroscopy
(75.70; 64.70; 76.80; S1.1; S1.2)
(a) Vol. **96**, No. 2, p. 573—579 (1986).

K. Sumiyama, Y. Kawawake, and Y. Nakamura
Amorphous Fe–Cu–Ag Alloys Produced by Vapor Quenching
(61.40 and 75.50; 76.80; S1.1; S1.2; S1.3)
(a) Vol. **96**, No. 2, p. K107—K110 (1986).

G. L. Whittle, A. M. Stewart, and A. B. Kaiser
Magnetic and Electrical Properties of $(FeTM)_{85}B_{15}$ (TM = Cr, Ni) Metallic Glasses
(75.50; 72.15; 76.80; S1.1)
(a) Vol. **97**, No. 1, p. 199—211 (1986).

I. Okamoto, T. Miyazaki, and M. Takahashi
Spin-Glass in Rapidly Quenched $Cu_{100-x}Fe_x$ ($0 \leqq x \leqq 9.0$) Alloys
(75.50; 75.20; S1.1; S1.2)
(a) Vol. **97**, No. 1, p. 223—230 (1986).

R. Moskalewicz and W. Zych
Application of the DTMG-DTA(M) Technique for the Curie Point and Crystallisation Temperature Determination of $Fe_{80-x}Mn_xB_{20}$ Amorphous Alloys
(75.50; S1.1)
(a) Vol. **97**, No. 1, p. K43—K48 (1986).

W. Pfeiler and R. Reihsner
Concentration Dependence of Short-Range Order in Some Concentrated Alloys
(61.55; 64.75; S1.1; S1.2; S1.3)
(a) Vol. **97**, No. 2, p. 377—385 (1986).

N. Mattern, A. G. Ilinskii, H. Hermann, and A. V. Romanova
X-Ray Diffraction Study on Amorphous and Liquid Fe–B Alloys
(61.25 and 61.40; 61.10; S1.1)
(a) Vol. **97**, No. 2, p. 397—401 (1986).

M. Krause, C. Michalk, B. Lippold, M. Benedix, and J. Suwalski
Slow Paramagnetic Relaxation in Oxalate (2,2′-Bipyridine) Iron
(75.20; 61.40; 76.80; S1.1)
(a) Vol. **97**, No. 2, p. 521—526 (1986).

A. Raoufi, S. Arajs, and K. V. Rao
Effects of Short-Range Order on the Electrical Resistivity of Plastically Deformed Pd–Ag and Pd–Ag–Fe Alloys
(72.15; S1.1; S1.3)
(a) Vol. **97**, No. 2, p. 571—575 (1986).

A. Chassé and P. Rennert
Determination of Adsorbate Geometries from Angle-Resolved X-Ray Photoelectron Spectra. Nitrogen on Fe(111)
(68.20; 79.60; S1.1)
(b) Vol. **138**, No. 1, p. 53—64 (1986).

M. Cinal and S. Olszewski
Magnetic Atoms in Metallic Compounds Approached on the Basis of the Quantum-Statistical ab Initio Theory
(75.90; S1.1; S1.2)
(b) Vol. **138**, No. 1, p. K17—K20 (1986).

P. Rochegude and J. Foct
Interstitial Atom Ordering in Binary Fe–N Solid Solutions Studied by Mössbauer Spectrometry
(64.75; 76.80; S1.1)
(a) Vol. **98**, No. 1, p. 51—62 (1986).

K. Benhaddane and P. Beauchamp
Core Structure of the $a\langle 100\rangle$ $\langle 001\rangle$ Edge Dislocation in a B2 Ordered Alloy
(61.70; S1.1)
(a) Vol. **98**, No. 1, p. 195—202 (1986).

K. Sumiyama, N. Ohshima, and Y. Nakamura
Magnetic Properties of Metastable α-Mn-Type $Mn_{1-y}Fe_y$ Alloys Produced by Vapor Quenching
(75.50; 61.10; 76.80; S1.1)
(a) Vol. **98**, No. 1, p. 229—238 (1986).

Y. D. Yao, T. Y. Yen, and J. Y. Chen
Electrical Resistitivity and Magnetization Studies of Some Commercial Steels
(72.15 and 75.50; S1.1)
(a) Vol. **98**, No. 1, p. 285—289 (1986).

Z. A. Matysina, A. O. Mekhrabov, A. Z. Menshikov, Z. M. Babaev, and S. Yu. Zaginaichenko
Impurities in Ni_3Fe Magnetic Alloys
(64.60; 65.50; 65.70; S1.1; S1.2)
(b) Vol. **138**, No. 2, p. 399—406 (1986).

A. Jezierski
Electronic Density of States in Binary Non-Crystalline Alloys with Different Chemical Short Range Ordering
(71.20; S1.1)
(b) Vol. **138**, No. 2, p. K119—K123 (1986).

A. A. Bahgat, E. E. Shaisha, S. F. El-Desouki, and A. I. Sabry
On the Determination of the Partial Pair Distribution Functions in Amorphous Alloys
(61.40; S1.1)
(a) Vol. **98**, No. 2, p. 351—357 (1986).

M. Tarabieh, M. Ahmad, and A. Zihlif
Initial Magnetic Susceptibility and Electrical Resistivity of $Fe_{92}Zr_8$ Amorphous Alloy
(72.15 and 75.50; S1.1)
(a) Vol. **98**, No. 2, p. 545—550 (1986).

A. G. Rustamov, R. Z. Sadykhov, L. M. Valiev, and A. O. Ismailov
Paramagnetic Susceptibility of the $Me_{1-x}Fe_xCr_2S_4$ (Me = Zn, Co) Composition System
(75.20; S1.1; S1.2)
(a) Vol. **98**, No. 2, p. K153—K156 (1986).

S1.2 Other 3d metals

Review Article

B. X. Liu
Ion Mixing and Metallic Alloy Phase Formation
(61.80 and 64.75; 61.70; S1.1; S1.2; S1.3; S4)
(a) Vol. **94**, No. 1, p. 11—34 (1986).

Original Papers and Short Notes

G. C. Fletcher
The Band Structure of Co_3Ga and CO_2MnGa and the Electrical Conductivity of Related Alloys
(71.25 and 72.15; 75.50; S1.2)
(b) Vol. **133**, No. 1, p. 241—248 (1986).

Sadhana Garg, H. C. Gupta, and B. B. Tripathi
The Width of the Scattering Resonance and the Electrical Transport Properties in Partially Filled d-Band Transition Metals
(72.15; S1.2)
(b) Vol. **133**, No. 1, p. K73—K76 (1986).

J. P. Zhang, H. Q. Ye, and K. H. Kuo
A High-Resolution Electron Microscopy Study of Domain Structure in Ni_3Nb. I. Translation Domain Boundaries
(61.70; 61.16; S1.2)
(a) Vol. **93**, No. 1, p. 35—44 (1986).

E. Biagi, S. Frosini, L. Masotti, and E. Borchi
The Propagation of Acoustic Waves in Polycrystals with Axial Texture
(62.65; 62.20; S1.2; S4; S5.11)
(a) Vol. **93**, No. 1, p. 151—162 (1986).

M. R. Nagy, M. S. Sakr, M. M. Mostafa, and R. Kamel
Recovery Behaviour after Stress Reductions during the Steady State Creep of Al–14 wt% Zn
(62.20; S1.2; S4)
(a) Vol. **93**, No. 1, p. 185—193 (1986).

S. K. Shadangi, M. Singh, S. C. Panda, and S. Bhan
On the Structure of Equilibrium and Metastable Phases in $Ni_{75}Sn_{25-x}Ge_x$ System
(61.50; 64.70; S1.2)
(a) Vol. **93**, No. 1, p. K1—K4 (1986).

V. N. Baier, V. M. Katkov, and V. M. Strakhovenko
Radiation Yield of High-Energy Electrons in Thick Crystals
(61.80 and 78.70; S1.1; S1.2; S5; S5.11; S5.12)
(b) Vol. **133**, No. 2, p. 583—592 (1986).

J. M. Pelletier, F. Fouquet, and J. Hillairet
Evolution of the Electrical Properties of Amorphous $Ni_{24}Zr_{76}$ and $Ni_{35}Ti_{65}$ as a Result of Structural Relaxation and Crystallization
(72.15; S1.2)
(a) Vol. **93**, No. 2, p. 703—710 (1986).

V. I. Shulga
Surface Semichanneling of Molecular Ions
(61.80; S1.2)
(b) Vol. **134**, No. 1, p. 87—95 (1986).

T. Stobiecki and M. Przybylski
Spontaneous Hall Effect and Magnetization of Fe–(B, Zr) and Co–Mo Amorphous Films
(75.70; 73.60; S1.1; S1.2)
(b) Vol. **134**, No. 1, p. 131—139 (1986).

B. A. Oli and J. S. Ononiwu
Effect of Non-Locality in the Calculation of Liquid Metal Resistivity of Transition Metals
(72.15; 71.25; S1.1; S1.2; S1.3)
(b) Vol. **134**, No. 1, p. 325—330 (1986).

P. L. Rossiter and R. H. Mair
The Electrical Properties of Rapidly Quenched Cu–Ni Alloys
(72.15; 75.90; S1.2)
(b) Vol. **134**, No. 1, p. 335—342 (1986).

B. V. Kuvaldin, E. V. Zolotoyabko, E. M. Iolin, and E. A. Raitman
Neutron Topography Study of Large Single Crystals by the Local Rocking Curve Method
(61.12 and 61.50; S1.2; S5.11)
(a) Vol. **94**, No. 1, p. 45—52 (1986).

E. Ratner, A. Apfelbaum, R. Brener, and M. Eizenberg
Interactions of Thin Al Films with Ni–Cr Alloy and Bilayer Films Deposited on Si
(68.55; 61.10; 79.20; S1.2; S4)
(a) Vol. **94**, No. 1, p. 61—69 (1986).

R. Butz and H. Wagner
Reactions at the Ti/SiO_2 Interface Studied by EELS
(68.48; 68.30; 61.80; S1.2; S10.1)
(a) Vol. **94**, No. 1, p. 71—76 (1986).

H.-J. Burmeister and M. Berndt
Inclusion Approximations Applied to the Model of Heterogeneous Plastic Deformation for Fatigued Ni Polycrystals. I. Description of the Method. Eigenstresses Due to the Different Grain Orientations
(62.20; S1.2)
(a) Vol. **94**, No. 1, p. 121—129 (1986).

H. Siethoff
Dynamical Recovery of Cadmium, Zinc, and Magnesium at Intermediate Temperatures
(62.20; S1; S1.2; S3)
(a) Vol. **94**, No. 1, p. 183—189 (1986).

YU. S. NECHAEV, S. A. VLADIMIROV, N. A. OLSHEVSKII, and V. S. KHLOMOV
On the Effect of High-Rate Strain on Diffusion Mass Transfer in Metals
(66.30; S1.1; S1.2; S1.3; S4)
(a) Vol. **94**, No. 2, p. 553—560 (1986).

M. S. SAKR, A. E. E. ABDEL-RAHIEM, and A. A. EL-DALY
Steady State Creep during Transformation in Zn–40 wt% Al and Zn–0.5 wt% Al Alloys
(62.20; 61.70; S1.2)
(a) Vol. **94**, No. 2, p. 561—568 (1986).

A. TUROS
Defect–Impurity Interactions in Ion-Implanted Metals
(61.70; S1.1; S1.2; S1.3; S4)
(a) Vol. **94**, No. 2, p. 809—822 (1986).

B. RAUSCHENBACH and C. BLOCHWITZ
Electron Microscopical Investigations of Ion-Implanted and Fatigued Nickel
(61.70; 61.16; S1.2)
(a) Vol. **94**, No. 2, p. 827—831 (1986).

B. RAUSCHENBACH and K. HOHMUTH
Synthesis of Compounds by High-Fluence Nitrogen Ion Implantation in Titanium
(61.70 and 68.55; S1.2)
(a) Vol. **94**, No. 2, p. 833—837 (1986).

R. Z. SADYKHOV, L. M. VALIEV, N. B. NABIEVA, and Z. M. NAMAZOV
Magnetic Properties of $Fe_{1+x}Cr_{2-x}S_4$ $(0 < x \leqq 0.5)$
(75.50; S1.1; S1.2)
(a) Vol. **94**, No. 2, p. K103—K106 (1986).

J. V. KORNEV, D. I. SEMENTSOV, and V. V. SIDORENKOV
High-Frequency Properties of Cobalt Thin Films with Surface Anisotropy of the Electrical Conductivity
(73.60; S1.2)
(a) Vol. **94**, No. 2, p. K147—K152 (1986).

L. DĄBROWSKI
Multiparticle Correlations in Alloys
(63.75 and 65.50; S1.2)
(b) Vol. **135**, No. 1, p. 61—65 (1986).

V. K. THAKUR and T. N. SINGH
Lattice Dynamics of F.C.C. Metals
(63.20; 65.40; S1.2; S1.3)
(b) Vol. **135**, No. 1, p. 67—73 (1986).

I. MERTIG, E. MROSAN, V. N. ANTONOV, VL. ANTONOV, and P. ZIESCHE
Effective Optical Masses of Copper and Silver
(71.25; S1.2; S1.3)
(b) Vol. **135**, No. 1, p. K13—K16 (1986).
Addendum
(b) Vol. **138**, No. 1, p. K75 (1986).

A. WĘGRZYN, T. JAGIELIŃSKI, S. OHNUMA, and T. MASUMOTO
Saturation Magnetostriction and Forced Volume Magnetostriction in Amorphous $(Fe_{1-x}Co_x)_{90}Zr_{10}$ and $(Fe_{1-x}Ni_x)_{90}Zr_{10}$ Alloys. Experimental Results
(75.80; S1.1; S1.2)
(a) Vol. **95**, No. 2, p. 599—603 (1986).

A. R. ALI and G. SAID
On the Migration of Point Defects in Irradiated Pure Ni and Ni–Cu Alloys
(61.70; 61.80; S1.2)
(a) Vol. **95**, No. 2, p. K109—K112 (1986).

P. V. GELD, V. V. SINGER, I. Z. RADOVSKII, and S. V. KONOVALOV
Variable Valence of Ce and Pr in Liquid Alloys with Fe, Co, and Ni
(71.25; 61.25; 72.15; 75.30; S1.1; S1.2; S1.4)
(b) Vol. **136**, No. 1, p. 153—158 (1986).

J. PABST and H. RÖLL
Induced Magnetic Form Factors for 3d Transition Metals and 3d–3d Alloys. II. Induced Spin Form Factors
(71.25 and 75.25; S1.1; S1.2)
(b) Vol. **136**, No. 1, p. 225—231 (1986).

B. N. SINGH and S. M. RAFIQUE
Propriety of Harrison's Method: Electronic Transport Property
(72.15; S1.2)
(b) Vol. **136**, No. 1, p. K65—K69 (1986).

J. KUČERA, B. MILLION, and J. RŮŽIČKOVÁ
Magnetic Anomalies of ^{59}Fe and ^{60}Co Diffusion in Iron and of ^{48}V Diffusion in Cobalt. Quasi-Chemical Model
(66.30; 75.50; S1.1; S1.2)
(a) Vol. **96**, No. 1, p. 177—184 (1986).

T. WIELINGA and J. C. LODDER
Reversal Mechanism in Perpendicular Media with Columnar Microstructure
(75.70; S1.2)
(a) Vol. **96**, No. 1, p. 255—264 (1986).

V. GRÖGER
Influence of Decomposition and Coarsening on the Electrical Resistivity of Cu–2wt%Be–0.11 wt%Co Aged at 320 °C
(72.15; 64.80; S1.2)
(a) Vol. **96**, No. 1, p. 325—335 (1986).

D. AUGIE, M. BELHADIJ and P. VIGIER
Internal Swains in Amorphous Ni–Ag Thin Films
(68.55; S1.2; S1.3)
(a) Vol. **96**, No. 1, p. K5—K8 (1986).

I. M. Ghauri, P. Feltham, and M. Z. Butt
Solid-Solution Hardening in α-Brasses at Low Temperatures
(62.20; S1.2)
(a) Vol. **96**, No. 1, p. K43—K47 (1986).

P. Wosicki
On the Functional Integral Method to the Hubbard Model. Application to Nickel
(75.10; S1.2)
(b) Vol. **136**, No. 2, p. 567—574 (1986).

J. Pabst and H. Röll
Induced Magnetic Form Factors for 3d Transition Metals and 3d–3d Alloys. II. Diamagnetic Form Factors
(71.25 and 75.20; S1.1; S1.2)
(b) Vol. **136**, No. 2, p. 637—642 (1986).

P. Blaudeck
Cluster Embedding in an Effective Shell and Calculation of the Density of States for Amorphous Copper
(71.20; 61.40; S1.2)
(b) Vol. **136**, No. 2, p. 673—678 (1986).

A. B. Kaiser
Low Temperature Resistivity Anomaly in Glassy Metals. Two-Level Systems and Incipient Localization versus Correlations
(72.15; S1.2; S3)
(b) Vol. **136**, No. 2, p. 779—784 (1986).

T. Soma, M. Obata, and H.-Matsuo Kagaya
Mean-Square Displacements and Lindemann's Law for Melting of Cu
(63.20; 64.60; S1.2)
(b) Vol. **136**, No. 2, p. K75—K79 (1986).

Wang Yuan-hang, Hu An, Yu Zheng, Li Xiang-jin, and Feng Duan
A Structural Study of Compositionally Modulated Nb–Ti Films
(68.55; S1.2)
(a) Vol. **96**, No. 2, p. 377—384 (1986).

S. Hashimoto and H. Iwasaki
Microdomain Model Analysis for Diffuse X-Ray Scattering from Quenched Cu–40 and 60 at% Au Alloys
(61.10; S1.2; S1.3)
(a) Vol. **96**, No. 2, p. 435—444 (1986).

T. Tyliszczak, A. Budkowski, and J. A. Sawicki
Study of Magnetism and Interface Processes between Thin Iron Film and Nickel Substrate by in-Situ Conversion Electron Mössbauer Spectroscopy
(75.70; 64.70; 76.80; S1.1; S1.2)
(b) Vol. **96**, No. 2, p. 573—579 (1986).

V. D. Das, M. S. M. Sastry, and M. Pattabi
Aging and Field Effect Studies of Cu and Cu/Ag Composite Discontinuous Thin Films
(73.60; S1.2; S1.3)
(a) Vol. **96**, No. 2, p. 677—681 (1986).

K. Sumiyama, Y. Kawawake, and Y. Nakamura
Amorphous Fe–Cu–Ag Alloys Produced by Vapor Quenching
(61.40 and 75.50; 76.80; S1.1; S1.2; S1.3)
(a) Vol. **96**, No. 2, p. K107—K110 (1986).

G. Neumann
On the Calculation of Self-Diffusion and Impurity Diffusion Coefficients with the Aid of Elastic Constants
(66.30; S1.2; S1.3)
(b) Vol. **137**, No. 1, p. 57—63 (1986).

T. Soma, N. Suzuki, and H.-Matsuo Kagaya
Specific Heat of α-Phase Cu–Al Alloys
(65.40; S1.2; S4)
(b) Vol. **137**, No. 1, p. K15—K19 (1986).

K. Z. Botros, S. S. Sheinin, and S. M. Saleh
On Distinguishing between Electron Microscope Images of Overlapping and Intrinsic Faults in F.C.C. Materials
(61.16 and 61.70; S1.2)
(a) Vol. **97**, No. 1, p. 143—151 (1986).

N. Laouini and F. Hila
Diffusion in Thin Films. Case of Au–Ni Couple
(68.60; 68.48; S1.2; S1.3)
(a) Vol. **97**, No. 1, p. 173—179 (1986).

R. Hähnel, W. Miekeley, and H. Wever
Diffusion Studies on the B8 Phase of the Ni/Sb System
(66.30; S1.2)
(a) Vol. **97**, No. 1, p. 181—190 (1986).

I. Okamoto, T. Miyazaki, and M. Takahashi
Spin-Glass in Rapidly Quenched $Cu_{100-x}Fe_x$ ($0 \leqq x \leqq 9.0$) Alloys
(75.50; 75.20; S1.1; S1.2)
(a) Vol. **97**, No. 1, p. 223—230 (1986).

F. Corni and G. M. Antonini
EXAFS Investigation of Interdiffusive Kinetics in Cu–Au Thin Films
(66.30; 61.55; 78.70; S1.2; S1.3)
(a) Vol. **97**, No. 1, p. K21—K24 (1986).

R. Krishnan and M. Tessier
Enhanced Magnetization in Multilayer Ni–Ag Films
(75.70; S1.2; S1.3)
(a) Vol. **97**, No. 1, p. K39—K41 (1986).

M. Delannoy-Coutris and G. Perrin
Fourth-Order Thermal Expansion for Cubic Solids. Application to Aluminum, Copper, Germanium, and Silicon
(65.70; 64.10; S1.2; S4; S5.11; S5.12)
(b) Vol. **138**, No. 1, p. 93—99 (1986).

Z. Kąkol and H. Figiel
Co Contribution to Anisotropy of R–Co Type Pseudobinaries
(75.50; S1.2)
(b) Vol. **138**, No. 1, p. 151—160 (1986).

A. Beckmann
Modelling of Angle Resolved Photoemission Spectra at Low Energy
(73.20 and 79.60; S1.2)
(b) Vol. **138**, No. 1, p. 295—308 (1986).

F. Daver, T. M. A. Khajil, and M. Tomak
Finite Mean Free Paths and the Electrical Resistivity of Liquid Simple Metals and Binary Alloys
(72.15; S1; S1.2; S1.3; S2; S3)
(b) Vol. **138**, No. 1, p. 373—376 (1986).

K. Helming and B. Rauschenbach
Determination of Implantation-Induced Textures by Electron Diffraction
(61.14 and 68.55; 61.70; S1.2; S1.62)
(b) Vol. **138**, No. 1, p. K1—K6 (1986).

M. Cinal and S. Olszewski
Magnetic Atoms in Metallic Compounds Approached on the Basis of the Quantum-Statistical ab Initio Theory
(75.90; S1.1; S1.2)
(b) Vol. **138**, No. 1, p. K17—K20 (1986).

I. Tomov
Orientation Dependence of the Effective Depth of X-Ray Penetration
(61.10; S1.2)
(a) Vol. **98**, No. 1, p. 43—50 (1986).

R. Hergt, H. Pfeiffer, and L. Fritzsch
Formation of Texture in Co/Cr Films for Perpendicular Magnetic Recording
(68.55; 75.70; S1.2)
(a) Vol. **98**, No. 1, p. 69—80 (1986).

L. Kertész, A. Szász, M. I. Zakharova, and A. G. Khundzhua
Study of Crystal and Electronic Structure of Ti–Cr Alloys during Decomposition of β-Solid Solution
(61.55; 64.75; 71.25; S1.2)
(a) Vol. **98**, No. 1, p. 107—115 (1986).

A. H. Wafik and I. A. Ibrahim
Effect of Plastic Deformation on the Irreversible Magnetization in Nickel–Chromium Wire
(75.60; S1.2)
(a) Vol. **98**, No. 2, p. K141—K145 (1986).

H. J. Blythe, E. Klugmann, and F. Dworschak
Magnetic Relaxation in Cobalt Following Low-Temperature Electron Irradiation
(61.80 and 75.60; S1.2)
(a) Vol. **98**, No. 2, p. K147—K151 (1986).

A. G. Rustamov, R. Z. Sadykhov, L. M. Valiev, and A. O. Ismailov
Paramagnetic Susceptibility of the $Me_{1-x}Fe_xCr_2S_4$ (Me = Zn, Co) Composition System
(75.20; S1.1; S1.2)
(a) Vol. **98**, No. 2, p. K153—K156 (1986).

Th. Moch, B. Lippold, and A. Handstein
Spin Glass Behaviour of a Sputtered $Cu_{94.6}Tb_{5.4}$ Alloy
(75.60; S1.2)
(a) Vol. **98**, No. 2, p. K161—K164 (1986).

S1.3 Noble metals

Review Article

B. X. Liu
Ion Mixing and Metallic Alloy Phase Formation
(61.80 and 64.75; 61.70; S1.1; S1.2; S1.3; S4)
(a) Vol. **94**, No. 1, p. 11—34 (1986).

Original Papers and Short Notes

A. I. Melker, S. N. Romanov, and N. L. Tarasenko
Defect Clusters and Subcascades. II. The Splitting Threshold and Subcascade Dimensions
(61.80; S1.3)
(b) Vol. **133**, No. 1, p. 111—118 (1986).

M. Scrocco
Electron-Energy-Loss Spectra of Ag Metal
(61.80; S1.3)
(b) Vol. **133**, No. 1, p. K35—K38 (1986).

J. Szczyrbowski, K. Schmalzbauer, and H. Hoffmann
Optical Properties and Dynamic Current Density for Polycrystalline Metallic Films
(73.60 and 78.20; 78.65; S1.3)
(a) Vol. **93**, No. 1, p. 329—339 (1986).

S. M. M. Rahman and D. K. Sarker
Temperature Dependence of Structural and Transport Properties of Less-Simple Liquid Metals
(62.10 and 66.10; 61.30; S1.3; S4)
(b) Vol. **133**, No. 2, p. 517—524 (1986).

K. Z. Botros and S. S. Sheinin
On Identifying Crystal Surfaces in Electron Microscope Images of Stacking Faults
(61.70 and 68.20; 61.16; S1.2; S1.3)
(a) Vol. **93**, No. 2, p. 515—521 (1986).

B. A. Oli and J. S. Ononiwu
Effect of Non-Locality in the Calculation of Liquid Metal Resistivity of Transition Metals
(72.15; 71.25; S1.1; S1.2; S1.3)
(b) Vol. **134**, No. 1, p. 325—330 (1986).

B. Rauschenbach
Ion Beam Mixing Effects in Cu/Au Multilayered Thin Films
(61.80 and 68.55; 65.50; S1.2; S1.3)
(a) Vol. **94**, No. 1, p. 409—412 (1986).

R. Veresegyházy, B. Pécz, and I. Mojzes
The Influence of a Gold Layer on the Thermal Decomposition of InAs
(68.48; S1.3; S7.12)
(a) Vol. **94**, No. 1, p. K11—K12 (1986).

G. Schubert and V. Christoph
Phonon Drag Contribution to the Electrical Resistivity of Alkali and Noble Metals
(72.15; S1.3; S2)
(b) Vol. **134**, No. 2, p. 771—777 (1986).

R. Ducastel and G. Pitsi
The Lorenz Number of Some Ag–Cr Alloys
(72.15; S1.3)
(b) Vol. **134**, No. 2, p. 799—803 (1986).

Wang Rong-Yao
A Further Study of Superconducting Critical Temperature of $L1_2$ Type Superconductors
(61.55 and 74.70; S1.2; S1.3; S1.4; S4)
(a) Vol. **94**, No. 2, p. 445—452 (1986).

B. Pécz, E. Jároli, G. Radnóczi, R. Veresegyházy, and I. Mojzes
Pyramidal Pit Formation at the Au/GaAs Interface during Heat Treatment
(68.48; S1.3; S7.12)
(a) Vol. **94**, No. 2, p. 507—513 (1986).

Yu. S. Nechaev, S. A. Vladimirov, N. A. Olshevskii, and V. S. Khlomov
On the Effect of High-Rate Strain on Diffusion Mass Transfer in Metals
(66.30; S1.1; S1.2; S1.3; S4)
(a) Vol. **94**, No. 2, p. 553—560 (1986).

A. Turos
Defect–Impurity Interactions in Ion-Implanted Metals
(61.70; S1.1; S1.2; S1.3; S4)
(a) Vol. **94**, No. 2, p. 809—822 (1986).

V. K. Thakur and T. N. Singh
Lattice Dynamics of F.C.C. Metals
(63.20; 65.40; S1.2; S1.3)
(b) Vol. **135**, No. 1, p. 67—73 (1986).

I. Mertig, E. Mrosan, V. N. Antonov, Vl. Antonov, and P. Ziesche
Effective Optical Masses of Copper and Silver
(71.25; S1.2; S1.3)
(b) Vol. **135**, No. 1, p. K13—K16 (1986).
Addendum
(b) Vol. **138**, No. 1, p. K75 (1986).

J. Mäki
Coincidence Patterns of Two Perpendicular Long-Period Domains in Ag_3Mg
(61.55; 61.16; S1.3; S3)
(a) Vol. **95**, No. 1, p. 51—55 (1986).

N. Wagner, O. Brümmer, R. Hofmann, M. Horx, M. Jacobs, and M. Suchy
Photoelectron Spectroscopic Studies of Anodic Surface Oxidation on Pt–Ir Alloys
(79.60; S1.3)
(a) Vol. **95**, No. 2, p. K149—K152 (1986).

C. Schobèr, G. Kurz, H. Wonn, V. V. Nemoshkalenko, and V. N. Antonov
The Gyromagnetic Factor of Conduction Electrons in Silver, Gold, Palladium, and Platinum
(71.25; S1.3)
(b) Vol. **136**, No. 1, p. 233—239 (1986).

D. Augie, M. Belhadji and P. Vigier
Internal Strains in Amorphous Ni–Ag Thin Films
(68.55; S1.2; S1.3)
(a) Vol. **96**, No. 1, p. K5—K8 (1986).

S. Hashimoto and H. Iwasaki
Microdomain Model Analysis for Diffuse X-Ray Scattering from Quenched Cu–40 and 60 at% Au Alloys
(61.10; S1.2; S1.3)
(a) Vol. **96**, No. 2, p. 435—444 (1986).

V. D. Das, M. S. M. Sastry, and M. Pattabi
Aging and Field Effect Studies of Cu and Cu/Ag Composite Discontinuous Thin Films
(73.60; S1.2; S1.3)
(a) Vol. **96**, No. 2, p. 677—681 (1986).

Zhang Ji-kang, G. N. Okeke, and C. H. B. Mee
Work Function Measurements on a Silver–Oxygen–Caesium Photocathode Prepared on an Ag(111) Single Crystal
(73.30 and 79.60; S1.3)
(a) Vol. **96**, No. 2, p. 691—696 (1986).

K. Sumiyama, Y. Kawawake, and Y. Nakamura
Amorphous Fe–Cu–Ag Alloys Produced by Vapor Quenching
(61.40 and 75.50; 76.80; S1.1; S1.2; S1.3)
(a) Vol. **96**, No. 2, p. K107—K110 (1986).

M. R. Nagy, M. I. El-Agrab, M. H. Mahmood, and A. A. Mohamed
Hardness and Creep Behaviour Associated with Isothermal Aging of Aluminium–Silver Base Alloys
(62.20; S1.3; S4)
(a) Vol. **96**, No. 2, p. K139—K143 (1986).

G. Neumann
On the Calculation of Self-Diffusion and Impurity Diffusion Coefficients with the Aid of Elastic Constants
(66.30; S1.2; S1.3)
(b) Vol. **137**, No. 1, p. 57—63 (1986).

Y. Bouazra

On Some Multi-Beam Effects in a Convergent Beam
(61.16; S1.2; S1.3; S4)
(a) Vol. **98**, No. 1, p. K5—K9 (1986).

F. A. Khwaja and M. S. K. Razmi

Calculation of Binding Energies and Elastic Constants of Cu, Ag, and Au Using a One-Parameter Model Potential
(62.20; S1.2; S1.3)
(b) Vol. **138**, No. 2, p. K95—K99 (1986).

K. V. S. Prasad and C. Bansal

Resistivity and Thermoelectric Power Measurements on AgCd Shape Memory Alloys
(64.60; 72.15; S1.3)
(a) Vol. **98**, No. 2, p. 453—464 (1986).

S1.4 Lanthanides

Original Papers and Short Notes

A. I. Mitsek, K. Yu. Guslienko, and S. V. Pavlovskii
Electron Structure and Phase Diagrams of Low-Temperature Rare-Earth Magnets
(75.30; S1.4)
(b) Vol. **135**, No. 1, p. 173—188 (1986).

C. Sugiura, M. Kitamura, and S. Muramatsu
Chlorine K X-Ray Absorption-Edge Structure of Metal Trichlorides
(78.70; S1.4)
(b) Vol. **135**, No. 1, p. K57—K61 (1986).

W. Kaczmarek, I. Gościańska, J. Pietrzak, H. Ratajczak, and J. Baszyński
Ferromagnetic Resonance in Amorphous Gd–Fe Films
(76.50; 75.70; S1.1; S1.4)
(a) Vol. **95**, No. 1, p. 221—227 (1986).

M. N. Ghatikar
Phase Shift Functions in Rare-Earth Ternary Compounds
(78.70; S1.4)
(b) Vol. **135**, No. 2, p. 487—491 (1986).

L. Palermo and X. A. da Silva
Magnetic Order of Singlet–Singlet Ions Coupled to Conduction Electrons. Band Structure Effects
(75.10; 71.25; S1.4)
(b) Vol. **136**, No. 1, p. 97—102 (1986).

G. Fischer and A. Herr
Magnetic Moments and Initial Susceptibilities of f-Electron States Experiencing a Crystal Field of Cubic Symmetry
(75.10; S1.4)
(b) Vol. **136**, No. 1, p. 103—114 (1986).

S. L. Kakani and U. N. Upadhyaya
Frequency Dependent Susceptibility of Rare Earth Ternary Superconductors
(74.10 and 75.30; S1.4)
(b) Vol. **136**, No. 1, p. 115—121 (1986).

W. Borgieł, W. Nolting, and G. Borstel
Self-Consistent CPA Treatment of the Conduction Electron Spin Polarization of Degenerate Ferromagnetic Semiconductors
(75.10; 71.25; S1.4; S8)
(b) Vol. **136**, No. 1, p. 131—144 (1986).

P. V. Geld, V. V. Singer, I. Z. Radovskii, and S. V. Konovalov
Variable Valence of Ce and Pr in Liquid Alloys with Fe, Co, and Ni
(71.25; 61.25; 72.15; 75.30; S1.1; S1.2; S1.4)
(b) Vol. **136**, No. 1, p. 153—158 (1986).

M. K. Borombaev, R. Z. Levitin, A. S. Markosyan, Z. Smetana, P. Svoboda, and V. Šíma
Magnetic and Crystallographic Properties of $Gd(Cu_{1-x}Co_x)_2$
(61.60 and 75.50; S1.2; S1.4)
(a) Vol. **98**, No. 1, p. 221—227 (1986).

F. López-Aguilar and J. Costa-Quintana
Electronic Structure of the Ce Systems. II. Results
(71.25; S1.4)
(b) Vol. **138**, No. 2, p. 589—598 (1986).

C. Weber, E. Sigmund, and M. Wagner
An Unconventional Jahn-Teller Phenomenon in a Valency-Mixing Surrounding
(71.70; 76.30; S1.4)
(b) Vol. **138**, No. 2, p. 661—671 (1986).

P. Simonnin, R. Tourbot, B. Boucher, M. Perrin, and J. Vanhaute
Thermal, Electrical, and Magnetic Properties of the Amorphous Alloys RE_xSi_{1-x} (RE = Gd or Tb; $x = 0.87, 0.59, 0.18$)
(75.50; 61.40; 64.75; 72.60; S1.4; S1.61)
(a) Vol. **98**, No. 2, p. 551—566 (1986).

S1.5. Actinides

Original Papers and Short Notes

S1.61. Carbides, silicides, germanides

Original Papers and Short Notes

V. M. Koleshko, V. F. Belitskii, and A. A. Khodin
Electrical Conductance and Phase Transformations in Dy and Tb Thin Films on Si Substrates Subjected to Heat Treatment
(68.55; 73.60; S1.4; S1.61)
(a) Vol. **93**, No. 1, p. 29—33 (1986).

A. I. Gusev and A. A. Rempel
Order–Disorder Phase Transition Channel in Niobium Carbide
(64.60; S1.61)
(a) Vol. **93**, No. 1, p. 71—80 (1986).

D. M. Vanderwalker
The Glass Transition Phases in the Formation of PtSi
(61.40; 68.48; S1.61)
(a) Vol. **94**, No. 1, p. 77—81 (1986).

N. V. Moseev, V. G. Chudinov, B. N. Goshchitskii, and V. I. Protasov
Molecular Dynamics Study of Atomic Configuration of Point Defects in Intermetallic Compound Mo_3Si
(61.70; S1.61)
(a) Vol. **94**, No. 1, p. 147—152 (1986).

W. Bażela, A. Szytuła, and J. Leciejewicz
Magnetic Properties of ZrMnSi, ZrMnGe, and HfMnSi Intermetallics
(75.20; 61.55; S1.61)
(a) Vol. **94**, No. 1, p. 207—211 (1986).

I. Bryłowska, K. Paprocki, and M. Subotowicz
Obtaining Hafnium Silicides by the Mixing Method
(61.80 and 68.55; 73.30; S1.61)
(a) Vol. **94**, No. 1, p. 423—426 (1986).

C. Kaschner, A. Witzmann, K. Gärtner, and G. Götz
Laser Annealing of Ion-Implanted NiSi Layers
(68.55; 61.80; S1.61)
(a) Vol. **94**, No. 2, p. 787—791 (1986).

O. Engström, H. Pettersson, and B. Sernelius
Photoelectric Yield Spectra of Metal–Semiconductor Structures
(79.60; 73.40; S1; S1.3; S1.61; S5.11)
(a) Vol. **95**, No. 2, p. 691—701 (1986).

S1.62. Nitrides, phosphides, arsenides

Original Papers and Short Notes

U. PITTERMANN and S. RIPPER
Crystallization Behaviour of Thin NiP Alloy Films Prepared by Electroless or Galvanic Deposition
(68.55; S1.62)
(a) Vol. **93**, No. 1, p. 131—142 (1986).

C. SUŁOWSKI, Z. WOKULSKI, and K. WOKULSKA
Superconducting Properties of TiN Whiskers
(74.70; S1.62)
(a) Vol. **95**, No. 1, p. K67—K70 (1986).

A. ARMIGLIATO, G. QUEIROLO, and G. VALDRÈ
Nitrogen Profiles of Ion-Implanted TiN Films by Electron Energy Loss Spectroscopy
(68.55; 61.80; S1.62; S5.11)
(a) Vol. **96**, No. 1, p. 89—99 (1986).
Erratum
(a) Vol. **98**, No. 2, p. 651 (1986).

G. A. GOVOR
The Effect of Titanium Additions on Phase Transitions in Manganese Arsenide Single Crystals
(75.50; S1.62)
(a) Vol. **97**, No. 2, p. K145—K148 (1986).

K. HEMACHANDRAN and A. R. CHETAL
Correlation between X-Ray Chemical Shift and Partial Charge in Some Minerals
(78.70; S1.62; S7)
(b) Vol. **138**, No. 1, p. 229—233 (1986).

K. HELMING and B. RAUSCHENBACH
Determination of Implantation-Induced Textures by Electron Diffraction
(61.14 and 68.55; 61.70; S1.2; S1.62)
(b) Vol. **138**, No. 1, p. K1—K6 (1986).

S1.63. Sulfides, selenides, tellurides

Original Papers and Short Notes

P. R. Sarode
EXAFS in Niobium Dichalcogenides Intercalated with First-Row Transition Metals
(61.50; 78.70; S1.63)
(a) Vol. **98**, No. 2, p. 391—397 (1986).

S2. Alkali metals and their alloys and compounds (for alkaline halides, see S9.11)

Original Papers and Short Notes

M. Kumari and N. Dass
On the Pressure Dpendence of Grüneisen Parameter in Solids
(63.90; 65.50; S2; S9.11; S9.12; S10.1)
(b) Vol. **133**, No. 1, p. 101—110 (1986).

M. Kato and Y. H. Ohtsuki
Stopping Power of Atoms near Metal Surfaces at Grazing Incidence Condition
(61.80; 71.35; S2)
(b) Vol. **133**, No. 1, p. 267—275 (1986).

D. J. González
Density Functional Pseudopotential Calculation of the Cohesive Properties of Ordered (CsCl-Type) Compounds of Nontransition Metals
(71.10; S2; S3; S4)
(b) Vol. **133**, No. 1, p. 277—284 (1986).

Tran Thoai Duy Bao
Static Polarizability and Dipolar Surface Plasmon of Small Metal Particles. II. Including the Lattice Structure
(71.45 and 73.20; S2)
(b) Vol. **133**, No. 1, p. 329—332 (1986).

R. Sharma, K. S. Sharma, and L. Dass
Linearised Screened Pseudopotential and Superconducting State Parameters of a Number of Metals
(74.20; 71.10; S1.2; S2; S3; S4)
(b) Vol. **133**, No. 2, p. 701—706 (1986).

T. Soma and H.-Matsuo Kagaya
Lattice Anharmonicity and Thermal Expansion of Alkali Metals
(65.70; S2)
(b) Vol. **134**, No. 1, p. K7—K9 (1986).

S. O. Cholakh, V. A. Pustovarov, and B. V. Shulgin
The Effects of Electron Excitation Energy Transfer to the Impurity Centre in LiH Crystals
(71.35 and 71.55; 78.55; S2)
(b) Vol. **134**, No. 2, p. 741—744 (1986).

G. Schubert and V. Christoph
Phonon Drag Contribution to the Electrical Resistivity of Alkali and Noble Metals
(72.15; S1.3; S2)
(b) Vol. **134**, No. 2, p. 771—777 (1986).

K. P. Zolnikov, S. G. Psakhe, and V. E. Panin
Partial Structure Factors of the Melt in the Long-Wave Limit. The Calculation of the State Diagram of the Li–Na System
(64.30 and 71.25; S2)
(b) Vol. **134**, No. 2, p. K115—K119 (1986).

T. Soma, H. Higuchi, and H.-Matsuo Kagaya
High-Temperature Dependence of the Thermal Expansion Coefficient for Alloy Systems Composed of Alkali Metal Elements
(65.70; S2)
(b) Vol. **135**, No. 1, p. K1—K3 (1986).

G. M. S. Srivastava, G. G. Agrawal, and J. Shanker
Thermodynamic Properties and the Equation of State of Alkali Hydride Crystals
(64.30 and 65.70; S2)
(b) Vol. **135**, No. 2, p. 529—534 (1986).

S. K. Lai
Density Functional Approach to the Surface Tension of Liquid Metals
(68.40; 71.10; S2)
(b) Vol. **136**, No. 2, p. 685—691 (1986).

M. Dębowska, A. Baranowski, K. Jerie, and W. Świątkowski
The Effect of Impregnation with Gold on Positron Lifetime Spectra in NaY Zeolite
(61.70; 68.20; 78.70; S2)
(a) Vol. **97**, No. 1, p. K35—K38 (1986).

K. N. Khanna and G. Shanker
One-Component Plasma Structure Factor of Liquid Alkali Metals
(61.25; S2)
(b) Vol. **137**, No. 2, p. K107—K110 (1986).

T. Soma, K. Hosaka, and H.-Matsuo Kagaya
Compression Effect on the Phase Diagrams at Low Temperatures of Alkali Alloys
(64.70; S2)
(b) Vol. **137**, No. 2, p. K111—K114 (1986).

C. Malinowska-Adamska
On the Stability of Anharmonic Crystals
(65.50; S1.2; S1.3; S2; S3; S4)
(b) Vol. **137**, No. 2, p. K115—K119 (1986).

O. V. Gritsenko and G. M. Zhidomirov
A New Correlation Energy Interpolation Functional for Many-Electron Systems in the Homogeneous Electron Gas Model. The Ratio of Paramagnetic and Ferromagnetic State Energies versus Gas Density
(75.10; 75.20; 75.50; S2)
(b) Vol. **138**, No. 1, p. 161—169 (1986).

S3. Alkaline earth metals and their alloys and compounds (for alkaline earth halides, see S9.12)

Original Papers and Short Notes

D. K. Papademetriou, D. Katsanos, and A. G. Doukas
Plasmon Bands in X-Ray Spectra of Polycrystalline Beryllium
(71.45 and 78.70; S3)
(b) Vol. **133**, No. 1, p. 223—228 (1986).

D. J. González
Density Functional Pseudopotential Calculation of the Cohesive Properties of Ordered (CsCl-Type) Compounds of Nontransition Metals
(71.10; S2; S3; S4)
(b) Vol. **133**, No. 1, p. 277—284 (1986).

R. Sharma, K. S. Sharma, and L. Dass
Linearised Screened Pseudopotential and Superconducting State Parameters of a Number of Metals
(74.20; 71.10; S1.2; S2; S3; S4)
(b) Vol. **133**, No. 2, p. 701—706 (1986).

H. Siethoff
Dynamical Recovery of Cadmium, Zinc, and Magnesium at Intermediate Temperatures
(62.20; S1; S1.2; S3)
(a) Vol. **94**, No. 1, p. 183—189 (1986).

J. Mäki
Coincidence Patterns of Two Perpendicular Long-Period Domains in Ag_3Mg
(61.55; 61.16; S1.3; S3)
(a) Vol. **95**, No. 1, p. 51—55 (1986).

P. H. Bisio and A. M. Monti
Small Vacancy Clusters in H.C.P. Metals
(61.70; S3)
(b) Vol. **135**, No. 2, p. 545—553 (1986).

G. Singh and R. P. S. Rathore
Two- and Three-Body Interactions and Phonons in Some F.C.C. Metals
(63.20; 62.20; S1; S3)
(b) Vol. **136**, No. 1, p. 57—61 (1986).

A. B. Kaiser
Low Temperature Resistivity Anomaly in Glassy Metals. Two-Level Systems and Incipient Localization versus Correlations
(72.15; S1.2; S3)
(b) Vol. **136**, No. 2, p. 779—784 (1986).

C. Malinowska-Adamska
On the Stability of Anharmonic Crystals
(65.50; S1.2; S1.3; S2; S3; S4)
(b) Vol. **137**, No. 2, p. K115—K119 (1986).

D. O. Northwood and I. O. Smith
Three Creep Behaviour of High Purity Polycrystalline Magnesium at 0.32 to $0.51 T_m$
(62.20; S3)
(a) Vol. **97**, No. 2, p. 441—450 (1986).

F. Daver, T. M. A. Khajil, and M. Tomak
Finite Mean Free Paths and the Electrical Resistivity of Liquid Simple Metals and Binary Alloys
(72.15; S1; S1.2; S1.3; S2; S3)
(b) Vol. **138**, No. 1, p. 373—376 (1986).

J. Ni and T. S. Kê
A Theory of Anomalously Amplitude-Dependent Internal Friction Peaks Due to the Interaction between Dislocation Kinks and Substitutional Solute Atoms in F.C.C. Metals
(61.70; 62.40; 66.30; S1.2; S3; S4)
(a) Vol. **98**, No. 2, p. 487—501 (1986).

S4. Other metals and their alloys

Review Article

B. X. Liu
Ion Mixing and Metallic Alloy Phase Formation
(61.80 and 64.75; 61.70; S1.1; S1.2; S1.3; S4)
(a) Vol. **94**, No. 1, p. 11—34 (1986).

Original Papers and Short Notes

M. Kotera, K. Murata, and K. Nagami
Monte-Carlo Studies of Energy Straggling of Electrons in Solids
(79.20; S4)
(b) Vol. **133**, No. 1, p. 203—209 (1986).

D. J. González
Density Functional Pseudopotential Calculation of the Cohesive Properties of Ordered (CsCl-Type) Compounds of Nontransition Metals
(71.10; S2; S3; S4)
(b) Vol. **133**, No. 1, p. 277—284 (1986).

A. R. Jani and H. K. Patel
Quantum Number Dependent Model Pseudopotential for Metallic Properties
(71.25; 72.15; S4)
(b) Vol. **133**, No. 1, p. K21—K24 (1986).

G. Hübner and H. Löffler
The Growth of Guinier-Preston Zones in an Al–12 at% Zn Alloy during the Ageing at RT in Dependence on the Homogenization Temperature
(61.50; S4)
(a) Vol. **93**, No. 1, p. 113—118 (1986).

E. Biagi, S. Frosini, L. Masotti, and E. Borchi
The Propagation of Acoustic Waves in Polycrystals with Axial Texture
(62.65; 62.20; S1.2; S4; S5.11)
(a) Vol. **93**, No. 1, p. 151—162 (1986).

M. R. Nagy, M. S. Sakr, M. M. Mostafa, and R. Kamel
Recovery Behaviour after Stress Reductions during the Steady State Creep of Al–14 wt% Zn
(62.20; S1.2; S4)
(a) Vol. **93**, No. 1, p. 185—193 (1986).

S. M. M. Rahman and D. K. Sarker
Temperature Dependence of Structural and Transport Properties of Less-Simple Liquid Metals
(62.10 and 66.10; 61.30; S1.3; S4)
(b) Vol. **133**, No. 2, p. 517—524 (1986).

R. Sharma, K. S. Sharma, and L. Dass
Linearised Screened Pseudopotential and Supercoducting State Parameters of a Number of Metals
(74.20; 71.10; S1.2; S2; S3; S4)
(b) Vol. **133**, No. 2, p. 701—706 (1986).

T. Soma, K. Wada, and H.-Matsuo Kagaya
Temperature Effect on the Elastic Stiffness Constants of Al
(62.20; S4)
(b) Vol. **133**, No. 2, p. K87—K90 (1986).

J. Dutkiewicz, J. Salawa, H. Löffler, and E. Korngiebel
The Effect of Small Additions of Si on the Structure of an Al–2 at% Zn–1 at% Mg Mother Alloy
(61.55; S4)
(a) Vol. **93**, No. 2, p. 429—441 (1986).

G. Wendrock and H. Löffler
Information about the Position of the Coherent Spinodal Curve in the Al–Zn System by Means of the Course of the SAXS Intensity at Elevated Temperatures
(61.55; 64.70; S1.2; S4)
(a) Vol. **93**, No. 2, p. 487—493 (1986).

A. E. Smith and D. F. Lynch
Calculated Low Energy Electron Total Current Spectra for the Al(111) Surface
(61.14; 68.20; S4)
(b) Vol. **134**, No. 1, p. 11—15 (1986).

M. Kotera, K. Murata, and K. Nagami
Monte-Carlo Studies of Energy Straggling of Electrons in Solids. II. Consideration of Electron Angular Scattering
(79.20; S4)
(b) Vol. **134**, No. 1, p. 203—210 (1986).

T. M. A. Khajil and M. Tomak
The Resistivity of Liquid Metals with Finite Mean-Free Path
(72.15; S4)
(b) Vol. **134**, No. 1, p. 321—324 (1986).

F. Völklein and E. Kessler
Temperature and Thickness Dependence of Electrical and Thermal Transport Coefficients of $Bi_{1-x}Sb_x$ Films in an Anisotropic, Non-Degenerate Two-Band Model
(72.10 and 73.60; S4)
(b) Vol. **134**, No. 1, p. 351—362 (1986).

E. Ratner, A. Appelbaum, R. Brener, and M. Eizenberg
Interactions of Thin Al Films with Ni–Cr Alloy and Bilayer Films Deposited on Si
(68.55; 61.10; 79.20; S1.2; S4)
(a) Vol. **94**, No. 1, p. 61—69 (1986).

C. M. Su and T. S. Kê
Further Experiments on the High-Temperature Internal Friction Peak in High-Purity Single Crystal Aluminium
(62.40; 61.70; S4)
(a) Vol. **94**, No. 1, p. 191—202 (1986).

Wang Rong-Yao
A Further of Superconducting Critical Temperature of $L1_2$ Type Superconductors
(61.55 and 74.70; S1.2; S1.3; S1.4; S4)
(a) Vol. **94**, No. 2, p. 445—452 (1986).

Yu. S. Nechaev, S. A. Vladimirov, N. A. Olshevskii, and V. S. Khlomov
On the Effect of High-Rate Strain on Diffusion Mass Transfer in Metals
(66.30; S1.1; S1.2; S1.3; S4)
(a) Vol. **94**, No. 2, p. 553—560 (1986).

S. Takamura and M. Kobiyama
Effect of Low-Temperature Deformation on Internal Friction Peaks of Irradiated Al Alloys
(61.70; 62.40; S4)
(a) Vol. **94**, No. 2, p. 595—600 (1986).

K. Iwasaki
Strain Amplitude Dependence of the Grain Boundary Internal Friction Peak of Pure Aluminium
(61.70; 62.40; S4)
(a) Vol. **94**, No. 2, p. 601—607 (1986).

A. Turos
Defect–Impurity Interactions in Ion-Implanted Metals
(61.70; S1.1; S1.2; S1.3; S4)
(a) Vol. **94**, No. 2, p. 809—822 (1986).

U. Schmidt, A. Zahra, R. Schülbe, and H. Löffler
On the Stability of Guinier-Preston Zones in an Al–1.87 at% Cu Alloy at Elevated Temperature
(61.55; 62.20; S4)
(a) Vol. **95**, No. 1, p. 73—80 (1986).

K. Takaishi, T. Kikuchi, K. Furuya, I. Hashimoto, H. Yamaguchi, E. Yagi, and M. Iwaki
Thermal Extraction of Krypton in Aluminium Using Mass Spectrometer
(66.30; 61.70; S4)
(a) Vol. **95**, No. 1, p. 135—139 (1986).

S. Takamura and M. Kobiyama
Dislocation Pinning in Al and Ag Alloys after Low-Temperature Deformation
(61.70; 62.20; 72.15; S1.3; S4)
(a) Vol. **95**, No. 1, p. 165—172 (1986).

K.-D. Durst, H. Kronmüller, F. T. Parker, and H. Oesterreicher
Temperature Dependence of Coercivity of Cellular Sm_2Co_{17}–$SmCo_5$ Permanent Magnets
(75.50; 75.60; S1.2; S4)
(a) Vol. **95**, No. 1, p. 213—219 (1986).

I. F. I. Mikhail
L-Holes and Their Effect on the Band Parameters of Bismuth
(71.25; S4)
(b) Vol. **136**, No. 2, p. 643—649 (1986).

Wang Rong-yao and Luo Qi-guang
Structure of A15 Compounds
(61.55; 74.70; S1; S4)
(a) Vol. **96**, No. 2, p. 397—406 (1986).

R. Tognato
Interfacial Free Energies per Unit Area of Liquid Flat Aggregates Embedded in a Crystal Matrix
(68.45; 61.20; S4)
(a) Vol. **96**, No. 2, p. 451—461 (1986).

T. S. Kê and B. S. Zhang
Contribution of Bamboo Boundaries to the Internal Friction Peak in Macrocrystalline High-Purity Aluminium
(61.70; 62.40; S4)
(a) Vol. **96**, No. 2, p. 515—525 (1986).

M. R. Nagy, M. I. El-Agrab, M. H. Mahmood, and A. A. Mohamed
Hardness and Creep Behaviour Associated with Isothermal Aging of Aluminium–Silver Base Alloys
(62.20; S1.3; S4)
(a) Vol. **96**, No. 2, p. K139—K143 (1986).

T. Soma, N. Suzuki, and H.-Matsuo Kagaya
Specific Heat of α-Phase Cu–Al Alloys
(65.40; S1.2; S4)
(b) Vol. **137**, No. 1, p. K15—K19 (1986).

J. Mucha, A. V. Krasnokutskii, V. A. Merisov, N. Azumai, and J. Rafałowicz
Thermal Conductivity Minimum of 6N Purity Aluminium Monocrystal
(72.15; S4)
(a) Vol. **97**, No. 1, p. 307—312 (1986).

M. A. Kenawy, M. R. Nagy, M. S. Sakr, and M. T. Mostafa
*The Effect of Precipation on the Electrical Resistivity in Al–2.5 wt% Cu
(72.15; 64.75; S4)
(a) Vol. **97**, No. 1, p. K73—K78 (1986).

M. Załużny and A. Łukasik
Theory of the Infrared Spectrum of Size-Quantized Bi Films
(78.30; 71.25; 78.65; S4)
(b) Vol. **137**, No. 2, p. 607—617 (1986).

C. Malinowska-Adamska
On the Stability of Anharmonic Crystals
(65.50; S1.2; S1.3; S2; S3; S4)
(b) Vol. **137**, No. 2, p. K115—K119 (1986).

S. Maciejewski

Electron Structure Dependence on Crystal Lattice Parameters for $Cd_{1-x}Hg_x$ ($0.30 \leqq x \leqq 0.70$) and $In_{1-x}Sn_x$ ($0.10 \leqq x \leqq 0.30$) Alloys

(71.25; S4)

(b) Vol. **138**, No. 2, p. K113—K117 (1986).

J. Ni and T. S. Kê

A Theory of Anomalously Amplitude-Dependent Internal Friction Peaks Due to the Interaction between Dislocation Kinks and Substitutional Solute Atoms in F.C.C. Metals

(61.70; 62.40; 66.30; S1.2; S3; S4)

(a) Vol. **98**, No. 2, p. 487—501 (1986).

V. I. Danilov, V. M. Drobin, and V. N. Trofimov

Ferromagnetic Behaviour of Non-Magnetic Samples

(75.50; S1.2; S4; S5; S12)

(a) Vol. **98**, No. 2, p. K137—K140 (1986).

S5. Nonmetallic elements

Review Article

YU. N. SHUNIN and K. K. SHVARTS
Calculation of the Electronic Structure in Disordered Semiconductors
(71.25; 71.10; 71.20; S5; S5.11)
(b) Vol. **135**, No. 1, p. 15—36 (1986).

Original Papers and Short Notes

V. N. BAIER, V. M. KATKOV, and V. M. STRAKHOVENKO
Nondipole Radiation of Electrons at Planar Channeling
(61.80 and 78.70; S5; S5.11)
(b) Vol. **133**, No. 1, p. 211—222 (1986).

V. K. BASHENOV, V. E. GORBACHEV, and D. I. MARVAKOV
Bond Charge Approximation for Valence Electron Density in Elemental Semiconductors
(71.25; S5; S5.11; S5.12)
(b) Vol. **133**, No. 1, p. 285—296 (1986).

V. N. BAIER, V. M. KATKOV, and V. M. STRAKHOVENKO
Radiation Yield of High-Energy Electrons in Thick Crystals
(61.80 and 78.70; S1.1; S1.2; S5; S5.11; S5.12)
(b) Vol. **133**, No. 2, p. 583—592 (1986).

P. ENDERS and R. SCHUCHARDT
Anomalous Composition Dependence of the Dielectric Constant in Bismuth–Antimony Alloys Due to Disorder-Induced Indirect Interband Polarization Processes
(71.25; 77.20; S5)
(b) Vol. **135**, No. 1, p. 207—216 (1986).

K. YASUDA, H. MASUDA, M. TAKEDA, and A. YOSHIDA
Annealing Effects in Amorphous Hydrogenated Carbon Films Prepared by Plasma Deposition from Butane Gas
(78.65; 68.55; S5)
(a) Vol. **95**, No. 1, p. 249—254 (1986).

V. S. VARICHENKO, A. M. ZAITSEV, and V. F. STELMAKH
Luminescence of Natural IIa Diamond Implanted with Nitrogen Ions
(78.55 and 78.60; 61.70; S5)
(a) Vol. **95**, No. 1, p. K25—K28 (1986).

R. RAMIRÉZ and M. C. BÖHM
An Efficient Technique for the Evaluation of Lattice Sums in Crystal Orbital (CO) Calculations
(71.10 and 71.25; S5; S7.1)
(b) Vol. **135**, No. 2, p. 661—667 (1986).

I. E. Bolotov, V. Yu. Kolosov, and V. B. Malkov
Electron Microscope Investigation of Crystals Based on Bend-Contour Arrangement. III. Formation of Subgrain Boundaries in Dislocation-Free Crystals of Selenium
(61.70; 61.16; S5)
(a) Vol. **95**, No. 2, p. 377—383 (1986).

P. Burgaud, J. C. Bernede, G. Safoula, and A. Ameziane
Effect of Chlorine on Some Properties of Chlorine-Doped Selenium Films
(73.60; 68.55; S5)
(a) Vol. **95**, No. 2, p. 721—733 (1986).

V. S. Varichenko, A. M. Zaitsev, and V. F. Stelmakh
Spatial Distribution of Impurities and Defects Introduced in Diamond by High Energy Ion Implantation
(61.70; S5)
(a) Vol. **95**, No. 2, p. K123—K126 (1986).

M. Kohyama, R. Yamamoto, and M. Doyama
Energies and Structures of (111) Coincidence Twist Boundaries in 3C-SiC, Diamond, and Silicon
(61.70; S5; S5.11; S6)
(b) Vol. **136**, No. 1, p. 31—36 (1986).

P. V. Smith, J. E. Szymanski, and J. A. D. Matthew
A Reformulation of the LUC-CNDO Approach to the Properties of Solids. Application to Diamond and Silicon
(71.10; 63.20; S5; S5.11)
(b) Vol. **136**, No. 1, p. 261—271 (1986).

Y. Kazumata, Y. Nakano, S. Yugo, and T. Kimura
Exchange Coupling Model in Neutron-Irradiated Graphite
(76.30; 61.80; 71.70; S5)
(b) Vol. **136**, No. 1, p. 325—329 (1986).

F. M. Muntyanu
Quantum Oscillations of Magnetoresistance and Thermomagnetic Power and the Fermi Surface of As–Sb Alloys
(72.15; 71.25; S5)
(b) Vol. **136**, No. 2, p. 749—756 (1986).

Chen Guanghua, Xu Xixiang, Zhang Fangqing, and Jiang Zhichen
Valence-Electron Properties of Glow-Discharge a-C:H Films Studied by Electron Energy Loss Spectroscopy
(71.45; 61.80; S5; S6)
(b) Vol. **137**, No. 1, p. K55—K57 (1986).

I. M. Chapnik
On the Relation between the Energy Gap Data of Semiconducting Compounds and the Position of the Components of the Compounds in the Periodic Table
(72.80; S5; S6; S7; S8; S10.1)
(b) Vol. **137**, No. 1, p. K95—K100 (1986).

B. A. Vaid and K. C. Sharma
Structural Disorder Model for Amorphous Semiconductors: a-Se
(61.40; 65.50; 78.20; S5)
(b) Vol. **137**, No. 2, p. 433—440 (1986).

V. I. Danilov, V. M. Drobin, and V. N. Trofimov
Ferromagnetic Behaviour of Non-Magnetic Samples
(75.50; S1.2; S4; S5; S12)
(a) Vol. **98**, No. 2, p. K137—K140 (1986).

S5.11. Silicon

Review Article

Yu. N. Shunin and K. K. Shvarts
Calculation of the Electronic Structure in Disordered Semiconductors
(71.25; 71.10; 71.20; S5; S5.11)
(b) Vol. **135**, No. 1, p. 15—36 (1986).

Original Papers and Short Notes

M. V. Kruglov, I. K. Solomin, and A. V. Lunev
Determination of Photoemission Generation Depth with Use of Experiments on the Dynamic Scattering of X-Rays
(79.60; 61.10; S5.11; S5.12)
(b) Vol. **133**, No. 1, p. 47—55 (1986).

P. K. Dixit, B. A. Vaid, and K. C. Sharma
The Structural Disorder Model: Pressure Induced Transitions in Semiconductors Si and Ge
(64.70; 65.50; S5.11; S5.12)
(b) Vol. **133**, No. 1, p. 71—74 (1986).

V. N. Baier, V. M. Katkov, and V. M. Strakhovenko
Nondipole Radiation of Electrons at Planar Channeling
(61.80 and 78.70; S5; S5.11)
(b) Vol. **133**, No. 1, p. 211—222 (1986).

V. K. Bashenov, V. E. Gorbachev, and D. I. Marvakov
Bond Charge Approximation for Valence Electron Density in Elemental Semiconductors
(71.25; S5; S5.11; S5.12)
(b) Vol. **133**, No. 1, p. 285—296 (1986).

D. Wruck and F. Spiegelberg
Metastable Species of Oxygen-Related Donors in Silicon
(71.55 and 78.30; S5.11)
(b) Vol. **133**, No. 1, p. K39—K43 (1986).

K. H. Heinig and H.-D. Geiler
Phenomenological Theory of Explosive Solid-Phase Crystallization of Amorphous Silicon. II. Dynamical Processes
(61.40 and 61.50; 64.70; S5.11)
(a) Vol. **93**, No. 1, p. 99—104 (1986).

H. S. Tan and S. C. Ng
Thickness Dependence of Threshold Energy Density for Pulsed-Laser Melting of Silicon
(64.70 and 79.20; S5.11)
(a) Vol. **93**, No. 1, p. 119—122 (1986).

E. Biagi, S. Frosini, L. Masotti, and E. Borchi
The Propagation of Acoustic Waves in Polycrystals with Axial Texture
(62.65; 62.20; S1.2; S4; S5.11)
(a) Vol. **93**, No. 1, p. 151—162 (1986).

V. A. Gurtov and A. I. Nazarov
Isothermal Annealing of Space Charge in MNOS Structures
(73.40; S5.11)
(a) Vol. **93**, No. 1, p. 347—351 (1986).

P. R. Vaya, J. Majhi, B. S. V. Gopalam, and C. Dattatreyan
Study on n-PbTe/p-Si Heterostructures
(73.40; S5.11; S8)
(a) Vol. **93**, No. 1, p. 353—360 (1986).

S. V. Bogdanov, A. B. Kravchenko, A. F. Plotnikov, and V. E. Shubin
Model of the Avalanche Multiplication in MIS Structures
(73.40; S5.11)
(a) Vol. **93**, No. 1, p. 361—367 (1986).

A. B. Kravchenko, A. F. Plotnikov, A. G. Solonko, and V. E. Shubin
Avalanche Spectroscopy of MIS Structure Generation Centres
(73.40; S5.11)
(a) Vol. **93**, No. 1, p. 369—376 (1986).

N. T. Tam and Tran Chot
Experimental Richardson Constant of Metal–Semiconductor Schottky Barrier Contacts
(73.30; S5.11; S7.11; S7.12; S8.11; S8.12)
(a) Vol. **93**, No. 1, p. K91—K95 (1986).

M. Kittler, W. Seifert, and K.-W. Schröder
Temperature-Dependent EBIC Diffusion-Length Measurements in Silicon
(72.20; S5.11)
(a) Vol. **93**, No. 1, p. K101—K104 (1986).

S. G. Gagarin, A. V. Mudryi, A. L. Pushkarchuk, A. G. Ulyashin, and Yu. A. Teterin
Theoretical Analysis of the Chemical Activity of Noble Gas Atoms in Silicon
(61.70 and 71.55; S5.11)
(b) Vol. **133**, No. 2, p. 499—503 (1986).

A. M. Taratin and S. A. Vorobiev
Computer Simulation of Proton Channeling Catch-Up in Bent Crystals
(61.80; S5.11)
(b) Vol. **133**, No. 2, p. 511—516 (1986).

V. N. Baier, V. M. Katkov, and V. M. Strakhovenko
Radiation Yield of High-Energy Electrons in Thick Crystals
(61.80 and 78.70; S1.1; S1.2; S5; S5.11; S5.12)
(b) Vol. **133**, No. 2, p. 583—592 (1986).

H. Conzelmann, A. Hangleiter, and J. Weber
Thallium-Related Isoelectronic Bound Excitons in Silicon. A Bistable Defect at Low Temperatures
(71.70; 71.35; 78.55; S5.11)
(b) Vol. **133**, No. 2, p. 655—668 (1986).

S. Wilke, J. Mašek, and B. Velický
A CBLM Approach to the Electronic Structure of Transition Metal Impurities in Silicon
(71.55; S5.11)
(b) Vol. **133**, No. 2, p. K111—K117 (1986).

E. Scheid and P. Chenevier
1 D Simulation of Oxidation-Enhanced and Oxidation Retarded Diffusion in Silicon and Validity of the Physical Model. Transition to the 2D Case
(66.30; S5.11)
(a) Vol. **93**, No. 2, p. 523—533 (1986).

S. Guimarães, R. Kögler, E. Landi, S. Solmi, and E. Wieser
Anomalous Enhanced Diffusion and Electrical Activation of Boron in Silicon after Rapid Isothermal Annealing
(66.30; 72.80; S5.11)
(a) Vol. **93**, No. 2, p. 549—557 (1986).

K. D. Glinchuk, N. M. Litovchenko, and V. Yu. Ptitsin
A Variety of Oxygen-Induced Recombination Centres in 450 to 600 °C Heat-Treated Silicon
(61.70; 71.55; S5.11)
(a) Vol. **93**, No. 2, p. 565—571 (1986).

H. Indusekhar, V. Kumar, and D. Sengupta
Investigation of Deep Defects Due to α-Particle Irradiation in n-Silicon
(71.55; 61.80; S5.11)
(a) Vol. **93**, No. 2, p. 645—653 (1986).

S. A. Abo-Namous and R. W. Fane
Post-Annealing Effects on Doping Efficiency in Ne-Sputtered a-Si
(73.60; S5.11)
(a) Vol. **93**, No. 2, p. 669—674 (1986).

W. Fertig and H. Pagnia
Photoemission of Electrons from Interface States in Silicon MOS Structures
(73.40; S5.11)
(a) Vol. **93**, No. 2, p. 717—722 (1986).

V. Yu. Balandin, A. V. Dvurechenskii, and L. N. Aleksandrov
Intermediate and Self-Sustaining Crystallization of a-Si Layers during Pulsed Electron Beam Annealing
(68.55; 61.40; S5.11)
(a) Vol. **93**, No. 2, p. K105—K109 (1986).

F.-G. Kirscht, V. I. Nikitenko, H. Richter, E.-A. Steinman, and E. B. Yakimov
Photoluminescence of Preannealed Plastically Deformed Silicon Crystals
(61.70 and 78.55; S5.11)
(a) Vol. **93**, No. 2, p. K143—K146 (1986).

L. I. Murin
On the Nature of Interstitial Defect with E_c — 0.16 eV Level in Irradiated Silicon
(61.70; 71.55; S5.11)
(a) Vol. **93**, No. 2, p. K147—K149 (1986).

N. I. Akulovich, V. M. Maksimenko, and V. I. Utenko
On Localization of Phosphorus in Neutron-Doped Silicon
(61.70; 72.80; S5.11)
(a) Vol. **93**, No. 2, p. K157—K159 (1986).

Ya. I. Latushko, L. F. Makarenko, V. P. Markevich, and L. I. Murin
Electrical and Optical Characterization of Thermal Donors in Silicon
(71.55; 78.50; S5.11)
(a) Vol. **93**, No. 2, p. K181—K184 (1986).

R. Petrova and S. Grancharov
Mobile Ion Gettering in MOS Structures with Polysilicon Gate Electrodes
(73.40; S5.11)
(a) Vol. **93**, No. 2, p. K215—K218 (1986).

D. A. Aronov, D. K. Isamukhamedova, and Yu. M. Yuabov
On the Theory of Majority-Carrier Silicon Solar-Cell MTIS Structures with a Schottky Barrier
(73.40; 73.30; S5.11)
(a) Vol. **93**, No. 2, p. K219—K221 (1986).

D. E. Ioannou, Y. J. Huang, P. K. McLarty, and S. M. Johnson
On the Barrier Height of Al/p-Si Schottky Diodes
(73.30; S5.11)
(a) Vol. **93**, No. 2, p. K223—K226 (1986).

H. Übensee, G. Paasch, and J.-P. Zöllner
Self-Consistent Theory of the Electronic Structure of Inversion Layers. II. Quantum Effects and the Electronic Properties of MIS Structures
(73.40; S5.11; S7.1; S10.1)
(b) Vol. **134**, No. 1, p. 367—381 (1986).

Chen Guanghua, Zhang Fangqing, and Xu Xixiang
Doping Effect on the Hydrogen Content of a-$Si_{1-x}C_x$:H Films
(61.70; S5.11)
(b) Vol. **134**, No. 1, p. K27—K29 (1986).

H. Wiesner, W. Cordts, and J. Monecke
On the Influence of the Non-Locality of Exchange and Correlation on the Total Energy and on the Band Gap of Silicon
(71.25; S5.11)
(b) Vol. **134**, No. 1, p. K63—K66 (1986).

B. V. Kuvaldin, E. V. Zolotoyabko, E. M. Iolin, and E. A. Raitman
Neutron Topography Study of Large Single Crystals by the Local Rocking Curve Method
(61.12 and 61.50; S1.2; S5.11)
(a) Vol. **94**, No. 1, p. 45—52 (1986).

M. Berti, A. V. Drigo, G. Lulli, P. G. Merli, and M. Vittori Antisari
On the Dynamic Annealing Mechanism in P^+-Implanted Silicon
(61.55 and 61.80; 61.16; S5.11)
(a) Vol. **94**, No. 1, p. 95—106 (1986).

A. M. Mazzone, M. Servidori, and G. Cembali
Monte Carlo Simulation of the Annealing Kinetics of Oxygen Recoils Formed in Through-Oxide Implants on Silicon
(61.70; S5.11)
(a) Vol. **94**, No. 1, p. 171—176 (1986).

F. Lukeš
Ellipsometry of Silicon with Natural Surface Film at 632.8 nm
(78.20; 78.65; S5.11)
(a) Vol. **94**, No. 1, p. 223—230 (1986).

S. A. Azimov, P. M. Karageorgy-Alkalaev, A. Yu. Leiderman, and V. S. Rubin
Computer Simulation of Current–Voltage Characteristics in Amorphous Silicon p–n Diode Structures under Impurity Photoexcitation
(73.40, S5.11)
(a) Vol. **94**, No. 1, p. 299—304 (1986).

P. Chakrabarti and B. B. Pal
Computer Simulation of Ion-Implanted DAR Impatt around 94 GHz under Steady State Condition
(73.40; S5.11; S7.11; S7.12)
(a) Vol. **94**, No. 1, p. 305—313 (1986).

F. Cembali, M. Servidori, E. Landi, and S. Solmi
Influence of Damage Depth Profile on the Characteristics of Shallow p^+/n Implanted Junctions
(73.40; 61.70; S5.11)
(a) Vol. **94**, No. 1, p. 315—319 (1986).

V. S. Vavilov, I. P. Akimchenko, and V. V. Krasnopevtsev
Ion Implantation as a Tool to Control Properties of Amorphous Hydrogenated Silicon
(61.70 and 73.60; 78.65; S5.11)
(a) Vol. **94**, No. 1, p. 343—350 (1986).

J. Hlávka
Excess Carrier Recombination Centre Distribution in Implanted and Annealed Silicon
(61.70; 72.40; S5.11)
(a) Vol. **94**, No. 1, p. 353—355 (1986).

K. Holldack, H. Kerkow, and W. Frentrup
Anomalous Diffusion of Boron in Silicon Due to Heavy Ion Bombardement
(61.80 and 66.30; S5.11)
(a) Vol. **94**, No. 1, p. 357—363 (1986).

D. S. Karpuzov
Profile Evolution in High-Dose Ion Implantation. A Computer Simulation Study
(61.70; 79.20; S5.11)
(a) Vol. **94**, No. 1, p. 365—370 (1986).

A. F. Khokhlov, A. I. Mashin, A. V. Ershov, and Yu. A. Mordvinova
The Properties of Amorphous Silicon Doped with the Isovalent Germanium Impurity
(73.60 and 78.65; 61.70; S5.11)
(a) Vol. **94**, No. 1, p. 379—384 (1986).

J. Kotliński, K. Mojejko-Kotlińska, M. Subotowicz, and I. Bryłowska
Deep Levels in Si(p) Implanted with C^+, N^+, O^+, and Ne^+ Ions Measured by DLTS Method
(61.70; S5.11)
(a) Vol. **94**, No. 1, p. 385—389 (1986).

K. Paprocki and I. Bryłowska
Recoil Implantation of Al into Disordered Silicon
(61.80; 61.70; S5.11)
(a) Vol. **94**, No. 1, p. 391—394 (1986).

P. V. Pavlov, D. I. Tetelbaum, V. D. Skupov, Yu. A. Semin, and G. V. Zorina
Abnormally Deep Structural Changes in Ion-Implanted Silicon
(61.50 and 61.70; S5.11)
(a) Vol. **94**, No. 1, p. 395—402 (1986).
Erratum
(a) Vol. **98**, No. 2, p. 649 (1986).

M. Posselt
Effect of Subcascades on Energy-Loss and Damage Profiles
(61.80; S5.11; S5.12)
(a) Vol. **94**, No. 1, p. 403—408 (1986).

G. Vízkelethy, M. Fried, G. Mezey, F. Pászti, and J. Gyulai
Simulations of Energetic Ions in Solids by Monte-Carlo Method (SEISM) and Its Comparison with Experiments
(61.70; S5.11)
(a) Vol. **94**, No. 1, p. 413—418 (1986).

P. Gaworzewski and E. Hild
Maximum Concentration of Thermal Donors Formed at 450 °C in Silicon in Dependence on the Oxygen Concentration
(61.70; S5.11)
(a) Vol. **94**, No. 1, p. K33—K38 (1986).

H. Lemke
Doping Properties of Silver in Silicon
(71.55; 61.70; S5.11)
(a) Vol. **94**, No. 1, p. K55—K59 (1986).

B. V. Klimkovich, N. A. Poklonski, and V. F. Stelmakh
The Temperature Dependence of the AC Hopping Conductivity of Lightly Compensated Semiconductors
(72.20; S5.11; S5.12)
(b) Vol. **134**, No. 2, p. 763—770 (1986).

H. WIESNER, W. CORDTS, and J. MONECKE
Self-Consistent Calculation of the Equilibrium Lattice Constant and of the Bulk Modulus of Silicon with Local and Non-Local Exchange-Correlation Functionals
(61.50 and 71.45; S5.11)
(b) Vol. **134**, No. 2, p. K93—K95 (1986).

V. P. POPOV, A. V. DVURECHENSKII, B. P. KASHNIKOV, and A. I. POPOV
Defects in Supersaturated Solid Solution of Arsenic in Silicon at Rapid Thermal Annealing
(61.70; 72.80; S5.11)
(a) Vol. **94**, No. 2, p. 569—572 (1986).

C. M. VAN VLIET and P. VASILOPOULOS
Electrical and Optical Bandgap Narrowing Due to Heavy Doping in Silicon
(71.25 and 71.55; S5.11)
(a) Vol. **94**, No. 2, p. 635—644 (1986).

D. BAITHER, R. KOEGLER, D. PANKNIN, and E. WIESER
Residual Defects in Implanted Silicon after Annealing with Incoherent Light
(61.70; 61.16; S5.11)
(a) Vol. **94**, No. 2, p. 767—772 (1986).

H. BARTSCH, G. ANDRÄ, and E. GLASER
TEM Cross-Sectional Imaging of Laser-Induced Crystallization after Ion Implantation
(68.55; 61.16; 61.80; S5.11)
(a) Vol. **94**, No. 2, p. 773—779 (1986).

E. GLASER, G. ANDRÄ, H. BARTSCH, K. DRENDA, and G. GÖTZ
Formation of Monocrystalline Layers by Explosive Crystallization of Ion-Implanted Amorphous Silicon
(68.55; 61.70; S5.11)
(a) Vol. **94**, No. 2, p. 781—786 (1986).

H. KERKOW, G. KREYSCH, G. WOLF, and K. HOLLDACK
Ion-Beam-Assisted Annealing of Amorphized Silicon
(68.55; 61.80; S5.11)
(a) Vol. **94**, No. 2, p. 793—796 (1986).

B. TILLACK, P. MÖCK, R. BANISCH, R. REINBOTH, H. H. RICHTER, M. VOELSKOW, J. MATTHÄI, and E. WIESER
Thick Monocrystalline Silicon on Oxidized Silicon Wafers Produced by a Zone Melting Process Using a Scanning Halogen Lamp
(68.55; S5.11)
(a) Vol. **94**, No. 2, p. 871—877 (1986).

G. FERENCZI, J. BODA, and T. PAVELKA
Isothermal Frequency Scan DLTS
(71.55; S5.11)
(a) Vol. **94**, No. 2, p. K119—K124 (1986).

H. Bergner, V. Brückner, F. Kerstan, L. Leine, M. Schulze, and W. Nowick
Thermally and Carrier Induced Changes in the Photoconductivity in a-Si in the Subnanosecond Time Domain
(72.40; S5.11)
(a) Vol. **94**, No. 2, p. K137—K141 (1986).

S. L. Agrawal and R. Swami
On the Inductive Behaviour of the Unijunction Transistor in Magnetic Field
(73.40; S5.11)
(a) Vol. **94**, No. 2, p. K159—K161 (1986).

S. Wilke, J. Mašek, and B. Velický
Electronic Structure of a-SiH_x with Arbitrary Hydrogen Concentration
(61.40 and 71.20; S5.11)
(b) Vol. **135**, No. 1, p. 309—319 (1986).

M. Bianconi and A. M. Mazzone
On the Concept of Cascade Volume in Spike Effects
(61.80; S5.11)
(b) Vol. **135**, No. 1, p. K5—K7 (1986).

J. Mašek and J. J. Mareš
Model of Inactive Nitrogen Incorporation in Amorphous Silicon: Nitridated Vacancy
(61.70 and 71.55; S5.11)
(b) Vol. **135**, No. 1, p. K33—K38 (1986).

V. V. Aristov, A. Yu. Nikulin, A. A. Snigirev, and P. Zaumseil
Experimental Investigation of X-Ray Bragg Diffraction on the Periodic Surface Relief of a Perfect Crystal
(61.10 and 68.20; S5.11)
(a) Vol. **95**, No. 1, p. 81—86 (1986).

A. M. Mazzone
Defect Distribution in Ion-Implanted Silicon. A Monte Carlo Simulation
(61.70; 61.80; S5.11)
(a) Vol. **95**, No. 1, p. 149—154 (1986).

I. E. Bondarenko, H. Blumtritt, J. Heydenreich, V. V. Kazmiruk, and E. B. Yakimov
Recombination Properties of Dislocation Slip Planes
(61.70; 71.55; S5.11)
(a) Vol. **95**, No. 1, p. 173—177 (1986).

J. L. Lindström, G. S. Oehrlein, and J. W. Corbett
A Study of the Annealing of the 830 cm^{-1} IR Band Observed in Electron-Irradiated Silicon
(61.70 and 78.30; S5.11)
(a) Vol. **95**, No. 1, p. 179—184 (1986).

H. Indusekhar and V. Kumar
Properties of Iron Related Quenched-In Levels in p-Silicon
(71.55; 61.70; 73.40; S5.11)
(a) Vol. **95**, No. 1, p. 269—278 (1986).

V. I. TURCHANIKOV, V. S. LYSENKO, and V. A. GUSEV
Isothermal DLTS Method Using Sampling Time Scanning
(71.55; 73.40; S5.11)
(a) Vol. **95**, No. 1, p. 283—289 (1986).

S. D. KOUIMTZI
AC Conductivity Measurements in Electron-Irradiated Al-Doped Si
(61.80 and 72.20; S5.11)
(a) Vol. **95**, No. 1, p. 317—321 (1986).

D. A. ARONOV, D. K. ISAMUKHAMEDOVA, and YU. M. YUABOV
The Influence of Inversion Layer on the Characteristics of the Contact-Adjoining Layer in Solar-Cell Silicon Schottky MTIS Structures
(73.40; S5.11)
(a) Vol. **95**, No. 1, p. 337—342 (1986).

W. R. FAHRNER, C. P. SCHNEIDER, and E. F. GOREY
Electrical Properties of 1 MeV Helium Implanted Metal–Oxide–Semiconductor Structures
(73.40; 61.70; S5.11)
(a) Vol. **95**, No. 1, p. 343—351 (1986).

K. SCHMALZ, V. V. BOLOTOV, H. RICHTER, and K. TITTELBACH-HELMRICH
DLTS Investigation on the Interaction of Radiation Defects with Thermal Defects in Czochralski p-Si
(61.70 and 61.80; 71.55; S5.11)
(a) Vol. **95**, No. 1, p. K15—K20 (1986).

JUH TZENG LUE and SEN-YEN SHAW
Laminant Distribution of Hydrogenated Bonds in Amorphous Silicon Deposited by RF Sputtering
(72.40; 79.60; S5.11)
(a) Vol. **95**, No. 1, p. K75—K80 (1986).

L. TULACH, H. FRANK, B. PÍNA, and M. JANŮ
Gold as Residual Impurity in Silicon Devices
(71.55; S5.11)
(a) Vol. **95**, No. 1, p. K87—K89 (1986).

D. A. ARONOV, B. N. ZAVERYUKHIN, R. A. MUMINOV, V. V. RUBINOV, and A. SH. SHAMAGDIEV
The Use of the Strong Exclusing Effect in Ionizing Radiation Detectors Based on "Pure" Semiconductors with Non-Injecting Contacts
(73.40; S5.11; S5.12)
(a) Vol. **95**, No. 1, p. K95—K100 (1986).

K. HÜBNER, E.-D. KLINKENBERG, and A. STERN
Theoretical Study of the Gradual Chemical Transition at the Si–SiO_2 Interface. I. Structural Models for the Si(111)–SiO_2 and Si(100)–SiO_2 Interregions
(68.48; S5.11; S10.1)
(b) Vol. **135**, No. 2, p. 475—485 (1986).

V. Rybka, V. Odzhaev, J. Červená, V. Hnatowicz, J. Kvítek, and H. Jelínková
Laser Annealing of Bismuth-Implanted (111) Silicon
(61.80; S5.11)
(a) Vol. **95**, No. 2, p. 511—515 (1986).

L. I. Fedina and A. L. Aseev
Study of Interaction of Point Defects with Dislocations in Silicon by Means of Irradiation in an Electron Microscope
(61.70; 61.16; 61.80; S5.11)
(a) Vol. **95**, No. 2, p. 517—529 (1986).

B. G. Svensson and J. L. Lindström
Annealing Studies of the 862 cm^{-1} Infrared Band in Silicon
(61.70; 78.30; S5.11)
(a) Vol. **95**, No. 2, p. 537—542 (1986).

K. T. Gabrielyan, J. Kuběna, and V. Holý
A Study of the Nature of Dislocation Loops in Growth Striations in Silicon by X-Ray Double Crystal Topography
(61.10 and 61.70; S5.11)
(a) Vol. **95**, No. 2, p. 579—588 (1986).

S. Guimarães, E. Landi, and S. Solmi
Enhanced Diffusion Phenomena during Rapid Thermal Annealing of Preamorphized Boron-Implanted Silicon
(61.70 and 66.30; S5.11)
(a) Vol. **95**, No. 2, p. 589—598 (1986).

A. V. Dvurechenskii, V. A. Dravin, I. A. Ryazantsev, A. Kh. Antonenko, and I. G. Landochkin
Transport Phenomena in Amorphous Silicon Doped by Ion Implantation of 3d Metals
(71.55; 72.80; S5.11)
(a) Vol. **95**, No. 2, p. 635—640 (1986).

H. Lemke
Defect Interactions in Cu-Doped Silicon Crystals
(61.70 and 71.55; S5.11)
(a) Vol. **95**, No. 2, p. 665—677 (1986).

O. Engström, H. Pettersson, and B. Sernelius
Photoelectric Yield Spectra of Metal–Semiconductor Structures
(79.60; 73.40; S1; S1.3; S1.61; S5.11)
(a) Vol. **95**, No. 2, p. 691—701 (1986).

Dao Khac An and F. Pavlyák
Use of SIMS for Determining the Segregation Coefficient of Boron at the Si–SiO_2 Interface in Steam Oxidation Following Ion Implantation
(64.75; 79.20; S5.11; S10.1)
(a) Vol. **95**, No. 2, p. K117—K121 (1986).

H. Lemke
On the Time Dependence of the Barrier Capacity during the Filling Process of Deep Impurity
(71.55; S5.11)
(a) Vol. **95**, No. 2, p. K153—K158 (1986).

A. Balasiński, A. Jakubowski, T. Brożek, and K. Iniewski
On γ-Radiation and Electron Beam Dose and Gate Oxide Thickness Dependence of Radiation Defects in MOS Capacitors
(61.80 and 73.40; S5.11)
(a) Vol. **95**, No. 2, p. K203—K206 (1986).

P. Bury, J. Ďurček, and K. Sakalauskas
On the Piezoactivity of Si MOS Structures
(73.40 and 77.60; S5.11)
(a) Vol. **95**, No. 2, p. K207—K209 (1986).

M. Kohyama, R. Yamamoto, and M. Doyama
Energies and Structures of (111) Coincidence Twist Boundaries in 3C-SiC, Diamond, and Silicon
(61.70; S5; S5.11; S6)
(b) Vol. **136**, No. 1, p. 31—36 (1986).

B. K. Ghosh and Bal K. Agrawal
Fluorinated Hydrogenated Amorphous Silicon Alloys. Vibrational Excitations of Chains
(63.50; S5.11)
(b) Vol. **136**, No. 1, p. 51—55 (1986).

T. Soma, T. Matsuoka, and H.-Matsuo Kagaya
Formation and Migration Energies of an Interstitial Atom in Si and Ge
(61.70; S5.11; S5.12)
(b) Vol. **136**, No. 1, p. 85—90 (1986).

K. Hübner, A. Stern, and E.-D. Klinkenberg
Theoretical Study of the Gradual Chemical Transition at the Si–SiO_2 Interface. II. Electronic Density of States Calculations
(71.20 and 73.40; 68.48; S5.11; S10.1)
(b) Vol. **136**, No. 1, p. 211—223 (1986).

A. A. Frantsuzov, S. A. Okhonin, and A. G. Pogosov
The Influence of Quantization on the Space-Charge Layer Capacitance of Si in Strong Accumulation
(73.40; 77.50; S5.11)
(b) Vol. **136**, No. 1, p. 241—249 (1986).

P. V. Smith, J. E. Szymanski, and J. A. D. Matthew
A Reformulation of the LUC-CNDO Approach to the Properties of Solids. Application to Diamond and Silicon
(71.10; 63.20; S5; S5.11)
(b) Vol. **136**, No. 1, p. 261—271 (1986).

A. A. KARPUSHIN, A. N. SOROKIN, I. YU. SEMENOVA, and M. TOMÁŠEK
Method of Generalized Bethe Lattices in Chemisorption Theory
(71.10 and 73.20; 68.45; S5.11)
(b) Vol. **136**, No. 1, p. 331—339 (1986).

J. T. LENKKERI
Nonlinear Effects in the Piezoresistivity of p-Type Silicon
(72.20; S5.11)
(b) Vol. **136**, No. 1, p. 373—385 (1986).

T. DRÜSEDAU and R. BINDEMANN
The Meyer-Neldel Rule and the Fundamental Pre-Exponential Factor in the Conductivity of a-Si:H
(72.80; S5.11)
(b) Vol. **136**, No. 1, p. K61—K64 (1986).

V. P. PARKHUTIK, V. A. LABUNOV, and G. G. CHIGIR
Kinetics and Mechanism of Transient Thermal Oxidation of Silicon
(68.55; 73.20; 73.40; 79.60; S5.11; S10.1)
(a) Vol. **96**, No. 1, p. 11—18 (1986).

D. R. MCKENZIE, P. S. TURNER, and D. J. H. COCKAYNE
Electron Imaging and Energy Loss Studies of the Crystallization of Hydrogenated Amorphous Silicon
(68.55; 61.16; 61.80; S5.11)
(a) Vol. **96**, No. 1, p. 67—73 (1986).

A. ARMIGLIATO, G. QUEIROLO, and G. VALDRÈ
Nitrogen Profiles of Ion-Implanted TiN Films by Electron Energy Loss Spectroscopy
(68.55; 61.80; S1.62; S5.11)
(a) Vol. **96**, No. 1, p. 89—99 (1986).
Erratum
(a) Vol. **98**, No. 2, p. 651 (1986).

M. B. DUTT, R. KUMAR, R. NATH, M. N. SEN, and K. K. LAROIA
Depth Profiling of Ion Implanted Silicon by Electrical Methods
(61.70; 66.30; 73.40; S5.11)
(a) Vol. **96**, No. 1, p. 121—127 (1986).

V. V. BOLOTOV, A. V. KARPOV, V. A. STUCHINSKII, and K. SCHMALZ
Accumulation of Radiation Defects in Oxygen-Rich n-Type Silicon Heat-Treated at Temperatures from 600 to 1000 °C
(61.80; 71.55; S5.11)
(a) Vol. **96**, No. 1, p. 129—134 (1986).

A. V. DVURECHENSKII, V. V. SUPRUNCHIK, and L. S. SMIRNOV
Divacancy-Type Defect with Low Symmetry in Silicon
(61.70; 71.55; 76.30; S5.11)
(a) Vol. **96**, No. 1, p. 155—159 (1986).

L. A. Kazakevich, V. I. Kuznetsov, and P. F. Lugakov
The Effect of Dislocations on the Activation Energy of Radiation Defect Thermal Ionization
(61.70 and 61.80; 71.55; 72.80; S5.11)
(a) Vol. **96**, No. 1, p. 243—247 (1986).

S. Alexandrova and A. Szekeres
Low-Temperature RF Plasma Treatment of Si–SiO_2 Structures as a Substitution for High-Temperature Anneals
(73.40; S5.11; S10.1)
(a) Vol. **96**, No. 1, p. 363—366 (1986).

F. Cembali, E. Gabilli, R. Lotti, M. Servidori, and A. Zani
Ambiguity in the Strain Profile Determination by Simulation of Double Crvstal X-Ray Rocking Curves
(61.10 and 61.80; S5.11)
(a) Vol. **96**, No. 1, p. K31—K34 (1986).
Erratum
(a) Vol. **97**, No. 2, p. K215 (1986).

D. V. Kazakovtsev and Y. B. Levinson
The Effect of Phonon–Phonon Scattering in the Substrate on Temperature Dynamics of a Phonon Film Injector
(63.20; 66.70; S5.11; S5.12; S7.12; S9.12)
(b) Vol. **136**, No. 2, p. 425—434 (1986).

K. Winer and F. Wooten
Vibrational Properties of Metastable Phases of Si and Ge
(63.50; 61.40; S5.11; S5.12)
(b) Vol. **136**, No. 2, p. 519—527 (1986).

A. M. Mazzone
Channeling of P^+ Ions at 1 MeV in Silicon. A Monte Carlo Simulation
(61.80; S5.11)
(b) Vol. **136**, No. 2, p. K81—K83 (1986).

G. D. Maksimović, D. R. Mašović, and Z. S. Popović
Application of the Modified Veljković-Slavić Pseudopotential for Calculation of the Electronic Charge Density of Si
(71.20; S5.11)
(b) Vol. **136**, No. 2, p. K101—K104 (1986).

D. E. Popov, S. E. Savushkin, and S. A. Vorobiev
Orientation and Energety Dependence of Electron–Atom Close Collisions at Molecular Channeling
(61.80; S5.11)
(b) Vol. **136**, No. 2, p. K109—K112 (1986).

S. Wilke, J. Mašek, and B. Velický
The Electronic Structure of Polysilane Alloys
(71.20; S5.11)
(b) Vol. **136**, No. 2, p. K113—K117 (1986).

I. M. Filippov, L. A. Kazakevich, P. F. Lugakov, and V. V. Shusha
The Charge Carrier Recombination Processes in Dislocated Silicon
(61.70 and 71.55; 72.80; S5.11)
(a) Vol. **96**, No. 2, p. 527—531 (1986).

R. Angelucci, S. Solmi, and Q. Zini
Influence of Precipitation on Phosphorus Diffusivity in Silicon
(66.30; 61.70; S5.11)
(a) Vol. **96**, No. 2, p. 541—546 (1986).

R. Herzog, G. Schramm, and T. Flade
Impurity Analysis of Silicon Epitaxial Layers by the 4.2 K Photoluminescence Spectroscopy
(73.60 and 78.55; 78.65; S5.11)
(a) Vol. **96**, No. 2, p. 629—636 (1986).

C. A. Londos
Charge-Dependent Defect Traces in the DLTS and MCTS Spectra of Silicon
(71.55; 61.80; S5.11)
(a) Vol. **96**, No. 2, p. 637—642 (1986).

S. Sinha and A. P. B. Sinha
Effect of Substrate Temperature on Bond Angle Distortion in a-Si:H Films. A Raman Scattering Study
(68.55; 78.30; S5.11)
(a) Vol. **96**, No. 2, p. K117—K120 (1986).

M. Kittler, W. Seifert, K. Schmalz, and K. Tittelbach-Helmrich
Comparison of EBIC and DLTS Measurements on Boron-Doped CZ Silicon Contaminated with Iron
(61.70 and 71.55; S5.11)
(a) Vol. **96**, No. 2, p. K133—K137 (1986).

V. P. Markevich and L. I. Murin
The Nature of Intrinsic Defects in n-Type Silicon γ-Irradiated at 77 K
(61.70 and 71.55; 61.80; S5.11)
(a) Vol. **96**, No. 2, p. K151—K155 (1986).

V. Belenkii, V. G. Veselago, M. S. Gilenko, A. Yu. Leiderman, L. P. Maksimov, V. D. Ovsyannikov, and P. M. Karageorgy-Alkalaev
On the Enhanced Magnetosensitivity of Semiconductor Structures with Isotype Junction
(73.40; S5.11)
(a) Vol. **96**, No. 2, p. K223—K228 (1986).

M. Kohyama, R. Yamamoto, and M. Doyama
Structures and Energies of Symmetrical ⟨011⟩ Tilt Grain Boundaries in Silicon
(61.70; S5.11)
(b) Vol. **137**, No. 1, p. 11—20 (1986).

K. THONKE, U. SCHALL, and R. SAUER
Photoluminescence Study of the Gallium Defect Spectrum at ≈0.928 eV (Ga3) in Irradiated Silicon
(71.55 and 78.55; 71.35; S5.11)
(b) Vol. **137**, No. 1, p. 291—303 (1986).

U. SCHALL, K. THONKE, and R. SAUER
Photoluminescence Study of the Gallium Defect Spectrum at ≈1.049 eV (Ga2) in Irradiated Silicon
(71.55 and 78.55; 71.35; S5.11)
(b) Vol. **137**, No. 1, p. 305—317 (1986).

H. HAVLOVÁ and L. SMRČKA
Quantum Oscillations of Thermogalvanomagnetic Coefficients in Two-Dimensional Electron Systems
(72.20 and 73.40; S5.11; S7.15)
(b) Vol. **137**, No. 1, p. 331—343 (1986).

A. M. TARATIN, YU. M. FILIMONOV, and S. A. VOROBIEV
Computer Simulation of the Volume Capture Effect for Relativistic Electrons in Crystals
(61.80 and 71.90; S5.11; S5.12)
(b) Vol. **137**, No. 1, p. K63—K67 (1986).

A. SAKAI, H. SAKA, and T. IMURA
Reduction of Dislocation Density in Si by Thermal Cyclic Annealing
(61.70; 61.10; S5.11)
(a) Vol. **97**, No. 1, p. 57—63 (1986).

M. BERTI, A. V. DRIGO, G. LULLI, P. G. MERLI, and M. VITTORI ANTISARI
Dose Rate Effects on the Dynamic Annealing Mechanism in P^+-Implanted Silicon
(61.55 and 61.80; S5.11)
(a) Vol. **97**, No. 1, p. 77—85 (1986).

F. H. M. SPIT and H. BAKKER
Diffusion of Donor Elements (^{125}Sb, ^{32}P, $^{74(73)}$As) in Polycrystalline Silicon
(66.30; S5.11)
(a) Vol. **97**, No. 1, p. 135—142 (1986).

H. SIETHOFF, H. G. BRION, K. AHLBORN, and W. SCHRÖTER
Dynamical Recovery of ⟨111⟩ Germanium
(62.20; S5.11; S5.12)
(a) Vol. **97**, No. 1, p. 153—162 (1986).

T. WILSON and P. D. PESTER
Optical Beam Induced Minority Carrier Distribution in Semiconductors
(72.40 and 73.25; S5.11)
(a) Vol. **97**, No. 1, p. 323—329 (1986).

F. SCHMIDL, M. SIEGEL, P. SEIDEL, and H.-J. KÖHLER
Some Aspects of the AC Josephson Effect at Nb/Amorphous Si/Nb Structures
(74.50; S1; S5.11)
(a) Vol. **97**, No 1, p. K87—K89 (1986).

J. Bollmann, H. Klose, and A. Mertens
Formation of Schottky Junctions in Silicon by Ion Implantation
(73.30; 61.70; S5.11)
(a) Vol. **97**, No. 1, p. K95—K99 (1986).

G. Paasch, K. Tarnay, J.-P. Zöllner, and H. Übensee
Electric Field versus Band Bending in MIS Structures: Quantum Interpolation Formula for Device Modeling
(73.40; S5.11; S7.11; S7.12)
(a) Vol. **97**, No. 1, p. K107—K111 (1986).

U. Pietsch, V. G. Tsirelson, and R. P. Ozerov
X-Ray Electron Charge Density Distribution in Silicon
(61.10; S5.11)
(b) Vol. **137**, No 2, p. 441—447 (1986).

J. Donecker and J. Kluge
Excitonic Absorption in Silicon in the Presence of Low Magnetic Fields
(78.20; 71.35; S5.11)
(b) Vol. **137**, No. 2, p. 591—595 (1986).

G. V. Gadiyak, A. A. Karpushin, I. V. Korolenko, Yu. N. Morokov, I. Yu. Semenova, A. N. Sorokin, and M. Tomášek
Calculation of the Electronic Structure of Disordered Hydrogen Adsorption on the Si(111) Surface
(73.20; 73.40; S5.11)
(b) Vol. **137**, No. 2, p. 633—639 (1986).

H. Wiesner, W. Cordts, and J. Monecke
On the Shaw Potential and Its Application to Self-Consistent Band Structure and Total Energy Calculations for Silicon
(71.25; S5.11)
(b) Vol. **137**, No. 2, p. K161—K166 (1986).

B. Ya. Farber, Yu. L. Iunin, and V. I. Nikitenko
Experimental Study of the Double Kink Formation Kinetics and Kink Mobility on the Dislocation Line in Si Single Crystals
(61.70; S5.11)
(a) Vol. **97**, No. 2, p. 469—478 (1986).

A. A. Bugai, V. A. Girii, V. M. Maksimenko, Yu. V. Pomozov, and V. I. Shakhovtsov
Investigation of the Production of Quenching Thermodefects in Silicon
(71.55 and 72.80; 76.30; S5.11)
(a) Vol. **97**, No. 2, p. 549—555 (1986).

E. N. Vidalko, G. P. Gaidar, V. A. Girii, V. V. Savyak, and V. I. Shakhovtsov
Mobility of Majority Carriers in γ-Irradiated Oxygen-Containing Crystals of n-Si and n-Ge
(72.80; 61.80; S5.11; S5.12)
(a) Vol. **97**, No. 2, p. 565—570 (1986).

D. M. Vanderwalker
Hydrides on Dislocations in Silicon
(61.70; S5.11)
(a) Vol. **97**, No. 2, p. K127—K130 (1986).

J. Bollmann, H. Klose, and A. Mertens
Big Channeling Effects at Low-Energy Ion Implantation in Silicon
(61.70; 61.80; S5.11)
(a) Vol. **97**, No. 2, p. K135—K139 (1986).

A. Fischer, D. Knoll, T. Grabolla, and F. Fischer
Redistribution of Implanted and Thermally Predeposited Dopants during High Pressure Oxidation
(66.30; S5.11)
(a) Vol. **97**, No. 2, p. K141—K144 (1986).

V. P. Markevich, L. F. Makarenko, and L. I. Murin
Some New Features of Thermal Donor Formation in Silicon at $T < 800$ K
(61.70 and 71.55; S5.11)
(a) Vol. **97**, No. 2, p. K173—K176 (1986).

M. Delannoy-Coutris and G. Perrin
Fourth-Order Thermal Expansion for Cubic Solids. Application to Aluminum, Copper, Germanium, and Silicon
(65.70; 64.10; S1.2; S4; S5.11; S5.12)
(b) Vol. **138**, No. 1, p. 93—99 (1986).

M. Höhne
A Silicon–Gold Pair in Silicon Investigated by EPR
(71.55; 76.30; S5.11)
(b) Vol. **138**, No. 1, p. 337—346 (1986).

H. Yamamoto
Resonant Tunneling through Amorphous Double-Barrier Structures
(73.40; S5.11; S6)
(b) Vol. **138**, No. 1, p. K71—K73 (1986).

P. Danesh, I. Savatinova, E. Anachkova, and St. Georgiev
Raman Scattering in Glow Discharge Si:H:Cl Films
(78.30; 61.40; 68.55; S5.11)
(a) Vol. **98**, No. 1, p. 127—132 (1986).

D. A. Aronov and B. R. Mamatkulov
The Field and Spectrum Dependences of Impurity Photoconductivity with Exclusion in Compensated Semiconductors
(72.40; 73.40; S5.11; S5.12)
(a) Vol. **98**, No. 1, p. 263—277 (1986).

J. Shirafuji, S. Nagata, and K. Shirakawa
γ-Irradiation Effect on Electronic Properties in Hydrogenated Amorphous Silicon
(61.40 and 73.60; 61.80; S5.11)
(a) Vol. **98**, No. 1, p. 311—318 (1986).

V. I. Nikolaichik and I. I. Khodos
Determination of the Sign of a Dissociated Dislocation in Weak Beam
(61.70; S5.11)
(a) Vol. **98**, No. 1, p. K19—K21 (1986).

F. P. Korshunov, I. G. Marchenko, and V. T. Troshchinskii
Pecularities of Minority Carrier Lifetime Variations in p–n–p–n Structures Fabricated upon Neutron-Transmutation Doped Silicon and Irradiated by Fast Electrons
(61.80 and 73.40; S5.11)
(a) Vol. **98**, No. 1, p. K39—K42 (1986).

K.-D. Ufert, W. Ecke, M. Manzel, W. Michalke, and J. Salm
Investigations of Interface States in Si/Si-Oxynitride/Al Structures Using DLTS
(73.40; S5.11; S7)
(a) Vol. **98**, No. 1, p. K69—K72 (1986).

S. Ashok, K. Giewont, and H. P. Vyas
On the Ubiquity of Ion Bombardment Modification of Silicon Schottky Barriers
(61.80 and 73.40; S5.11)
(a) Vol. **98**, No. 1, p. K99—K104 (1986).

M. Kohyama, R. Yamamoto, and M. Doyama
Reconstructed Structures of Symmetrical ⟨011⟩ Tilt Grain Boundaries in Silicon
(61.70; S5.11)
(b) Vol. **138**, No. 2, p. 387—397 (1986).

V. A. Terekhov, S. N. Trostyanskii, E. P. Domashevskaya, O. A. Golikova, M. M. Mezdrogina, K. L. Sorokina, and M. M. Kazanin
Density of States and Photoconductivity of Hydrogenated Amorphous Silicon
(71.20; 72.40; 73.60; 78.70; S5.11)
(b) Vol. **138**, No. 2, p. 647—653 (1986).

L. Reggiani, R. Brunetti, and E. Normantas
Comparison between Monte-Carlo and Davydov Approaches to Hot-Carrier Transport in Semiconductors
(72.10; S5.11; S5.12)
(b) Vol. **138**, No. 2, p. 717—725 (1986).

F. L. Edelman, J. Heydenreich, D. Hoehl, J. Matthäi, I. Melnik, A. Rzhanov, M. Voelskov, and P. Werner
Structure of Annealed Polycrystalline Silicon Films. I. Recrystallization
(68.55; 64.70; S5.11)
(a) Vol. **98**, No. 2, p. 383—390 (1986).

K. Watanabe, Y. Kikuchi, and H. Yamaguchi
Screening Effect on Lattice Images of Silicon
(61.14; S5.11)
(a) Vol. **98**, No. 2, p. 409—416 (1986).

Lu Zhiheng
Similarity between Spinodal Decomposition and Clustering in Si Highly Doped by As
(64.75; S5.11)
(a) Vol. **98**, No. 2, p. 471—475 (1986).

F. Cembali, M. Servidori, S. Solmi, Z. Šourek, U. Winter, and P. Zaumseil
Structural and Electrical Characterization of Boron Implanted in Preamorphized Silicon Layers
(61.70; 61.10; 72.80; S5.11)
(a) Vol. **98**, No. 2, p. 511—516 (1986).

K. Holldack and H. Kerkow
Flux Dependence of Damage Accumulation in Silicon during Ion Bombardment
(61.80; S5.11)
(a) Vol. **98**, No. 2, p. 527—534 (1986).

S. Alexandrova, A. Szekeres, W. Füssel, and H. Flietner
RF Plasma Annealing Effects at the Wet Oxidized Si/SiO_2 Interface
(68.48 and 73.40; S5.11)
(a) Vol. **98**, No. 2, p. 645—648 (1986).

R. Tognato
On the Formation of Swirl Defects in Silicon and Germanium
(61.70; S5.11; S5.12)
(a) Vol. **98**, No. 2, p. K133—K136 (1986).

E. Caselli, R. Cabanillas, R. Wainschenker, and M. Cutella
High Temperature Drift Mobilities in High Resistivity Silicon
(72.20; S5.11)
(a) Vol. **98**, No. 2, p. K179—K183 (1986).

O. K. Malinovskaya, A. F. Plotnikov, and V. A. Tolokonnikov
Determination of Energy Barriers for the Photocharging of Trapping Sites Contained in ZnSe Layer of Ni–ZnSe–SiO_2–Si Structure by Electrons and Holes
(73.40; S5.11; S8.12; S10.1)
(a) Vol. **98**, No. 2, p. K191—K195 (1986).

S5.12. Germanium

Original Papers and Short Notes

M. V. Kruglov, I. K. Solomin, and A. V. Lunev
Determination of the Photoemission Generation Depth with Use of Experiments on the Dynamic Scattering of X-Rays
(79.60; 61.10; S5.11; S5.12)
(b) Vol. **133**, No. 1, p. 47—55 (1986).

P. K. Dixit, B. A. Vaid, and K. C. Sharma
The Structural Disorder Model: Pressure Induced Transitions in the Semiconductors Si and Ge
(64.70; 65.50; S5.11; S5.12)
(b) Vol. **133**, No. 1, p. 71—74 (1986).

V. K. Bashenov, V. E. Gorbachev, and D. I. Marvakov
Bond Charge Approximation for Valence Electron Density in Elemental Semiconductors
(71.25; S5; S5.11; S5.12)
(b) Vol. **133**, No. 1, p. 285—296 (1986).

V. I. Gatalskaya
Cyclotron Resonance Anisotropy of Charge Carriers in Irradiated Germanium
(76.40; 71.25; S5.12)
(b) Vol. **133**, No. 1, p. K31—K33 (1986).

C. Trallero Giner and M. Antón
Quantum Theory of Free-Carrier Absorption in Quasi-Two-Dimensional Semiconducting Structures
(78.20; 71.38; 78.65; S5.12; S7.12)
(b) Vol. **133**, No. 2, p. 563—572 (1986).

V. N. Baier, V. M. Katkov, and V. M. Strakhovenko
Radiation Yield of High-Energy Electrons in Thick Crystals
(61.80 and 78.70; S1.1; S1.2; S5; S5.11; S5.12)
(b) Vol. **133**, No. 2, p. 583—592 (1986).

V. I. Gatalskaya
On the Mechanism of Charge Carrier Scattering in Germanium with Point Defects under Cyclotron Resonance
(72.20; 76.40; S5.12)
(b) Vol. **133**, No. 2, p. K137—K141 (1986).

S. Bumelienė, J. Požela, and A. Tamaševičius
Period Multiplying and Chaoting Response in Driven n-Ge with Repulsive Defect Centres
(72.80; S5.12)
(b) Vol. **134**, No. 1, p. K71—K74 (1986).

I. B. Khaibullin, G. G. Zakirov, M. M. Zaripov, T. Lohner, L. Pogány, G. Mezey, M. Fried, E. Kótai, F. Pászti, A. Manuaba, and J. Gyulai
Effect of Heavy Ion Implantation and Laser Annealing on the Structural Properties of Germanium
(61.70; 61.80; S5.12)
(a) Vol. **94**, No. 1, p. 371—377 (1986).

M. Posselt
Effect of Subcascades on Energy-Loss and Damage Profiles
(61.80; S5.11; S5.12)
(a) Vol. **94**, No. 1, p. 403—408 (1986).

A. A. Manenkov, G. N. Mikhailova, S. Yu. Sokolov, and S. G. Tikhodeev
Phonon Wind Generation in Ge by CO_2 Laser Induced Heating of Electron–Hole Drops
(71.35; 71.38; 72.30; 78.70; S5.12)
(b) Vol. **134**, No. 2, p. 631—639 (1986).

B. V. Klimkovich, N. A. Poklonski, and V. F. Stelmakh
The Temperature Dependence of the AC Hopping Conductivity of Lightly Compensated Semiconductors
(72.20; S5.11; S5.12)
(b) Vol. **134**, No. 2, p. 763—770 (1986).

D. A. Aronov, B. N. Zaveryukhin, R. A. Muminov, V. V. Rubinov, and A. Sh. Shamagdiev
The Use of the Strong Exclusing Effect in Ionizing Radiation Detectors Based on "Pure" Semiconductors with Non-Injecting Contacts
(73.40; S5.11; S5.12)
(a) Vol. **95**, No. 1, p. K95—K100 (1986).

H. J. Juretschke
Anomalous Asymmetries in X-Ray π-Modes near n-Beam Interactions
(61.10; S5.12)
(b) Vol. **135**, No. 2, p. 455—466 (1986).

T. Soma, T. Matsuoka, and H.-Matsuo Kagaya
Formation and Migration Energies of an Interstitial Atom in Si and Ge
(61.70; S5.11; S5.12)
(b) Vol. **136**, No. 1, p. 85—90 (1986).

D. V. Kazakovtsev and Y. B. Levinson
The Effect of Phonon–Phonon Scattering in the Substrate on Temperature Dynamics of a Phonon Film Injector
(63.20; 66.70; S5.11; S5.12; S7.12; S9.12)
(b) Vol. **136**, No. 2, p. 425—434 (1986).

K. Winer and F. Wooten
Vibrational Properties of Metastable Phases of Si and Ge
(63.50; 61.40; S5.11; S5.12)
(b) Vol. **136**, No. 2, p. 519—527 (1986).

I. G. Kirnas, P. G. Litovchenko, E. E. Petrosyan, and N. N. Pashchuk
Peculiar Features of the Electric Field Distribution in One-Open-Face Coaxial Germanium Detectors of Nuclear Radiation
(72.20; 73.40; S5.12)
(a) Vol. **96**, No. 2, p. 657—662 (1986).

M. Dressel and H. W. Helberg
AC Conductivity of Deformed Germanium Single Crystals at $T = 4.2$ K
(72.20; 61.70; S5.12)
(a) Vol. **96**, No. 2, p. K199—K202 (1986).

T. Rother and W. D. Kraeft
On the Thermodynamics of Strongly Degenerate Two-Component Plasmas
(65.50 and 71.45; S5.12; S7.12)
(b) Vol. **137**, No. 1, p. 103—108 (1986).

A. M. Taratin, Yu. M. Filimonov, and S. A. Vorobiev
Computer Simulation of the Volume Capture Effect for Relativistic Electrons in Crystals
(61.80 and 71.90; S5.11; S5.12)
(b) Vol. **137**, No. 1, p. K63—K67 (1986).

H. Siethoff, H. G. Brion, K. Ahlborn, and W. Schröter
Dynamical Recovery of ⟨111⟩ Germanium
(62.20; S5.11; S5.12)
(a) Vol. **97**, No. 1, p. 153—162 (1986).

A. P. Medvid, J. B. Litaunieks, A. P. Krivich, and V. E. Prudents
Magnetoconcentration Effect in Inhomogeneous Electric Field
(72.20; S5.12)
(a) Vol. **97**, No. 1, p. K91—K94 (1986).

E. N. Vidalko, G. P. Gaidar, V. A. Girii, V. V. Savyak, and V. I. Shakhovtsov
Mobility of Majority Carriers in γ-Irradiated Oxygen-Containing Crystals of n-Si and n-Ge
(72.80; 61.80; S5.11; S5.12)
(a) Vol. **97**, No. 2, p. 565—570 (1986).

M. Delannoy-Coutris and G. Perrin
Fourth-Order Thermal Expansion for Cubic Solids. Application to Aluminum, Copper, Germanium, and Silicon
(65.70; 64.10; S1.2; S4; S5.11; S5.12)
(b) Vol. **138**, No. 1, p. 93—99 (1986).

A. I. Chumakov, G. V. Smirnov, M. V. Kruglov, and I. K. Solomin
Experimental Determination of Electron Escape Weight Functions by Using the Methods of "Calibrated Amorphous Layers" and "Total External Reflection"
(79.60; S5.12)
(a) Vol. **98**, No. 1, p. 11—24 (1986).

D. A. Aronov and B. R. Mamatkulov
The Field and Spectrum Dependences of Impurity Photoconductivity with Exclusion in Compensated Semiconductors
(72.40; 73.40; S5.11; S5.12)
(a) Vol. **98**, No. 1, p. 263—277 (1986).

L. Reggiani, R. Brunetti, and E. Normantas
Comparison between Monte-Carlo and Davydov Approaches to Hot-Carrier Transport in Semiconductors
(72.10; S5.11; S5.12)
(b) Vol. **138**, No. 2, p. 717—725 (1986).

R. Tognato
On the Formation of Swirl Defects in Silicon and Germanium
(61.70; S5.11; S5.12)
(a) Vol. **98**, No. 2, p. K133—K136 (1986).

S5.2. Solidified Gases

Original Papers and Short Notes

D. I. PUSHKAROV
On the Defecton Statistics in Quantum Crystals
(67.80; S5.2)
(b) Vol. **133**, No. 2, p. 525—531 (1986).

R. BAHADUR, N. BHARDWAJ, and C. M. KACHHAVA
Excitation Spectrum and Virial Coefficients of Superfluid He^4
(67.40; S5.2)
(b) Vol. **134**, No. 2, p. 671—674 (1986).

D. K. NEB, R. K. SINGH, and S. P. SANYAL
On Stability of Crystal Structures of Rare Gas Solids
(61.50; S5.2)
(b) Vol. **134**, No. 2, p. K97—K100 (1986).

Y. SHIBATA, K. ISHI, and S.-I. TAKAHASHI
Lattice Dynamics of Solid N_2O
(63.20; 65.40; S5.2)
(b) Vol. **135**, No. 1, p. 85—91 (1986).

S6. Group IV compounds
(for transition metal compounds, see S1.61)

Original Papers and Short Notes

N. Saito and N. Tanaka
Preparation and Properties of Magnetron Sputtered Fluorinated and Hydrogenated Amorphous Si–C Films
(73.60 and 78.65; 68.55; S6)
(a) Vol. **93**, No. 1, p. 341—345 (1986).

H. C. Gupta, G. S. N. Reddy, and B. B. Tripathi
Phonon Density of States in Amorphous $Si_{0.5}Ge_{0.5}$
(63.50; S6)
(b) Vol. **134**, No. 1, p. 17—20 (1986).

Yu. M. Azhnyuk, V. V. Artamonov, M. Ya. Valakh, and A. P. Litvinchuk
Raman Scattering from Polaritons and Plasmaritons in 6H-SiC
(78.30; 71.36; 71.45; S6)
(b) Vol. **135**, No. 1, p. 75—84 (1986).

M. Kohyama, R. Yamamoto, and M. Doyama
Energies and Structures of (111) Coincidence Twist Boundaries in 3C-SiC, Diamond, and Silicon
(61.70; S5; S5.11; S6)
(b) Vol. **136**, No. 1, p. 31—36 (1986).

Chen Guanghua, Xu Xixiang, Zhang Fangquing, and Jiang Zhichen
Valence-Electron Properties of Glow-Discharge a-C:H Films Studied by Electron Energy Loss Spectroscopy
(71.45; 61.80; S5; S6)
(b) Vol. **137**, No. 1, p. K55—K57 (1986).

I. M. Chapnik
On the Relation between the Energy Gap Data of Semiconducting Compounds and the Position of the Components of the Compounds in the Periodic Table
(72.80; S5; S6; S7; S8; S10.1)
(b) Vol. **137**, No. 1, p. K95—K100 (1986).

M. S. Saidov, Kh. A. Shamuratov, M. A. Kadyrov, and S. I. Vlaskina
Investigation of Crystal Structure of Cubic Silicon Carbide Layers
(68.55; S6)
(a) Vol. **97**, No. 2, p. 347—351 (1986).

H. Yamamoto
Resonant Tunneling through Amorphous Double-Barrier Structures
(73.40; S5.11; S6)
(b) Vol. **138**, No. 1, p. K71—K73 (1986).

S7. Group V compounds (for transition metal compounds, see S1.62)

Original Papers and Short Notes

J. Weszka, M. Renucci, and A. Zwick
Some Aspects of Raman Scattering in Cd_3As_2 Single Crystals
(78.30; S7)
(b) Vol. **133**, No. 1, p. 57—64 (1986).

M. Goiran, A. Zwick, R. Carles, J. Leotin, E. Arushanov, and M. von Ortenberg
Cadmium Antimonide Lattice Excitations
(78.30; S7)
(b) Vol. **133**, No. 1, p. K1—K4 (1986).

P. Roura, A. Lousa, A. Cornet, J. L. Morenza, and J. R. Morante
Photoluminescence Measurements in Small Grain Size Polycrystalline Zn_3P_2
(78.55; S7)
(a) Vol. **94**, No. 1, p. K47—K50 (1986).

V. K. Gueorguiev, L. I. Popova, I. N. Petrov, and I. G. Stoev
Anisotropic Etchants Inhibiting Properties of Silicon-Nitride Compound Layers Produced by Ion Implantation
(68.55; S7)
(a) Vol. **94**, No. 2, p. 865—869 (1986).

P. A. Pundur, J. G. Shavalgin, and V. A. Gritsenko
On the Nature of Deep Centres Responsible for the Memory Effect and Luminescence of a-SiN_x with $x \leqq 4/3$
(71.55; 78.55; 78.60; S7)
(a) Vol. **94**, No. 2, p. K107—K112 (1986).

K. Sierański and J. Szatkowski
Electrical Conductivity of Zn_3P_2
(72.80; S7)
(a) Vol. **94**, No. 2, p. K133—K135 (1986).

M. Mondal and K. P. Ghatak
Effect of Carrier Degeneracy on the Screening Length in Degenerate Tetragonal Semiconductors
(71.25; S7)
(b) Vol. **135**, No. 1, p. 239—251 (1986).

I. L. Baginskii and E. G. Kostsov
Experimental Determination of the Mechanism of Transient-Single-Injection Current Flow in Dielectrics
(73.40 and 77.55; S7)
(a) Vol. **95**, No. 1, p. 291—301 (1986).

A. M. Zaitsev, V. B. Shipilo, E. M. Shishonok, and A. A. Melnikov
Cathodoluminescence of Annealed Cubic Boron Nitride
(78.60; 61.70; S7)
(a) Vol. **95**, No. 1, p. K29—K32 (1986).

K.-D. UFERT, W. ECKE, M. MANZEL, W. MICHALKE, and J. SALM
Investigations of Interface States in Si/Si-Oxynitride/Al Structures Using DLTS
(73.40; S5.11; S7)
(a) Vol. **98**, No. 1, p. K69—K72 (1986).

L. GIER, A. SCHARMANN, and D. SCHALCH
Hopping Conduction and Defect States in Reactively Sputtered Silicon Nitride Thin Films
(73.60; 71.20; S7)
(a) Vol. **98**, No. 2, p. 605—610 (1986).

S7.1. III–V compounds

Review Article

J. Shanker and W. N. Bhende
Higher-Order Elastic Constants and Thermoelastic Properties of Ionic Solids
(62.20 and 63.90; 64.30; 65.70; S7.1; S9.11; S9.12; S10.1)
(b) Vol. **136**, No. 1, p. 11—30 (1986).

Original Papers and Short Notes

T. Jung
Kinetics and Mechanism of Plasma Nitridation of Thin Aluminium Films
(68.55; S7.1)
(a) Vol. **93**, No. 2, p. 479—485 (1986).

U. Pietsch
Bond Charges and Electronic Charge Transfer in Ternary Semiconductors
(61.50 and 71.50; S7.1; S7.15; S7.16)
(b) Vol. **134**, No. 1, p. 21—27 (1986).

H. Übensee, G. Paasch, and J.-P. Zöllner
Self-Consistent Theory of the Electronic Structure of Inversion Layers. II. Quantum Effects and the Electronic Properties of MIS Structures
(73.40; S5.11; S7.1; S10.1)
(b) Vol. **134**, No. 1, p. 367—381 (1986).

O. V. Romanov and A. M. Yafyasov
Manifestation of Structure and Occupation of Energy Bands in Surface Effects of Semiconductors
(73.20; 73.25; S7.1; S8.15)
(a) Vol. **94**, No. 1, p. 235—242 (1986).

A. M. Zaitsev, A. A. Melnikov, V. B. Shipilo, and E. M. Shishonok
Cathodoluminescence of Electron Irradiated Cubic Boron Nitride
(61.80 and 78.60; S7.1)
(a) Vol. **94**, No. 2, p. K125—K127 (1986).

P. K. Dixit, B. A. Vaid, and K. C. Sharma
Structural Disorder Model. Pressure-Induced Transitions in Sn and Semiconducting Compounds
(64.60; 65.50; S4; S7.1; S8.1)
(b) Vol. **135**, No. 2, p. 523—528 (1986).

R. Ramiréz and M. C. Böhm
An Efficient Technique for the Evaluation of Lattice Sums in Crystal Orbital (CO) Calculations
(71.10 and 71.25; S5; S7.1)
(b) Vol. **135**, No. 2, p. 661—667 (1986).

J. Kelber and Ch. Schnittler
Carrier Density in Layer Structures of Microelectronics. A Modified Local Density Approximation for Semiconductor Layers
(71.20 and 73.40; S7.1)
(b) Vol. **135**, No. 2, p. 783—793 (1986).

T. Jung and M. Schmidt
Properties of AlN_xO_y/GaAs Structures Made by Plasma Nitridation of Evaporated Aluminium Films
(68.55 and 73.60; 79.20; S7.1; S7.12)
(a) Vol. **96**, No. 1, p. 75—81 (1986).

Ya. D. Lebedev and M. D. Shagalov
Electrophysical Properties of Gallium Nitride in the Model of a Nonhomogeneous Semiconductor
(72.40 and 78.55; 71.55; S7.1)
(a) Vol. **97**, No. 2, p. 539—548 (1986).

H.-R. Höche, H. S. Leipner, and G. Stadermann
Line Defects in $A^{III}B^{V}$ Semiconductors
(61.70; S7.1)
(a) Vol. **98**, No. 2, p. 503—510 (1986).

S7.11. Phosphides

Original Papers and Short Notes

J. M. Peransin, B. E. F. Da Silva, and J. F. Bresse
Characterization of Au–n-InP Schottky Diodes by EBIC
(73.40; S7.11)
(a) Vol. **94**, No. 2, p. 713—718 (1986).

V. Rakovics, T. Görög, and E. Lendvay
Semitransparent LPE Reactor and Its Application for the Growth of the InP/InGaAsP Heterosystem
(68.45 and 68.55; S7.11; S7.15)
(a) Vol. **94**, No. 2, p. 727—730 (1986).

M. Somogyi
Dopant Concentration Determination at Frequencies above 10 kHz on GaP–Electrolyte System
(73.40; 68.45; S7.11; S7.16)
(a) Vol. **94**, No. 2, p. 731—734 (1986).

M. Garrigues and S. K. Krawczyk
Characterization of the InP Surface by Photoluminescence Imaging
(78.55; S7.11)
(a) Vol. **94**, No. 2, p. K163—K166 (1986).

W. Siegel, G. Kühnel, H. Koi, and W. Gerlach
Electrical Properties of n-Type and p-Type InP Grown by the Synthesis, Solute Diffusion Technique
(72.80; 71.55; S7.11)
(a) Vol. **95**, No. 1, p. 309—316 (1986).

Mao Er-wang, Zhang Hung-rei, Wang Wei-yuan, Ye Liang, and Fang Guo-ping
Inelastic Scattering of Electrons at the InP–SiO_2 Interface
(73.40; 71.38; S7.11; S10.1)
(b) Vol. **135**, No. 2, p. 825—830 (1986).

T. Soma and H.-Matsuo Kagaya
Pressure-Derivatives of Elastic Constants and Mode Grüneisen Parameters under Pressure for Tetrahedral Compounds
(62.20 and 65.50; S7.11; S8.11; S8.12; S8.13)
(b) Vol. **135**, No. 2, p. K103—K106 (1986).

H. Weinert, A. Glismann, and J. Voigt
"Memory" Effect of Bound Exciton Luminescence in GaP:N
(78.55; 71.35; S7.11)
(b) Vol. **135**, No. 2, p. K133—K138 (1986).

B. Akamatsu, A. M. Jean-Louis, A. C. Papadopoulo and P. Henoc
Electrical Activity of Dislocations Introduced in Optically Degraded InP
(61.70; 61.16; 78.60; S7.11)
(a) Vol. **95**, No. 2, p. 557—564 (1986).

U. Pietsch
Erratum to: X-Ray Structure Amplitudes for GaAs and InP
(61.10; S7.11; S7.12)
(a) Vol. **95**, No. 2, p. 749 (1986).

P. GALL, A. KALLEL, N. LAURET, J. P. PEYRADE, B. BROUSSEAU, and J. MAZZASCHI
An Experimental Evidence of the Destruction of Complex Defects by the Introduction of Dislocations in InP
(61.70; 78.55; S7.11)
(a) Vol. **97**, No. 1, p. K25—K27 (1986).

A. S. POPOV and D. B. KUSHEV
Deep Levels in Fe-Doped InP and GaP Formed by Iron Complexes
(71.55; S7.11)
(a) Vol. **97**, No. 1, p. K63—K67 (1986).

G. PAASCH, K. TARNAY, J.-P. ZÖLLNER, and H. ÜBENSEE
Electric Field versus Band Bending in MIS Structures: Quantum Interpolation in Formula for Device Modeling
(73.40; S5.11; S7.11; S7.12)
(a) Vol. **97**, No. 1, p. K107—K111 (1986).

W. SIEGEL, G. KÜHNEL, and H. A. SCHNEIDER
Symmetrical and Asymmetrical Potential Barriers of Grain Boundaries in p-Type GaP. Determination of Barrier Heights by *C–U* Measurements
(73.40; S7.11)
(a) Vol. **97**, No. 2, p. 609—618 (1986).

N. T. NGHI, D. D. KHANG, and B. T. NHI
Effect of Re-Use of Gallium Melt on Electrical Characteristics of p-GaP LPE Layers
(72.80; S7.11)
(a) Vol. **97**, No. 2, p. K201—K204 (1986).

MAO ER-WANG, ZHANG HUNG-REI, WANG WEI-YUAN, YE LIANG, and FANG GUO-PING
Localization Effect of Electrons in the InP–SiO_2 Interface
(73.40; 71.38; S7.11)
(b) Vol. **138**, No. 2, p. 735—743 (1986).

G. J. PAPAIOANNOU, P. C. EUTHYMIOU, C. CARABATOS, B. LEPLEY, and A. BATH
On the Electron-Irradiation Induced Defects in GaP:Zn
(61.80 and 71.55; S7.11)
(a) Vol. **98**, No. 2, p. K125—K127 (1986).

S7.12. Arsenides

Review Article

V. V. Rossin and V. G. Sidorov
Reabsorption of Recombination Radiation in Semiconductors with High Internal Quantum Efficiency
(73.40 and 78.60; 72.40; 78.20; S7.12)
(a) Vol. **95**, No. 1, p. 15—40 (1986).

Original Papers and Short Notes

Wu Xiaoguang, F. M. Peeters, and J. T. Devreese
Screening of the Electron–Phonon Interaction in GaAs Heterostructures
(71.38; 73.40; S7.12; S7.15)
(b) Vol. **133**, No. 1, p. 229—240 (1986).

E. Wendler, W. Wesch, and G. Götz
Near-Edge Optical Absorption Behaviour in Weakly Damaged Ion-Implanted GaAs
(61.70 and 78.50; S7.12)
(a) Vol. **93**, No. 1, p. 207—212 (1986).

M. Mondal and K. P. Ghatak
Effect of Quantizing Magnetic Field on the Gate Capacitance of MOS Structures of Kane-Type Semiconductors
(73.40; S7.12)
(a) Vol. **93**, No. 1, p. 377—383 (1986).

N. T. Tam and Tran Cнот
Experimental Richardson Constant of Metal–Semiconductor Schottky Barrier Contacts
(73.30; S5.11; S7.11; S7.12; S8.11; S8.12)
(a) Vol. **93**, No. 1, p. K91—K95 (1986).

C. Trallero Giner and M. Antón
Quantum Theory of Free-Carrier Absorption in Quasi-Two-Dimensional Semiconducting Structures
(78.20; 71.38; 78.65; S5.12; S7.12)
(b) Vol. **133**, No. 2, p. 563—572 (1986).

D. S. Domanevskii, V. V. Krasovskii, M. V. Prokopenya, V. A. Vilkotskii, and S. V. Zhokhovets
The Origin of the Minority Impurity States in Heavily Doped Semiconductors
(71.55; 78.55; S7.11; S7.12)
(b) Vol. **133**, No. 2, p. 693—700 (1986).

H. Burghardt, T. Frauenheim, and C. Hamann
The Influence of an Electric Field on the Mobility in Semiconductors
(72.10; 72.20; S7.12)
(b) Vol. **133**, No. 2, p. 755—767 (1986).

V. Milanović, D. Tjapkin, and R. Šašić
The Influence of Nonparabolicity on Self-Consistent Treatment for GaAs–$Al_xGa_{1-x}As$ Heterojunction
(73.40; 71.25; S7.12; S7.15)
(b) Vol. **133**, No. 2, p. 775—784 (1986).

V. Riede, H. Neumann, H. Sobotta, R. Schwabe, W. Seifert, and S. Schwetlick
The Localized Vibrational Mode of Nitrogen in GaAs
(78.30; S7.12)
(a) Vol. **93**, No. 2, p. K151—K156 (1986).

J. Betko and K. Měřínský
Temperature Dependences of Electrical Parameters of Semi-Insulating GaAs Determined from Galvanomagnetic Measurements
(72.80; S7.12)
(a) Vol. **93**, No. 2, p. K205—K209 (1986).

W. Siegel, G. Kühnel, S. Felix, and H. A. Schneider
Contacts for Electrical Characterization of Semi-Insulating GaAs
(73.40; S7.12)
(a) Vol. **93**, No. 2, p. K211—K214 (1986).

R. Strehlow, W. Kühn, and M. Hanke
Electronic Structure of Neutral and Charged Point Defects. A Charge Self-Consistent Empirical Tight-Binding Study
(71.10 and 71.55; S7.11; S7.12)
(b) Vol. **134**, No. 1, p. 257—265 (1986).

J. L. Farvacque
Improvement of Debye-Hückel Evaluation of Free Carrier Screening Effects in Semiconductors
(72.10; S7.12; S7.13)
(b) Vol. **134**, No. 1, p. 343—349 (1986).

K. D. Glinchuk, V. I. Guroshev, and A. V. Prokhorovich
Photoluminescence in Chromium-Doped Semi-Insulating GaAs after Irradiation by 2.2 MeV Electrons and Annealing
(78.55; 61.80; S7.12)
(a) Vol. **94**, No. 1, p. 203—206 (1986).

P. Chakrabarti and B. B. Pal
Computer Simulation of Ion-Implanted DAR Impatt around 94 GHz under Steady State Condition
(73.40; S5.11; S7.11; S7.12)
(a) Vol. **94**, No. 1, p. 305—313 (1986).

H. Bachert, W. Pittroff, I. Rechenberg, S. Stoeff, and K. Vogel
The Influence of Mg Dopping on the Optical and Electrical Properties of GaAs/(AlGa)As Injection Lasers
(66.30 and 68.55; 73.40; 78.45; S7.12; S7.15)
(a) Vol. **94**, No. 1, p. 321—327 (1986).

S. S. Kubakaddi, B. G. Mulimani, and V. M. Jali
Electronic Thermal Conductivity in Quasi-Two-Dimensional Semiconductor Quantum Well Structures
(72.15; S7.12)
(b) Vol. **135**, No. 2, p. K163—K167 (1986).

B. Tasche und H. Richtering
Spin–Lattice Relaxation of ^{115}In and ^{75}As in InAs between 300 and 1000 K and Diffusion of Copper in InAs
(66.30 and 76.60; S7.12)
(a) Vol. **95**, No. 2, p. 565—571 (1986).

A. M. Narsale and B. M. Arora
C–V–ω and *G–V–ω* Characteristics of Native Oxide–GaAs Interfaces of $\langle\bar{1}\bar{1}\bar{1}\rangle$ and $\langle 100\rangle$ Orientations
(73.40; S7.12)
(a) Vol. **95**, No. 2, p. 743—748 (1986).

U. Pietsch
Erratum to: X-Ray Structure Amplitudes for GaAs and InP
(61.10; S7.11; S7.12)
(a) Vol. **95**, No. 2, p. 749 (1986).

B. I, Sysoev, N. N. Bezryadin, and Yu. K. Shlyk
Electrophysical Properties of In_2Te_3–InAs Heterojunctions
(73.40; S7.12; S8)
(a) Vol. **95**, No. 2, p. K169—K173 (1986).

J. Braciník and J. Štelina
Sound Generation in Photoresistive Semiconductors
(62.65 and 77.60; S7.12)
(a) Vol. **95**, No. 2, p. K179—K182 (1986).

M. Procop and B. Sandow
Au–Ge Based Ohmic Contacts on GaAs
(73.40 and 79.20; S7.12)
(a) Vol. **95**, No. 2, p. K211—K215 (1986).

B. A. Danilchenko, V. N. Poroshin, M. I. Slutskii, and M. Asche
Investigation of Nonequilibrium Phonons in GaAs
(63.20; S7.12)
(b) Vol. **136**, No. 1, p. 63—68 (1986).

V. M. Fomin and E. P. Pokatilov
Optical Properties of Multi-Layer Structures. I. Polaritons
(71.36; 73.40; 78.20; S7.12; S7.15)
(b) Vol. **136**, No. 1, p. 187—199 (1986).

T. Jung and M. Schmidt
Properties of AlN_xO_y/GaAs Structures Made by Plasma Nitridation of Evaporated Aluminium Films
(68.55 and 73.60; 79.20; S7.1; S7.12)
(a) Vol. **96**, No. 1, p. 75—81 (1986).

A. L. Golovin and U. Pietsch
X-Ray Investigations of Submicrometer Layer Heterostructures
(61.10; 68.55; S7.12; S9.15)
(a) Vol. **96**, No. 2, p. K111—K115 (1986).

J. Jiménez, J. Bonnafé, M. A. González, P. Hernández, and J. C. Merino
Photoconductivity Properties of Vanadium Doped Bulk GaAs
(72.40; S7.12)
(a) Vol. **96**, No. 2, p. K207—K212 (1986).

T. Rother and W. D. Kraeft
On the Thermodynamics of Strongly Degenerate Two-Component Plasmas
(65.50 and 71.45; S5.12; S7.12)
(b) Vol. **137**, No. 1, p. 103—108 (1986).

P. Kleinert
Ionized-Impurity-Mediated Free-Carrier Absorption in n-GaAs Slabs
(63.20 and 71.45; 78.30; S7.12)
(b) Vol. **137**, No. 1, p. 125—135 (1986).

M. Tomak and V. E. Godwin
Electron Energy Levels in GaAs/$Ga_{1-x}Al_xAs$ Heterojunctions (I)
(71.55 and 73.40; S7.12; S7.15)
(b) Vol. **137**, No. 1, p. 183—186 (1986).

L. Wendler and R. Haupt
Damping of Polaritons in Bilayer Systems. Long-Range Surface Phonon-Polaritons
(68.30 and 71.36; S7.11; S7.12; S9; S9.12)
(b) Vol. **137**, No. 1, p. 269—290 (1986).

P. Trung Dung and M. Láznička
Origin and Formation Mechanism of Macroscopic Defects in GaAs Films Grown by Molecular Beam Epitaxy
(68.55; 68.70; S7.12)
(a) Vol. **97**, No. 1, p. 103—109 (1986).

B. Diegner, H. Weinert, R. Pickenhain, and W. Hörig
An Experimental Study on Deep Level Incorporation and Background Doping in Hydride VPE GaAs
(73.60; 68.55; S7.12)
(a) Vol. **97**, No. 1, p. 313—321 (1986).

P. Sircar
Excimer Laser Annealed Gold and Silver Ohmic Contacts on n^+-GaAs
(73.40; S7.12)
(a) Vol. **97**, No. 1, p. K69—K72 (1986).

S. Shigetomi and T. Matsumori
Electrical Properties of p-Type Layer in Si-Implanted GaAs
(72.80; 61.70; S7.12)
(a) Vol. **97**, No. 1, p. K83—K86 (1986).

H. Weinert, B. Diegner, and J. Kugler
Photoluminescence Study on the Influence of Substrate Doping on the Properties of Nominally Undoped VPE-GaAs
(61.70 and 78.55; S7.12)
(a) Vol. **97**, No. 2, p. K177—K182 (1986).

U. Pietsch, V. G. Tsirelson, and R. P. Ozerov
X-Ray Electron Density Distribution of GaAs
(61.10; S7.12)
(b) Vol. **138**, No. 1, p. 47—52 (1986).

S. Benakki, A. Goltzene, C. Schwab, Wang Guangyu, and Zou Yuanxi
Electron Paramagnetic Resonance Analysis of Strain-Induced Defects in Semi-Insulating GaAs
(61.70 and 76.30; S7.12)
(b) Vol. **138**, No. 1, p. 143—149 (1986).

K. Ishida and H. Kawano
Different Etch Pit Shapes Revealed by Molten KOH Etching on the (001) GaAs Surface and Their Dependence on the Burgers Vectors
(61.70; S7.12)
(a) Vol. **98**, No. 1, p. 175—181 (1986).

K. Somogyi and I. Gyuró
A Method for Studying Parameter Variations in VPE Technology
(68.55; S7.12)
(a) Vol. **98**, No. 1, p. K1—K4 (1986).

H.-Matsuo Kagaya and T. Soma
Thermal Expansion Coefficient of III–V Covalent Crystals
(65.70; 63.20; S7.12; S7.13)
(b) Vol. **138**, No. 2, p. K91—K93 (1986).

F. Garofalo and A. M. Mazzone
Channeling in GaAs. A Monte Carlo Simulation of Low Energy Implants
(61.80; S7.12)
(a) Vol. **98**, No. 2, p. 517—520 (1986).

S7.13. Antimonides

Original Papers and Short Notes

W. Kraak, G. Nachtwei, and R. Herrmann
Quantum Hall Effect in the Inversion Layer of p-Type InSb Bicrystals under High Hydrostatic Pressure
(73.40; S7.13)
(b) Vol. **133**, No. 1, p. 403—408 (1986).

K. P. Maheshwari and G. Tarey
Resonant Excitation of Helicon Waves by Two Microwave Beams in a Solid State Plasma
(72.30; S7.13)
(b) Vol. **133**, No. 1, p. 417—423 (1986).

R. Herrmann, U. Preppernau, and M. Gliński
Thermopower Measurements on the Two-Dimensional Electron Gas of a Grain Boundary in an InSb Bicrystal
(73.40; S7.13)
(b) Vol. **133**, No. 1, p. K57—K59 (1986).

W. Braune, N. Kubicki, N. Pruss, and R. Herrmann
Subband Characteristics of the Two-Dimensional Inversion Carrier System in p-InSb Bicrystals
(73.40; S7.13)
(b) Vol. **133**, No. 2, p. K149—K151 (1986).

E. I. Kafediiska, I. Mojzes, R. Veresegyházy, and B. Pécz
Mass-Spectrometric Study of the Dissociation of GaSb Covered with Contact Layers
(68.48; S7.13)
(a) Vol. **93**, No. 2, p. K125—K128 (1986).

S. I. Radautsan and O. V. Kulikova
Photocapacitance Effect in $Zn_xCd_{1-x}Te$–InSb Structures
(73.40; 72.40; S7.13; S8.15)
(a) Vol. **93**, No. 2, p. K201—K203 (1986).

J. L. Farvacque
Improvement of Debye-Hückel Evaluation of Free Carrier Screening Effects in Semiconductors
(72.10; S7.12; S7.13)
(b) Vol. **134**, No. 1, p. 343—349 (1986).

G. Diener and U. Schlink
Interpretation of the Hall Curve of n-Inversion Layers at InSb Grain Boundaries
(73.40; S7.13)
(b) Vol. **134**, No. 1, p. K79—K82 (1986).

S. A. Studenikin and E. M. Skok
Electron Gas Heating under Spin-Flip Raman Scattering Conditions in n-InSb
(72.20; S7.13)
(b) Vol. **134**, No. 2, p. 745—750 (1986).

W. KRAAK, U. SCHALLER, and R. HERRMANN
Metal–Insulator Transition and Metastable State in High-Purity n-InSb under High Hydrostatic Pressure
(71.30; 71.55; 72.20; S7.13)
(b) Vol. **134**, No. 2, p. 779—788 (1986).

J.-P. ZÖLLNER, H. ÜBENSEE, G. PAASCH, T. FIEDLER, and G. GOBSCH
A Novel Self-Consistent Theory of the Electronic Structure of Inversion Layers in InSb MIS Structures
(73.40; 71.25; S7.13)
(b) Vol. **134**, No. 2, p. 837—845 (1986).

H.-MATSUO KAGAYA and T. SOMA
Volume-Dependence of Phonon Frequencies for III–V Covalent Crystals
(63.20; 63.90; S7.12; S7.13)
(b) Vol. **134**, No. 2, p. K105—K110 (1986).

G. GOBSCH, G. PAASCH, and H. ÜBENSEE
Self-Consistent Calculation of the Electronic Structure of n-Inversion Layers Adjacent to the Grain Boundary in InSb Bicrystals
(71.25 and 73.40; S7.13)
(b) Vol. **135**, No. 1, p. 283—297 (1986).

R. HERRMANN, W. KRAAK, G. NACHTWEI, and TH. SCHURIG
Quantum Properties of the Two-Dimensional Electron Gas in the n-Inversion Layers of InSb Grain Boundaries under High Hydrostatic Pressure
(73.40; S7.13)
(b) Vol. **135**, No. 1, p. 423—435 (1986).

M. MONDAL and K. P. GHATAK
Effect of Varying the Orientation of a Quantizing Magnetic Field on the Effective Electron Mass in Stressed Kane-Type Semiconductors
(71.25; S7.13)
(b) Vol. **135**, No. 1, p. K21—K25 (1986).

S. HUANT, L. DMOWSKI, M. BAJ, and L. C. BRUNEL
Pressure Dependence of the Electronic Effective Mass and Effective *g*-Factor in the Narrow Gap Semiconductor InSb
(71.25; S7.13)
(b) Vol. **135**, No. 2, p. K129—K132 (1986).

I. GORCZYCA
Calculation of the Impurity Levels in Semiconductors by the Effective-Mass-Pseudopotential Method with Fitting Parameters
(71.55; S7.13)
(b) Vol. **136**, No. 1, p. K45—K48 (1986).

P. K. BHAT, J. J. DUBOWSKI, and D. F. WILLIAMS
Emission of Neutral and Charged Species from InSb Induced by Pulsed Laser Evaporation
(68.55 and 79.20; S7.13)
(a) Vol. **96**, No. 1, p. K9—K14 (1986).

A. V. Vdovin and E. M. Skok
Precise Measurement of the Free Electron g-Factor in InSb
(71.25; 76.30; S7.13)
(b) Vol. **136**, No. 2, p. 603—613 (1986).

B. Chalupa, R. Michalec, L. Horalik, and P. Mikula
The Study of Neutron Acoustic Effect by Neutron Diffraction on InSb Single Crystal
(61.12; S7.13)
(a) Vol. **97**, No. 2, p. 403—409 (1986).

V. N. Davydov and A. S. Petrov
Electric Field Effect on Characteristics of Indium Antimonide MOS Structures
(73.20 and 73.40; S7.13)
(a) Vol. **98**, No. 1, p. 253—261 (1986).

H.-Matsuo Kagaya and T. Soma
Thermal Expansion Coefficient of III–V Covalent Crystals
(65.70; 63.20; S7.12; S7.13)
(b) Vol. **138**, No. 2, p. K91—K93 (1986).

Th. Schurig and R. Herrmann
Simple Calculation of Pressure Dependent Electronic Properties of n-Inversion Layers in InSb Bicrystals
(73.40; S7.13)
(b) Vol. **138**, No. 2, p. K131—K136 (1986).

S7.15. Mixed crystals

Original Papers and Short Notes

Wu Xiaoguang, F. M. Peeters, and J. T. Devreese
Screening of the Electron–Phonon Interaction in GaAs Heterostructures
(71.38; 73.40; S7.12; S7.15)
(b) Vol. **133**, No. 1, p. 229—240 (1986).

U. Pietsch and D. Marlow
The Thermal Expansion Coefficient of $Ga_xIn_{1-x}As_yP_{1-y}$ Epitaxial Layers Grown on InP Substrate
(65.70; 68.25; S7.15)
(a) Vol. **93**, No. 1, p. 143—149 (1986).

U. Pietsch
Thermal Expansion Coefficients of Ternary $A^{III}B^{V}$ Solid Solutions
(65.70; S7.15)
(b) Vol. **133**, No. 2, p. 483—489 (1986).

V. Milanović, D. Tjapkin, and R. Šašić
The Influence of Nonparabolicity on Self-Consistent Treatment for GaAs–Al_xGa_{1-x}As Heterojunction
(73.40; 71.25; S7.12; S7.15)
(b) Vol. **133**, No. 2, p. 775—784 (1986).

U. Pietsch
Bond Charges and Electronic Charge Transfer in Ternary Semiconductors
(61.50 and 71.50; S7.1; S7.15; S7.16)
(b) Vol. **134**, No. 1, p. 21—27 (1986).

A. K. Giri and G. B. Mitra
Theoretical Prediction of Elastic Constants of Mixed Systems
(62.20; S7.15; S9.15)
(b) Vol. **134**, No. 1, p. K11—K16 (1986).

H. Bachert, W. Pittroff, I. Rechenberg, S. Stoeff, and K. Vogel
The Influence of Mg Doping on the Optical and Electrical Properties of GaAs/(AlGa)As Injection Lasers
(66.30 and 68.55; 73.40; 78.45; S7.12; S7.15)
(a) Vol. **94**, No. 1, p. 321—327 (1986).

P. Krispin and J. Maege
The Ge-Related DX Level in Sn/Ge-Doped Al_xGa_{1-x}As Heterojunctions Grown by LPE
(73.40; 71.55; S7.15)
(a) Vol. **94**, No. 1, p. 329—335 (1986).

M. de Dios Leyva, J. Sabin, and J. López Gondar
Band and Branch Inversion Phenomena in GaAs–Al_xGa_{1-x}As Superlattices
(71.25; S7.12; S7.15)
(b) Vol. **134**, No. 2, p. 615—620 (1986).

E. GEHRKE, K. WANDEL, and H. WEINERT
Experimental Investigations on the Influence of Surface and Interface Recombination on the Minority Carrier Lifetime in LPE-GaP:N, Si, VPE-GaP:N, Te, and VPE-$GaAs_{0.12}P_{0.88}$:N, Te
(73.40 and 78.55; S7.11; S7.15)
(a) Vol. **96**, No. 2, p. 683—689 (1986).

M. TOMAK and V. E. GODWIN
Electron Energy Levels in GaAs/$Ga_{1-x}Al_xAs$ Heterojunctions (I)
(71.55 and 73.40; S7.12; S7.15)
(b) Vol. **137**, No. 1, p. 183—186 (1986).

H. HAVLOVÁ and L. SMRČKA
Quantum Oscillations of Thermogalvanomagnetic Coefficients in Two-Dimensional Electron Systems
(72.20 and 73.40; S5.11; S7.15)
(b) Vol. **137**, No. 1, p. 331—343 (1986).

V. E. GODWIN and M. TOMAK
Electron Energy Levels in GaAs/$Ga_{1-x}Al_xAs$ Heterojunctions. II. Optical Properties
(71.55 and 73.40; 78.50; S7.12; S7.15)
(b) Vol. **137**, No. 2, p. 619—622 (1986).

S. S. KUBAKADDI, B. G. MULIMANI, and V. M. JALI
Thermopower in Quasi-Two-Dimensional Semiconductor Quantum Well Structures
(72.20; 73.40; S7.15)
(b) Vol. **137**, No. 2, p. 683—689 (1986).

G. HOLZ and CH. SCHNITTLER
Calculation of the Electronic Properties of a High Electron Mobility Transistor. I. Equilibrium Conditions
(73.40; 71.25; S7.12; S7.15)
(a) Vol. **97**, No. 2, p. 635—644 (1986).

J. PASTRŇÁK, F. KAREL, J. OSWALD, and O. PETŘÍČEK
The Electric Field Dependence of Photoconductivity Spectra in AlGaAs/GaAs DH Laser Diodes in Relation with Laser Parameters
(72.40 and 73.40; 78.45; S7.12; S7.15)
(a) Vol. **97**, No. 2, p. 657—665 (1986).

A. POPOV and R. ATANASOVA
The Origin of Nonradiative Centers in Degraded LEDs from $GaAs_{0.6}P_{0.4}$
(73.40 and 78.60; 71.55; S7.15)
(a) Vol. **98**, No. 2, p. 629—632 (1986).

S7.16. Higher compounds

Review Article

F. HENNEBERGER
Optical Bistability at the Absorption Edge of Semiconductors
(78.20; 71.35; S7.16; S8.16)
(b) Vol. **137**, No. 2, p. 371—432 (1986).
Erratum
(b) Vol. **138**, No. 2, p. 744 (1986).

Original Papers and Short Notes

M. MONDAL and K. P. GHATAK
On the Generalized Einstein Relation in Ultrathin Films of Ternary Chalcopyrite Semiconductors in the Presence of Crossed Electric and Magnetic Fields
(73.60; S7.16)
(b) Vol. **133**, No. 1, p. K67—K72 (1986).

U. PIETSCH
Bond Charges and Electronic Charge Transfer in Ternary Semiconductors
(61.50 and 71.50; S7.1; S7.15; S7.16)
(b) Vol. **134**, No. 1, p. 21—27 (1986).

M. SOMOGYI
Dopant Concentration Determination at Frequencies above 10 kHz on GaP–Electrolyte System
(73.40; 68.45; S7.11; S7.16)
(a) Vol. **94**, No. 2, p. 731—734 (1986).

S. MARCINKEVIČIUS, G. BABONAS, B. V. KORZUN, and I. G. TREGUB
Optical Activity of Semiconductors with Chalcopyrite Structure
(78.20; S7.16; S8.16)
(b) Vol. **135**, No. 2, p. 615—621 (1986).

R. MADELON, J. HUREL, and E. PAUMIER
Exciton Absorption Spectrum in a $ZnSiP_2$ Crystal
(71.35; 78.40; S7.16)
(b) Vol. **136**, No. 2, p. 679—684 (1986).

N. V. K. PRABHAKAR and S. P. SANYAL
Anharmonic Properties of the Ternary Semiconductor $CuGe_2P_3$
(62.20; 64.30; S7.16)
(b) Vol. **137**, No. 2, p. 453—458 (1986).

S8. Chalcogenides (for transition metal compounds, see S1.63)

Review Articles

K. Lischka
Deep Level Defects in Narrow Gap Semiconductors
(71.55; 72.20; 72.40; S8; S8.15)
(b) Vol. **133**, No. 1, p. 17—46 (1986).

K. V. S. Rama Rao and S. Ramaprabhu
Nuclear Quadrupole Resonance Studies on Inorganic Solids
(76.60; S8; S9; S10; S11)
(a) Vol. **93**, No. 1, p. 17—27 (1986).

J. Shanker, G. G. Agrawal, and N. Dutt
Electronic Polarizabilities and Photoelastic Behaviour of Ionic Crystals
(77.20 and 77.30; 78.20; S8; S9)
(b) Vol. **138**, No. 1, p. 9—30 (1986).

Original Papers and Short Notes

F. M. Gashimzade and E. V. Tagirov
Selection Rules for Matrix Elements of Electron–Phonon Scattering Processes in Crystals with TlSe Structure
(71.38; S8)
(b) Vol. **133**, No. 1, p. 179—182 (1986).

O. P. Singh and V. P. Gupta
Electronic Properties of Iron Chalcogenides FeS, FeS_2, $FeSe_2$, and $FeTe_2$
(71.25; 76.80; S8)
(b) Vol. **133**, No. 1, p. 249—252 (1986).

S. Asano, Y. Nakao, N. Yamashita, and I. Matsuyama
Luminescence and Non-Radiative $^3T_{1u} \rightarrow {}^3A_1$ Transition in the BaS:Bi^{3+} Phosphor
(78.55; 71.55; 78.50; S8)
(b) Vol. **133**, No. 1, p. 333—344 (1986).

K. I. Andronik, V. F. Banar, V. G. Kantser, and A. S. Sidorenko
Kondo-Type Behaviour of Conductivity in Tl-Doped PbTe Single Crystals
(72.80; S8)
(b) Vol. **133**, No. 1, p. K61—K65 (1986).

P. J. S. Foot and B. A. Nevett
Electronic Conduction in p- and n-Type $NiPS_3$
(72.20; 78.30; S8)
(a) Vol. **93**, No. 1, p. 283—292 (1986).

G. Valiukonis, D. A. Guseinova, G. Krivaitė, and A. Šileika
Optical Spectra and Energy Band Structure of Layer-Type $A^{IV}B^{VI}$ Compounds
(71.25 and 78.20; S8)
(b) Vol. **135**, No. 1, p. 299—307 (1986).

U. Berg and T. Chassé
Angular-Resolved X-Ray Photoelectron Spectra and Electronic Structure of SnS_2
(71.25; 79.60; S8)
(b) Vol. **135**, No. 2, p. 633—637 (1986).

A. M. Elkorashy
Indirect Forbidden Fundamental Absorption Edge in Germanium Selenide Single Crystals
(78.20; S8)
(b) Vol. **135**, No. 2, p. 707—713 (1986).

N. M. Gasanly, N. N. Melnik, V. I. Tagirov, and A. A. Yushin
Special Features of First- and Second-Order Raman and Infrared Spectra of $GaS_{1-x}Se_x$ Mixed Crystals
(78.30; 63.20; S8)
(b) Vol. **135**, No. 2, p. K107—K112 (1986).

S. Mróz and A. Mróz
Quantitative AES Analysis of PbTe and SnTe Samples with and without Matrix Effect Correction
(68.20; 79.20; S8)
(a) Vol. **95**, No. 2, p. 407—412 (1986).

D. Degenhardt, P. Rabe, and R. Haensel
EXAFS Investigations of TiS_2 and Cu_xTiS_2 Single Crystals ($x = 0.35, 0.4$)
(61.60; 78.70; S8; S8.16)
(a) Vol. **95**, No. 2, p. 439—445 (1986).

A. Klimakow, M. Schenk, and B. Christ
Incorporation of Tl in Monocrystalline n-PbS
(61.70; 61.60; 66.30; S8)
(a) Vol. **95**, No. 2, p. 573—578 (1986).

A. M. Karguppikar and A. G. Vedeshwar
Transport Properties of Thin Iron Pyrite Films
(73.60; S8)
(a) Vol. **95**, No. 2, p. 717—720 (1986).

A. Simanovskis, S. V. Stolyarova, and J. Teteris
Effect of Environment on Thermal Diffusion in $Ag–As_2S_3$
(66.30; S1.3; S8)
(a) Vol. **95**, No. 2, p. K105—K107 (1986).

B. I. Sysoev, N. N. Bezryadin, and Yu. K. Shlyk
Electrophysical Properties of In_2Te_3–InAs Heterojunctions
(73.40; S7.12; S8)
(a) Vol. **95**, No. 2, p. K169—K173 (1986).

M. C. AL-EDANI and K. S. DUBEY
Lattice Thermal Conductivity of Gadolinium Monosulphide above Room Temperature in the Frame of Two-Mode Conduction of Phonons
(63.20 and 66.70; S8)
(b) Vol. **137**, No. 2, p. 449—452 (1986).

F. LUKEŠ, E. SCHMIDT, J. HUMLÍČEK, P. DUB, and F. KOSEK
Absorption Edge of SnSe
(78.20; 71.25; S8)
(b) Vol. **137**, No. 2, p. 569—579 (1986).

C. JULIEN, E. HATZIKRANIOTIS, and K. KAMBAS
Electrical Transport Properties of Impurity-Doped In_2Se_3
(72.20; 71.55; S8)
(a) Vol. **97**, No. 2, p. 579—585 (1986).

D. BALTRŪNAS, S. MOTIEJŪNAS, and E. I. ROGACHEVA
Effect of the Deviation from Stoichiometry on the Mössbauer Parameters of SnTe
(64.75 and 76.80; S8)
(a) Vol. **97**, No. 2, p. K131—K133 (1986).

M. STÖLZER, M. STORDEUR, H. SOBOTTA, and V. RIEDE
IR Transmission Investigations of $(Bi_{1-x}Sb_x)_2Te_3$ Single Crystals
(78.20; 71.25; 78.30; S8)
(b) Vol. **138**, No. 1, p. 259—266 (1986).

M. SOUIRI, A. GOLTZENE, and C. SCHWAB
Correlations between Electron Paramagnetic Resonance and Bonding Parameters. Evidence of a Lattice Relaxation
(71.45 and 71.70; 76.30; S7; S8; S9; S10)
(b) Vol. **138**, No. 1, p. 321—330 (1986).

V. S. DNEPROVSKII, V. D. EGOROV, D. S. KHECHINASHVILI, and H. X. NGUYEN
Room Temperature Self-Screening of Excitons in GaSe
(71.35 and 78.40; S8)
(b) Vol. **138**, No. 1, p. K39—K42 (1986).

D. DĂDÂRLAT, M. CHIRTOC, and R. M. CÂNDEA
A Simple Detection Method in Photothermal Deflection Measurements on Thin-Film Semiconductors
(73.60; S8)
(a) Vol. **98**, No. 1, p. 279—283 (1986).

KH. A. MADY, M. M. EL-NAHAS, A. M. FARID, and H. S. SOLIMAN
Electrical Resistivity of Amorphous Antimony Trisulphide Films
(73.60; S8)
(a) Vol. **98**, No. 1, p. K83—K86 (1986).

S8.1. II–VI compounds

Original Paper

S8.11. Suldifes

Original Papers and Short Notes

O. P. Singh and V. P. Gupta
Modified PVV Theory and X-Ray K Absorption Edge Studies in Zinc Chalcogenides (Ionicity Determination and d Electron Effect on It)
(71.25; 78.70; S8.11; S8.12; S8.13)
(b) Vol. **133**, No. 1, p. K17—K19 (1986).

O. Goede, W. Heimbrodt, and F. Sittel
Photoacoustic Spectroscopy of Isovalent Impurities in II–VI Semiconductors
(78.20; 71.55; 78.50; S8.11)
(a) Vol. **93**, No. 1, p. 277—282 (1986).

H.-J. Fitting and A. von Czarnowski
Electron Breakdown in Polar Insulating and Semiconducting Layers
(73.60; 71.38; 79.70; S8.11; S10.1)
(a) Vol. **93**, No. 1, p. 385—396 (1986).

N. T. Tam and Tran Chot
Experimental Richardson Constant of Metal–Semiconductor Schottky Barrier Contacts
(73.30; S5.11; S7.11; S7.12; S8.11; S8.12)
(a) Vol. **93**, No. 1, p. K91—K95 (1986).

V. A. Kiselev, B. V. Novikov, A. E. Cherednichenko, and E. A. Ubushiev
Effect of Thin Space-Charge Layers on Exciton Reflectance
(71.35 and 78.40; S8.11)
(b) Vol. **133**, No. 2, p. 573—581 (1986).

W. Dułak and H. Męczyńska
Exciton Spectra of Thin CdS Films Prepared by Spray Pyrolysis
(78.65; 71.35; S8.11)
(a) Vol. **93**, No. 2, p. K171—K172 (1986).

O. Goede, W. Heimbrodt, T. Lau, G. Matzkeit, and B. Selle
Optical and RBS Studies of ZnS:Te Thin Films
(78.65; 61.70; S8.11)
(a) Vol. **94**, No. 1, p. 259—265 (1986).

K. A. Dmitrenko, S. G. Shevel, L. V. Taranenko, and A. V. Marintchenko
The Temperature Dependence (4.2 to 293 K) of the Resonance Energies of Excitonic Transitions in II–VI Compounds
(71.35 and 78.40; S8.11; S8.12; S8.13)
(b) Vol. **134**, No. 2, p. 605—613 (1986).

H.-Matsuo Kagaya and T. Soma
Temperature Dependence of Specific Heat and Debye Temperature of Tetrahedral Compounds
(65.40; S7.11; S8.11; S8.12; S8.13)
(b) Vol. **134**, No. 2, p. K101—K104 (1986).

N. S. Bogdanyuk and S. S. Ostapenko
Structure Peculiarities and Annealing Mechanism of the Red Flash-Like Luminescent Centre in CdS Single Crystals Irradiated with 1.2 MeV Electron Beam
(78.60; 61.80; 71.55; S8.11)
(a) Vol. **96**, No. 2, p. 621—628 (1986).

V. Z. Polinger and S. I. Boldyrev
The Multimode Jahn-Teller Effect in the Luminescence Spectrum of the Tetrahedral $^2T_2 \otimes (e + t_2 + t_2)$ Impurity System $ZnS:Cu^{2+}$
(63.20 and 71.70; 78.55; S8.11)
(b) Vol. **137**, No. 1, p. 241—253 (1986).

J. Kreissl
A Quantitative EPR Analysis of ZnS:Mn Powders of Different Crystal Structure
(76.30; 61.70; S8.11)
(a) Vol. **97**, No. 1, p. 191—198 (1986).

B. P. Chandra, P. K. Dubey, and S. C. Datt
Mechano- and Electroluminescence of Heavily Doped ZnS:Mn Phosphors
(78.60; S8.11)
(a) Vol. **97**, No. 1, p. K59—K61 (1986).

V. F. Gremenok, N. N. Koren, V. V. Kindyak, V. V. Moiseenko, and K. P. Grigorev
Properties of Heterojunctions Based on CdS
(73.40; S8.11)
(a) Vol. **97**, No. 2, p. K205—K209 (1986).

I. Haddad, M. Kretzschmar, H. Rossmann, and F. Henneberger
Increasing Absorption Bistability of CdS at Room Temperature
(78.20; S8.11)
(b) Vol. **138**, No. 1, p. 235—243 (1986).

M. Grinberg
The Pseudopotential Approach to the Shallow Donor Problem in II–VI Semiconductors
(71.55; S8.11; S8.12)
(b) Vol. **138**, No. 1, p. 275—284 (1986).

H.-Matsuo Kagaya and T. Soma
Mode Grüneisen Parameters of II–VI Tetrahedral Compounds
(63.20; S8.11; S8.12; S8.13)
(b) Vol. **138**, No. 1, p. K13—K16 (1986).

C. Gourdon and P. Lavallard
Comments on "Transmission and Damping of Excitonic Polaritons in CdS" by I. Broser, K.-H. Pantke, and M. Rosenzweig
(71.35 and 71.36; S8.11)
(b) Vol. **138**, No. 1, p. K29—K30 (1986).

S8.12. Selenides

Original Papers and Short Notes

O. P. SINGH and V. P. GUPTA
Modified PVV Theory and X-Ray K Absorption Edge Studies in Zinc Chalcogenides (Ionicity Determination and d Electron Effect on It)
(71.25; 78.70; S8.11; S8.12; S8.13)
(b) Vol. **133**, No. 1, p. K17—K19 (1986).

A. OPANOWICZ
On Thermoluminescence in Conductive Semiconductors
(78.60; S8.12)
(a) Vol. **93**, No. 1, p. K65—K67 (1986).

N. T. TAM and TRAN CHOT
Experimental Richardson Constant of Metal–Semiconductor Schottky Barrier Contacts
(73.30; S5.11; S7.11; S7.12; S8.11; S8.12)
(a) Vol. **93**, No. 1, p. K91—K95 (1986).

I. B. ERMOLOVICH and V. V. MILENIN
Nature of Deep Luminescence Centres in CdSe and $CdSe_xTe_{1-x}$
(78.55; 71.70; S8.12; S8.15)
(b) Vol. **133**, No. 2, p. 611—620 (1986).

K. A. DMITRENKO, S. G. SHEVEL, L. V. TARANENKO, and A. V. MARINTCHENKO
The Temperature Dependence (4.2 to 293 K) of the Resonance Energies of Excitonic Transitions in II–VI Compounds
(71.35 and 78.40; S8.11; S8.12; S8.13)
(b) Vol. **134**, No. 2, p. 605—613 (1986).

H.-MATSUO KAGAYA and T. SOMA
Temperature Dependence of Specific Heat and Debye Temperature of Tetrahedral Compounds
(65.40; S7.11; S8.11; S8.12; S8.13)
(b) Vol. **134**, No. 2, p. K101—K104 (1986).

F. J. BRYANT, W. E. HAGSTON, and A. KRIER
The Electroluminescence of Rare-Earth Implanted ZnS and ZnSe Diodes
(78.55 and 78.60; 73.40; S8.11; S8.12)
(a) Vol. **94**, No. 2, p. 701—712 (1986).

S. KISHIDA, K. MATSUURA, H. NAGASE, H. MORI, F. TAKEDA, and I. TSURUMI
The Photosensitive Optical Absorption Bands in Zn-Treated and Neutron-Irradiated ZnSe Single Crystals
(78.50; 61.70; S8.12)
(a) Vol. **95**, No. 1, p. 155—164 (1986).

V. A. KISELEV, B. V. NOVIKOV, A. S. BATYREV, E. A. UBUSHIEV, and A. E. CHEREDNICHENKO
Localization of Excitons in Space-Charge Layers
(71.35; 73.30; 78.40; S8.11; S8.12)
(b) Vol. **135**, No. 2, p. 597—604 (1986).

S8.13. Tellurides

Original Papers and Short Notes

O. P. Singh and V. P. Gupta
Modified PVV Theory and X-Ray K Absorption Edge Studies in Zinc Chalcogenides (Ionicity Determination and d Electron Effect on It)
(71.25; 78.70; S8.11; S8.12; S8.13)
(b) Vol. **133**, No. 1, p. K17—K19 (1986).

K. A. Dmitrenko, S. G. Shevel, L. V. Taranenko, and A. V. Marintchenko
The Temperature Dependence (4.2 to 293 K) of the Resonance Energies of Excitonic Transitions in II–VI Compounds
(71.35 and 78.40; S8.11; S8.12; S8.13)
(b) Vol. **134**, No. 2, p. 605—613 (1986).

H.-Matsuo Kagaya and T. Soma
Temperature Dependence of Specific Heat and Debye Temperature of Tetrahedral Compounds
(65.40; S7.11; S8.11; S8.12; S8.13)
(b) Vol. **134**, No. 2, p. K101—K104 (1986).

T. Soma and H.-Matsuo Kagaya
Pressure-Derivatives of Elastic Constants and Mode Grüneisen Parameters under Pressure for Tetrahedral Compounds
(62.20 and 65.50; S7.11; S8.11; S8.12; S8.13)
(b) Vol. **135**, No. 2, p. K103—K106 (1986).

J. W. Tomm, Karin Herrmann, A. E. Yunovich, and I. V. Koleshnikov
YAG-Laser Excited Photoluminescence in p-CdTe
(78.55; S8.13)
(a) Vol. **95**, No. 2, p. K131—K134 (1986).

A. M. Gue and A. Mazel
a ⟨100⟩ Dislocation Loops in Electron-Irradiated Cadmium Telluride
(61.80; S8.13)
(a) Vol. **96**, No. 2, p. K145—K146 (1986).

J. Mendolia and D. Lemoine
Electrical and Optical Properties of Evaporated Thin Films of a-CdTe
(73.60; 61.40; 72.40; S8.13)
(a) Vol. **97**, No. 2, p. 601—607 (1986).

H.Matsuo Kagaya and T. Soma
Mode Grüneisen Parameters of II–VI Tetrahedral Compounds
(63.20; S8.11; S8.12; S8.13)
(b) Vol. **138**, No. 1, p. K13—K16 (1986).

S8.15. Mixed crystals

Review Article

K. Lischka
Deep Level Defects in Narrow Gap Semiconductors
(71.55; 72.20; 72.40; S8; S8.15)
(b) Vol. **133**, No. 1, p. 17—46 (1986).

Original Papers and Short Notes

G. Ambrazevičius and G. Babonas
Temperature Dependence of Photoluminescence Excitation Spectra in $Cd_{1-x}Mn_xTe$
(78.55; S8.15)
(b) Vol. **133**, No. 1, p. K45—K48 (1986).

D. Shaw
Erratum to: The Chemical Diffusion of In in $Hg_{0.8}Cd_{0.2}Te$
(66.30; S8.15)
(a) Vol. **93**, No. 1, p. 398 (1986).

I. B. Ermolovich and V. V. Milenin
Nature of Deep Luminescence Centres in CdSe and $CdSe_xTe_{1-x}$
(78.55; 71.70; S8.12; S8.15)
(b) Vol. **133**, No. 2, p. 611—620 (1986).

M. M. El-Zaidia, A. A. Ammar, and A. A. El-Rhmany
Isothermal Crystallization and Kinetics of Amorphous $As_{34.2}Ge_xTe_{65.8-x}$
(64.70; 72.80; S8.15)
(a) Vol. **93**, No. 2, p. K111—K116 (1986).

N. R. Stratieva and E. M. Konstantinova
Structural and Photoelectrical Properties of $Zn_xCd_{1-x}S$ Layers Prepared by the $Zn_xCd_{1-x}S + Cr_2O_3$
(68.55 and 73.60; S8.15)
(a) Vol. **93**, No. 2, p. K191—K193 (1986).

S. I. Radautsan and O. V. Kulikova
Photocapacitance Effect in $Zn_xCd_{1-x}Te$–InSb Structures
(73.40; 72.40; S7.13; S8.15)
(a) Vol. **93**, No. 2, p. K201—K203 (1986).

K. P. Möllmann and M. Happ
Inhomogeneous Electric-Field Quenching of Persistent Photoconductivity in Strongly Indium Doped Lead Tin Telluride
(72.40; S8.15)
(b) Vol. **134**, No. 1, p. K75—K78 (1986).

O. A. Aleksandrova, M. I. Kamchatka, M. S. Miropolskii, and V. V. Passynkov
Diffusion of Native Defects in $(Pb_{1-x}Sn_x)_{1-y}Te_y$ during Liquid Phase Epitaxy
(66.30; 61.70; S8.15)
(a) Vol. **94**, No. 1, p. 139—145 (1986).

T. V. Saunina, D. B. Chesnokova, and D. A. Yaskov
The Equilibrium of Native Defects in the Crystals of $(Pb_{1-x}Sn_x)_{1-y}Te_y$
(61.70; 72.20; S8.15)
(a) Vol. **94**, No. 1, p. 161—170 (1986).

O. V. Romanov and A. M. Yafyasov
Manifestation of Structure and Occupation of Energy Bands in Surface Effects of Semiconductors
(73.20; 73.25; S7.1; S8.15)
(a) Vol. **94**, No. 1, p. 235—242 (1986).

K. Mochizuki and M. Takakusaki
Growth of In-Doped ZnS_xSe_{1-x} Single Crystals and Their Photoluminescence
(78.55; 61.70; 72.80; S8.15)
(a) Vol. **94**, No. 1, p. 243—250 (1986).

Ma Ke-Jun
Photoconductivity in $Mn_xCd_{1-x}Se$
(72.40; S8.15)
(a) Vol. **94**, No. 1, p. K75—K78 (1986).

J. Mycielski and A. M. Witowski
Effect of Chemical and Spin Disorder on Shallow Impurity States in Semiconductor Alloys
(71.55; 71.70; 75.30; S8.15)
(b) Vol. **134**, No. 2, p. 675—689 (1986).

G. Gobsch, J.-P. Zöllner, and G. Paasch
Two-Dimensional Electronic Subband Structure of Grain Boundaries in $Hg_{1-x}Mn_xTe$
(71.25; S8.15)
(b) Vol. **134**, No. 2, p. K149—K153 (1986).

Quian Dingrong
Below-Gap Absorption and Precipitation in n-Type $Hg_{1-x}Cd_xTe$
(78.30; 61.70; 63.20; S8.15)
(a) Vol. **94**, No. 2, p. 573—579 (1986).

N. N. Ablyazov, A. G. Areshkin, V. G. Melekhin, L. G. Suslina, and D. L. Fedorov
Fluctuation-Induced Broadening of Exciton Reflection Spectra in $A^{II}B^{VI}$ Solid Solutions
(71.35 and 78.40; 71.25; S8.15)
(b) Vol. **135**, No. 1, p. 217—225 (1986).

M. C. Desjardins-Deruelle, J. P. Lascaray, D. Coquillat, and R. Triboulet
Magneto-Optical Measurements in $Zn_{1-x}Mn_xTe$ with High Manganese Concentration
(71.35 and 71.70; 75.60; 78.20; S8.15)
(b) Vol. **135**, No. 1, p. 227—232 (1986).

D. GENZOW, M. MOCKER, and E. NORMANTAS
The Influence of Uniaxial Strain on the Band to Band Recombination in $Pb_{1-x}Sn_xTe$
(71.25; 73.60; S8.15)
(b) Vol. **135**, No. 1, p. 261—270 (1986).

S. KUŹMIŃKI and A. T. SZAYNOK
Surface Photovoltage Spectroscopy of $Cd_{0.85}Mn_{0.15}Te$
(73.20 and 73.25; 72.40; S8.15)
(a) Vol. **95**, No. 1, p. 279—282 (1986).

M. K. AGARWAL, P. K. GARG, and L. T. TALELE
Energy Band Location and Redox Analysis of Molybdenum Sulphoselenides (MoS_xSe_{2-x}, $0 \leqq x \leqq 2$) Photoelectrodes
(73.40; S8.15)
(a) Vol. **95**, No. 1, p. 303—307 (1986).

W. HOERSTEL and T. OTTO
Density of Trap Tunneling States in Diffused $PbS_{1-x}Se_x$ p–n Junctions
(73.40; 71.55; S8.15)
(a) Vol. **95**, No. 1, p. 361—366 (1986).

H. M. A. SCHEIJPEN and F. A. P. BLOM
Anisotropy in the Shubnikov-de Haas Effect in $Hg_{1-x}Mn_xSe$
(71.25 and 71.70; 72.20; 75.30; S8.15)
(b) Vol. **135**, No. 2, p. 605—614 (1986).

V. E. MASHCHENKO and A. I. ZIBOROV
Recombination Radiation and Electroreflection of Excitons in $Zn_xCd_{1-x}Se$ Solid Solutions
(71.35; 78.20; 78.55; S8.15)
(b) Vol. **135**, No. 2, p. 623—631 (1986).

W. HELMBRODT and O. GOEDE
Energy Transfer Processes between Te_n Centres in ZnS:Te and CdS:Te
(71.35 and 71.55; 78.55; S8.11; S8.15)
(b) Vol. **135**, No. 2, p. 795—804 (1986).

K. P. GHATAK and M. MONDAL
Effect of Alloy Composition on the Thermoelectric Power in Inversion Layers of Ternary Semiconductors in the Presence of a Classically Large Magnetic Field
(73.40; S8.15)
(b) Vol. **135**, No. 2, p. 819—824 (1986).

KE-JUN MA and W. GIRIAT
Dependence of Energy Gap on Temperature and Composition in $Mn_xCd_{1-x}S$
(71.25; 72.40; 78.20; S8.15)
(a) Vol. **95**, No. 2, p. K135—K138 (1986).

F. EL AKKAD, A. FELIMBAN, and F. SALLAM
Indium Surface Barriers on $Mg_xZn_{1-x}Te$
(73.40; S8.15)
(a) Vol. **95**, No. 2, p. K183—K186 (1986).

S. A. Azimov, T. M. Razykov, V. I. Vyaly, and B. Kh. Kadyrov
Physical Properties of $(ZnSe)_x(CdTe)_{1-x}$ Multicomponent System Films Fabricated by CVD in Hydrogen Flow
(68.55 and 71.25; 73.60; S8.15)
(a) Vol. **96**, No. 1, p. 281—284 (1986).

V. B. Orletskii, V. Z. Yurtsenyuk, and K. D. Tovstyuk
Localized Defect States in $PbSe_xTe_{1-x}$
(73.60; 71.55; S8.15)
(a) Vol. **96**, No. 1, p. K61—K63 (1986).

W. Giriat
Optical Absorption in $Mn_xCd_{1-x}S$
(78.40; S8.15)
(b) Vol. **136**, No. 2, p. K129—K131 (1986).

M. Zvára, P. Hlídek, P. Höschl, V. Koubele, Z. Kučera, V. Prosser, and V. Szöcs
Landau Levels and Phonon Oscillations in Photovoltage Response of $Cd_xHg_{1-x}Te$ Diodes
(72.40; 73.40; S8.15)
(b) Vol. **136**, No. 2, p. K145—K150 (1986).

I. Rückmann, K. Jarašiūnas, and J. Çesnulevičius
Laser-Induced Probe-Beam Defocusing at the Band Edge of CdS_xSe_{1-x} Mixed Crystal at Room Temperature
(78.20; S8.15)
(a) Vol. **96**, No. 2, p. 603—610 (1986).

E. Müller and W. Gebhardt
Position and Lifetime of Photoluminescence in $Cd_{1-x}Mn_xTe$ and $Zn_{1-x}Mn_xTe$. Exchange Dependent Effects
(78.55; S8.15)
(b) Vol. **137**, No. 1, p. 259—267 (1986).

V. D. Vulchev, L. D. Borisova, and S. K. Dimitrova
Preparation and Properties of the $Pb_{1-x}Cr_xTe$ System
(72.80; 64.75; S8.15)
(a) Vol. **97**, No. 1, p. K79—K82 (1986).

M. de Dios Leyva and J. López-Gondar
Electronic Structure of $PbTe–Pb_{1-x}Sn_xTe$ Superlattices
(71.25; S8.15)
(b) Vol. **138**, No. 1, p. 253—257 (1986).

E. Chrzanowski, W. Doliński, S. Mróz, and E. Woźniak
Investigation of $Pb_{1-x}Sn_xTe$ Crystal Surfaces by Quantitative AES, LEED, and Kikuchi Pattern Observation
(68.20; 61.14; 79.20; S8.15)
(a) Vol. **98**, No. 2, p. 359—366 (1986).

S8.16. Higher compounds

Review Article

F. Henneberger
Optical Bistability at the Absorption Edge of Semiconductors
(78.20; 71.35; S7.16; S8.16)
(b) Vol. **137**, No. 2, p. 371—432 (1986).
Erratum
(b) Vol. **138**, No. 2, p. 744 (1986).

Original Papers and Short Notes

S. G. Abdullaeva and N. T. Mamedov
Band Structure of $TlGaSe_2$ Ternary Layered Crystals
(71.25; S8.16)
(b) Vol. **133**, No. 1, p. 171—177 (1986).

M. Colocci, F. Fermi, R. Querzoli, and A. Vinattieri
Time-Resolved Luminescence of $CdGaInS_4$ under Pulsed Laser Excitation at 30 K
(78.55; S8.16)
(b) Vol. **133**, No. 1, p. 379—388 (1986).

S. G. Guseinov, G. D. Guseinov, N. Z. Gasanov, and S. B. Kyazimov
Special Features of Exciton Absorption Spectra of $A^{III}B^{III}X_2^{VI}$-Type Layer-Semiconductor Crystals
(71.35 and 78.40; S8.16)
(b) Vol. **133**, No. 1, p. K25—K30 (1986).

N. N. Belevich, A. I. Lukomskii, L. A. Makovetskaya, and A. P. Chernyakova
Absorption Edge of Stoichiometric and Nonstoichiometric $CuInSe_2$ Crystals
(78.30; 71.55; S8.16)
(b) Vol. **133**, No. 1, p. K49—K52 (1986).

N. N. Belevich and L. A. Makovetskaya
Semiconductor–Metal Transition and Variable Range Hopping Conductivity in $CuInSe_2$
(72.60; 71.30; S8.16)
(b) Vol. **133**, No. 1, p. K53—K56 (1986).

A. Serpi, M. Tapiero, and J. P. Zielinger
Investigation of Photoelectronic Processes in $CdIn_2S_4$ by Photoinduced Current Transient Spectroscopy
(72.40; 71.55; S8.16)
(a) Vol. **93**, No. 1, p. 241—249 (1986).

L. K. Samanta, D. K. Ghosh, and G. C. Bhar
Some Potential Mixed Chalcopyrite Crystals for Nonlinear Laser Devices
(78.20; S8.16)
(a) Vol. **93**, No. 1, p. K51—K54 (1986).

G. Massé
Cation Vacancy Acceptor in $AgGaSe_2$ and Some I–III–VI_2 Compounds
(71.55; 78.55; S8.16)
(a) Vol. **93**, No. 2, p. K177—K179 (1986).

A. Parisini, P. P. Lottici, C. Razzetti, and R. Bacewicz
Resonance Effects in the Raman Spectrum of $CdGa_2Se_4$
(78.30; 63.20; S8.16)
(b) Vol. **134**, No. 1, p. 81—86 (1986).

S. I. Gubarev
Valence Band Splittings in Wurtzite-Type Semimagnetic Semiconductors
(71.25 and 71.70; 78.20; S8.16)
(b) Vol. **134**, No. 1, p. 211—222 (1986).

J. M. Wesselinowa
The s–d Model Applied to $CdCr_2Se_4$ and $HgCr_2Se_4$
(75.10 and 75.50; S8.16)
(b) Vol. **134**, No. 1, p. K39—K42 (1986).

A. N. Georgobiani, V. B. Gutan, I. M. Tiginyanu, V. V. Ursaki, and T. F. Filina
Excitonic Luminescence in $CdGa_2S_4$ Single Crystals
(78.55 and 78.60; 71.35; S8.16)
(b) Vol. **134**, No. 1, p. K47—K49 (1986).

K. K. Mamedov, A. M. Abdullaev, and E. M. Kerimova
Heat Capacities of $TlInS_2$ and $TlInSe_2$ Crystals at Low Temperatures
(65.40; 64.70; S8.16)
(a) Vol. **94**, No. 1, p. 115—119 (1986).

G. D. Guseinov, V. A. Aliev, and E. F. Bagirzade
Recombination Levels in Non-Stoichiometric $TlGaSe_2$
(71.55; 72.40; S8.16)
(a) Vol. **94**, No. 2, p. 653—657 (1986).

K. R. Allakhverdiev, Sh. G. Gasymov, and T. G. Mamedov
Effect of Hydrostatic Pressure on the Electrical Conductivity and Hall Coefficient of $TlGaTe_2$ Crystals
(72.60; S8.16)
(a) Vol. **94**, No. 2, p. K143—K146 (1986).

D. K. Rao, J. J. B. Prasad, D. Sridevi, K. V. Reddy, and J. Sobhanadri
Properties of p-$CuInSe_2$/Al Schottky Devices
(73.30; S8.16)
(a) Vol. **94**, No. 2, p. K153—K158 (1986).

D. Crandles and E. Fortin
The Trap Distribution and the Luminescence Band in $ZnIn_2S_4$
(71.55; 78.55; S8.16)
(a) Vol. **95**, No. 1, p. K47—K51 (1986).

G. D. Guseinov, E. F. Bagirzade, S. N. Mustafaeva, N. Z. Gasanov, and S. G. Guseinov
Ionic and Electric Relaxation of the Conductivity of Intercalated $TlGaSe_2$ Single Crystals
(72.40; 72.80; S8.16)
(a) Vol. **95**, No. 1, p. K71—K74 (1986).

A. I. Dmitriev, G. V. Lashkarev, V. B. Orletskii, and K. D. Tovstyuk
The Properties of the Band Spectrum Parameters in the Band-Inversion Region of Narrow-Gap $Pb_{1-x}Sn_xTe$
(71.25; S8.16)
(b) Vol. **135**, No. 2, p. 587—596 (1986).

S. Marcinkevičius, G. Babonas, B. V. Korzun, and I. G. Tregub
Optical Activity of Semiconductors with Chalcopyrite Structure
(78.20; S7.16; S8.16)
(b) Vol. **135**, No. 2, p. 615—621 (1986).

D. Degenhardt, P. Rabe, and R. Haensel
EXAFS Investigations of TiS_2 and Cu_xTiS_2 Single Crystals ($x = 0.35; 0.4$)
(61.60; 78.70; S8; S8.16)
(a) Vol. **95**, No. 2, p. 439—445 (1986).

J. P. Aicardi and G. Aguero
Excitonic and Edge Luminescence of Silver Thiogallate ($AgGaS_2$)
(78.55 and 78.60; 71.35; 71.55; S8.16)
(a) Vol. **95**, No. 2, p. 679—689 (1986).

K. Kambas, A. Anagnostopoulos, B. Ploss, and J. Spyridelis
On the Optical Absorption Edge in the $CdInGaS_4$ Layered Compound at Low Temperatures
(78.40; S8.16)
(a) Vol. **95**, No. 2, p. K139—K144 (1986).

G. D. Guseinov, S. N. Mustafaeva, R. G. Guseinova, and E. G. Abdullaev
The Nature of Electric Memory in Lithium-Intercalated Chain-Structure $TlInSe_2$ Single Crystals
(72.80; 77.30; S8.16)
(a) Vol. **95**, No. 2, p. K159—K164 (1986).

Ch. I. Abilov
Electronic Conduction in $PbCr_2Te_4$
(72.80; S8.16)
(a) Vol. **95**, No. 2, p. K175—K178 (1986).

N. Frangis, G. van Tendeloo, C. Manolikas, J. van Landuyt, and S. Amelinckx
On the Structure of Layered β-$CuIn_5Se_8$. Electron Diffraction and High Resolution Electron Microscopy Study
(61.55; 64.70; S8.16)
(a) Vol. **96**, No. 1, p. 53—65 (1986).

P. N. Metlinskii, S. I. Radautsan, I. M. Tiginyanu, V. G. Tyrziu, and V. V. Ursaki
Deep Levels in $HgGa_2Se_4$ Single Crystals
(71.55; 72.40; 78.55; S8.16)
(a) Vol. **96**, No. 1, p. 295—299 (1986).

A. N. Y. SAMAAN, S. M. WASIM, A. E. HILL, D. G. ARMOUR, and R. D. TOMLINSON
Preparation and Properties of $CuInS_2$ Thin Films Deposited by RF Sputtering
(68.55 and 73.60; S8.16)
(a) Vol. **96**, No. 1, p. 317—324 (1986).

G. D. GUSEINOV, V. A. ALIEV, and E. F. BAGIRZADE
Impurity Photoconductivity in $TlInSe_2$ Single Crystals
(72.40; S8.16)
(a) Vol. **96**, No. 1, p. K73—K78 (1986).

H. NEUMANN
Bulk Modulus–Volume Relationship in Ternary Chalcopyrite Compounds
(62.20; S8.16)
(a) Vol. **96**, No. 2, p. K121—K125 (1986).

SH. NUROV, V. M. BURLAKOV, E. A. VINOGRADOV, N. M. GASANLY, and B. M. DZHAVADOV
Vibrational Spectra of $TlInS_2$, $TlIn_{0.95}Ga_{0.05}S_2$, and $TlIn(S_{0.8}Se_{0.2})_2$ Crystals in the Vicinity of Phase Transitions
(64.60 and 78.30; S8.16)
(b) Vol. **137**, No. 1, p. 21—32 (1986).

O. P. SINGH and V. P. GUPTA
Comments on Levin's Modifications in the Phillips and Van Vechten Theory. d-Electron Effect on Ionicity
(71.25; 61.50; S8.16; S9; S10.1)
(b) Vol. **137**, No. 1, p. 97—101 (1986).

O. S. RAJORA and A. E. CURZON
Crystal Structure and Staging of Cu_xNbSe_2 and Ag_xNbSe_2
(61.60; S8.16)
(a) Vol. **97**, No. 1, p. 65—76 (1986).

D. P. SINGH, K. D. KUNDRA and K. C. NAGPAL
X-Ray Study of Semiconducting Alloys Ga_2Te_3–In_2Se_3
(61.60; S8.16)
(a) Vol. **97**, No. 1, p. K5—K7 (1986).

O. B. TAGIEV and E. N. IBRAHIMOVA
Optical Properties of $EuGa_2S_4$ near the Fundamental Absorption Edge
(78.40; S8.16)
(a) Vol. **97**, No. 1, p. K49—K52 (1986).

O. UEMURA, A. SHISHIDO, and T. SATOW
The X-Ray Diffraction of Molten AgTlTe
(61.30; S8.16)
(a) Vol. **97**, No. 2, p. K113—K117 (1986).

A. A. Lavrentev, A. N. Gusatinskii, A. V. Soldatov, and V. Yu. Slivka
X-Ray Spectroscopic Investigation of the Electronic Energy Structure and Chemical Bond of the Tl_3AS_4 (A = P, As, Ta) Type Semiconductors
(71.25; 78.70; S8.16)
(b) Vol. **138**, No. 1, p. K49—K54 (1986).

B. Cendlewska, J. Szymaszek, and J. J. Wnuk
Upper Critical Field Deduced from Critical Current Data
(74.60; S8.16)
(a) Vol. **98**, No. 1, p. 305—309 (1986).

L. Artus, Y. Bertrand, C. Ance, and A. Lopez-Soler
Low-Temperature Reflectivity Spectra of $AgGaS_2$ and $AgGaSe_2$ Compounds
(78.20; S8.16)
(b) Vol. **138**, No. 2, p. 633—636 (1986).

E. J. K. B. Banda
Optical Absorption of $FePSe_3$ in the near Infrared, Visible, and near Ultraviolet Regions
(78.30 and 78.40; S8.16)
(b) Vol. **138**, No. 2, p. K125—K129 (1986).

V. K. Gandotra, K. V. Ferdinand, C. Jagadish, A. Kumar, and P. C. Mathur
Effect of Excess Copper on the Electrical Properties of Polycrystalline Thin Films of $CuInSe_2$
(73.60; S8.16)
(a) Vol. **98**, No. 2, p. 595—603 (1986).

I. V. Bodnar and A. I. Lukomskii
The Concentration Dependence of the Band Gap for $CuGa_xIn_{1-x}S_2$ and $AgGa_xIn_{1-x}S_2$ Solid Solutions
(71.25 and 78.40; S8.16)
(a) Vol. **98**, No. 2, p. K165—K169 (1986).

S9. Halides

Review Articles

K. V. S. RAMA RAO and S. RAMAPRABHU
Nuclear Quadrupole Resonance Studies on Inorganic Solids
(76.60; S8; S9; S10; S11)
(a) Vol. **93**, No. 1, p. 17—27 (1986).

J. SHANKER, G. G. AGRAWAL, and N. DUTT
Electronic Polarizabilities and Photoelastic Behaviour of Ionic Crystals
(77.20 and 77.30; 78.20; S8; S9)
(b) Vol. **138**, No. 1, p. 9—30 (1986).

Original Papers and Short Notes

J. P. SPOONHOWER and R. J. DERI
Photocarrier Kinetics in $IrBr_6^{3-}$-Doped AgBr
(72.40; 71.55; S9)
(b) Vol. **133**, No. 1, p. 301—313 (1986).

C. K. JAYASANKAR, E. ANTIC-FIDANCEV, M. LEMAITRE BLAISE, and P. PORCHER
Optical Properties of EuF_3 and GdF_3 and Paramagnetic Susceptibility of EuF_3
(71.70; 75.20; 78.55; S9)
(b) Vol. **133**, No. 1, p. 345—353 (1986).

K. SHAHI, W. WEPPNER, and A. RABENAU
Defects and the First-Order Phase Transition in AgI
(61.70 and 64.70; 66.30; S9)
(a) Vol. **93**, No. 1, p. 171—176 (1986).

J. F. HIGGS, W. Y. ZENG, and A. ANDERSON
Dynamical Model for the Vibrations of the Hydrogen Halide Crystals
(63.20; S9)
(b) Vol. **133**, No. 2, p. 475—482 (1986).

Z. WILAMOWSKI, B. OCZKIEWICZ, P. KACMAN, and J. BLINOWSKI
Asymmetry of the EPR Absorption Line in CdF_2 with Y
(76.30; S9)
(b) Vol. **134**, No. 1, p. 303—306 (1986).

M. PRASAD and S. K. GUPTA
Electrical Conductivity of $CuI_{1.0016}$ and $CuI_{1.0046}$ at Different Pressure
(72.80; S9)
(a) Vol. **94**, No. 1, p. 283—288 (1986).

A. M. SAMUEL, MEERA RAO, R. S. TIWARI, and O. N. SRIVASTAVA
On the Energetics of MX_2 Type Short Period Polytypes with Special Reference to CdI_2 Structures
(61.50; S9)
(a) Vol. **94**, No. 1, p. K5—K10 (1986).

H. B. Gon
Growth of AgI Single Crystals in Silicate Gel and DC Conductivity of AgI
(66.30; 64.70; S9)
(a) Vol. **94**, No. 1, p. K61—K64 (1986).

M. Prasad and S. K. Gupta
Hall Effect Measurement of a $CuI_{1.0046}$ Pellet
(72.80; S9)
(a) Vol. **94**, No. 1, p. K65—K68 (1986).

I. Kosacki, K. Hibner, A. P. Litvinchuk, and M. Ya. Valakh
Lattice Dynamics and Ionic Conductivity of PbF_2:6% LiF Superionic Crystals
(66.30 and 78.30; 61.70; S9)
(b) Vol. **134**, No. 2, p. 495—501 (1986).

A. A. Kaminskii, A. A. Kornienko, and M. I. Chertanov
Parametrization of Electric-Dipole Intensities in f^N Systems Due to Electron-Correlation Effects
(71.70; 78.50; S9; S11)
(b) Vol. **134**, No. 2, p. 717—729 (1986).

V. E. Mashchenko, V. F. Kharsik, and S. V. Brezhneva
Secondary Emission of Excitons in CuCl Polycrystals
(71.35; 78.55; S9)
(b) Vol. **135**, No. 1, p. 201—206 (1986).

B. Hönerlage, F. Tomasini, J. Y. Bigot, M. Frindi, and J. B. Grun
Nonlinear Optical Effects Studied by Time-Resolved Reflection Measurements
(71.35 and 78.40; S9)
(b) Vol. **135**, No. 1, p. 271—282 (1986).

R. A. Collins, S. C. Edwards, F. J. Johnson, A. G. Jones, and L. D. McMahon
Time-Dependent Reverse Negative Differential Resistance in Thin-Film MIM Devices
(73.40; S9; S10.1)
(a) Vol. **95**, No. 1, p. 353—359 (1986).

M. T. Sprackling and M. E. Bahrololoom
Strain-Ageing in Silver Iodobromide
(61.70; 62.20; S9)
(a) Vol. **96**, No. 1, p. 217—224 (1986).

N. V. Unnikrishnan and R. D. Singh
Temperature Dependence of Lux–Ampere Characteristics of CdI_2 Single Crystals
(72.40; S9)
(a) Vol. **96**, No. 2, p. K219—K222 (1986).

K. Srinivasan and S. Lakshmi Raghavan
Strain Derivatives of Dielectric Constants of MgO and Thallous Halides
(77.20; 62.20; S9; S10.1)
(b) Vol. **137**, No. 1, p. 33—38 (1986).

O. P. SINGH and V. P. GUPTA
Comments on Levin's Modifications in the Phillips and Van Vechten Theory. d-Electron Effect on Ionicity
(71.25; 61.50; S8.16; S9; S10.1)
(b) Vol. **137**, No. 1, p. 97—101 (1986).

L. WENDLER and R. HAUPT
Damping of Polaritons in Bilayer Systems. Long-Range Surface Phonon-Polaritons
(68.30 and 71.36; S7.11; S7.12; S9; S9.12)
(b) Vol. **137**, No. 1, p. 269—290 (1986).

M. FRINDI, B. HÖNERLAGE, and R. LEVY
Time-Resolved Degenerate Four-Wave Mixing in CuCl under Nanosecond Pulsed Excitation
(78.20; 71.35; S9)
(b) Vol. **138**, No. 1, p. 267—274 (1986).

M. SOUIRI, A. GOLTZENE, and C. SCHWAB
Correlations between Electron Paramagnetic Resonance and Bonding Parameters. Evidence of a Lattice Relaxation
(71.45 and 71.70; 76.30; S7; S8; S9; S10)
(b) Vol. **138**, No. 1, p. 321—330 (1986).

A. KESSLER and U. KLEIN
Spontaneous and Field Induced Current Peaks at the Order–Disorder Phase Transition of NH_4Cl
(77.80; S9)
(a) Vol. **98**, No. 1, p. K93—K98 (1986).

J. Y. BIGOT, M. FRINDI, F. FIDORRA, R. LEVY, and B. HÖNERLAGE
Nonlinear Propagation of Nanosecond Laser Pulses through Polycrystalline CuCl Layers
(78.20; 71.35; S9)
(b) Vol. **138**, No. 2, p. 621—631 (1986).

A. KESSLER and U. KLEIN
Conductivity and Polarization of NH_4Cl between 150 and 500 K
(66.90 and 77.30; 61.70; S9)
(a) Vol. **98**, No. 2, p. 617—627 (1986).

S9.11. Alkali halides

Review Article

J. Shanker and W. N. Bhende
Higher-Order Elastic Constants and Thermoelastic Properties of Ionic Solids
(62.20 and 63.90; 64.30; 65.70; S7.1; S9.11; S9.12; S10.1)
(b) Vol. **136**, No. 1, p. 11—30 (1986).

Original Papers and Short Notes

M. Kumari and N. Dass
On the Pressure Dependence of Grüneisen Parameter in Solids
(63.90; 65.50; S2; S9.11; S9.12; S10.1)
(b) Vol. **133**, No. 1, p. 101—110 (1986).

L. A. Bugaev, I. I. Gegusin, V. N. Datsyuk, A. A. Novakovich, and R. V. Vedrinskii
Multiple Scattering Approach to the XANES Theory of Alkali Halide Crystals. I. Crystalline Potential in the X-Ray Absorption Spectra Problem
(71.10; 78.70; S9.11)
(b) Vol. **133**, No. 1, p. 195—202 (1986).

T. Yanagihara and H. Yamaguchi
Potential Energies of a Silver Atom Adsorbed on Flat and Stepped Surfaces of Sodium Chloride Crystals
(68.40; S9.11)
(a) Vol. **93**, No. 1, p. 67—70 (1986).

R. Capelletti, R. Fava, M. Manfredi, and M. Solzi
Decay Scheme of 439 and 478 nm Emission Bands in the KCl:Eu^{2+} System
(78.55; 71.55; 78.40; S9.11)
(a) Vol. **93**, No. 1, p. 263—269 (1986).

A. Dhoble and M. P. Verma
Thermodynamic Analysis of the Anderson-Grüneisen Parameters
(65.50; 63.90; S9.11)
(b) Vol. **133**, No. 2, p. 491—497 (1986).

Yu. M. Annenkov, S. B. Nurmagambetov, and A. B. Chormonov
Nature of the Centres of Positron Annihilation in Intensively Irradiated Alkali Halides
(61.70 and 78.70; S9.11)
(b) Vol. **133**, No. 2, p. 505—509 (1986).

K. D. Nierzewski, B. Macalik, and H. Opyrchał
Effect of Low Temperature γ-Irradiation on I–V Dipoles in KCl:Eu^{2+} Crystals
(61.70 and 61.80; 78.50; S9.11)
(b) Vol. **133**, No. 2, p. K91—K93 (1986).

M. KAMADA and K. TSUTSUMI
Energy Spectra of Thermally Stimulated Exoelectrons from NaCl and NaCl:Cu X-Irradiated at 77 K
(79.75; S9.11)
(a) Vol. **94**, No. 1, p. 275—282 (1986).

R. C. BHUNIYA and S. E. RAO
Colour Centres and Luminescence in KCl:Ni Crystals
(61.70 and 78.60; 78.50; S9.11)
(b) Vol. **134**, No. 2, p. 561—568 (1986).

I. I. GEGUSIN, V. N. DATSYUK, A. A. NOVAKOVICH, L. A. BUGAEV, and R. V. VEDRINSKII
Multiple Scattering Approach to the XANES Theory of Alkali Halide Crystals. III. Comparison with the Experimental X-Ray Spectra
(71.10; 78.70; S9.11)
(b) Vol. **134**, No. 2, p. 641—650 (1986).

E. I. DMITRIEVA, N. N. KRISTOFFEL, I. A. PARFIANOVICH, L. G. POPOV, and V. N. SALOMATOV
The $F_H(F^-)$ Center Energy Structure in KCl
(71.55; S9.11)
(b) Vol. **134**, No. 2, p. K125—K128 (1986).

G. PRASAD and K. V. RAO
F-Band Absorption and Thermoluminescence of X-Ray Irradiated KCl Single Crystals Excited with Laser Light
(78.50 and 78.60; S9.11)
(a) Vol. **94**, No. 2, p. 659—663 (1986).

A. SHIROMANY, G. M. S. SRIVASTAVA, and J. SHANKER
Generalized Huggins-Mayer Potentials and Properties of Alkali Halides under the Effect of High Pressure
(64.30 and 77.20; 65.50; S9.11)
(b) Vol. **135**, No. 1, p. 119—126 (1986).

U. PURI and V. KUMAR
Pressure-Induced Polymorphic Phase Transitions of Potassium and Rubidium Halides
(64.60; 62.20; S9.11)
(b) Vol. **135**, No. 1, p. 127—131 (1986).

H. OPYRCHAŁ, K. D. NIERZEWSKI, B. MACALIK, and M. MLADENOVA
γ-Ray Induced Coloration of $KCl:Eu^{2+}$ Crystals
(61.70; 78.50; 78.60; S9.11)
(b) Vol. **135**, No. 1, p. 141—148 (1986).

E. RZEPKA, S. LEFRANT, and L. TAUREL
Optical Study of the Defect Stability in Tl-Doped Potassium Iodide
(61.70; 78.30; 78.50; 78.60; S9.11)
(b) Vol. **135**, No. 1, p. 149—154 (1986).

A. DHOBLE and M. P. VERMA
On the Temperature Variation of Anderson-Grüneisen Parameters of NaCl
(65.50 and 65.70; S9.11)
(b) Vol. **136**, No. 2, p. 497—504 (1986).

P. G. BARANOV, V. A. VETROV, N. G. ROMANOV, and V. TOPA
ODMR in Triplet Excited States of ns^2 Ions in Alkali Halide Crystals
(71.55 and 76.70; 78.55; S9.11)
(b) Vol. **136**, No. 2, p. 699—707 (1986).

E. RADZHABOV
Optical Transitions of Chalcogen-Vacancy Centers in Ionic Crystals
(71.55; 78.50; 78.55; S9.11)
(b) Vol. **136**, No. 2, p. K139—K143 (1986).

U. SEIPOLD and H. A. ULLNER
Martensitic Character of the Pressure Induced B1–B2 Phase Transition of Potassium Bromide
(64.70; 66.70; S9.11)
(a) Vol. **96**, No. 2, p. 469—474 (1986).

M. R. DALMAU, A. PAZ, and J. A. GORRI
The Electric Charge of Dislocations in Alkali Halides
(61.70; S9.11)
(a) Vol. **96**, No. 2, p. 533—540 (1986).

R. B. PODE
Formation of the Z_1 Colour Centre by Mechanical Bleaching in γ-Irradiated KCl:Ca
(61.70; 78.55; S9.11)
(a) Vol. **96**, No. 2, p. K147—K149 (1986).

L. N. KANTOROVICH
Multipole Theory of the Polarization of Solids by Point Defects. III. Transition Energy Calculations, the Approximation of Non-Point Charges
(71.55 and 77.30; S9.11)
(b) Vol. **137**, No. 1, p. 229—240 (1986).

A. MEMON and D. B. TANNER
Long Wavelength Optical Phonons in Mixed Alkali Halide Powder Crystals
(63.20 and 78.30; S9.11)
(b) Vol. **137**, No. 1, p. K9—K13 (1986).

M. M. KIDIBAEV, A. A. ALYBAKOV, and V. A. GUBANOVA
Absorption Bands of U^{4+} and U^{3+} Ions in X-Irradiated LiF:U, Me Crystals
(78.50; 61.80; S9.11)
(b) Vol. **137**, No. 1, p. K73—K75 (1986).

A. AUSMEES, M. ELANGO, A. KIKAS, and J. PRUULMANN
Monte Carlo Simulation of Electron–Phonon Scattering in the XUV-Induced Electron Emission of NaCl
(71.38 and 79.60; S9.11)
(b) Vol. **137**, No. 2, p. 495—500 (1986).

S9.12. Alkaline earth halides

Review Article

J. SHANKER and W. N. BHENDE
Higher-Order Elastic Constants and Thermoelastic Properties of Ionic Solids
(62.20 and 63.90; 64.30; 65.70; S7.1; S9.11; S9.12; S10.1)
(b) Vol. **136**, No. 1, p. 11—30 (1986).

Original Papers and Short Notes

M. KUMARI and N. DASS
On the Pressure Dependence of Grüneisen Parameter Solids
(63.90; 65.50; S2; S9.11; S9.12; S10.1)
(b) Vol. **133**, No. 1, p. 101—110 (1986).

J. KURIATA, N. GUSKOS, and P. PASTUSIAK
Dynamic Effects in the Temperature Dependence of the "Intrinsic" $\bar{b}_2$ Parameter for $MeF_2:Gd^{3+}$
(76.30; S9.12)
(b) Vol. **133**, No. 2, p. K133—136 (1986).

R. ALCALÁ, R. CASES, and P. J. ALONSO
EPR of Nickel Centers in $SrCl_2:Ni$
(76.30; S9.12)
(b) Vol. **135**, No. 1, p. K63—K67 (1986).

W. PASTUSIAK and J. KURIATA
On the Microscopic Interpretation of Tetragonal Second-Order Spin–Lattice Coefficients of $4f^7$ Ions in the Dynamic Relativistic Crystal Field Approach
(71.70; S9.12)
(b) Vol. **135**, No. 1, p. K69—K73 (1986).

E. RADZHABOV and P. FIGURA
Optical Properties of Oxygen-Vacancy Centers in Fluorite
(78.40 and 78.50; S9.12)
(b) Vol. **136**, No. 1, p. K55—K59 (1986).

D. V. KAZAKOVTSEV and Y. B. LEVINSON
The Effect of Phonon–Phonon Scattering in the Substrate on Temperature Dynamics of a Phonon Film Injector
(63.20; 66.70; S5.11; S5.12; S7.12; S9.12)
(b) Vol. **136**, No. 2, p. 425—434 (1986).

G. DAS, H. C. PADHI, and O. AIKALA
Directional Compton Profiles of CaF_2
(78.70 S9.12)
(b) Vol. **136**, No. 2, p. 623—627 (1986).

A. Hamaïdia, A. Hachimi, J. Margerie, and J. F. Hémidy
EPR of Colour Centres in Alkaline Metal Doped Calcium and Strontium Fluorides
(61.70; 76.30; 78.50; S9.12)
(b) Vol. **137**, No. 1, p. 47—56 (1986).

L. Wendler and R. Haupt
Damping of Polaritons in Bilayer Systems. Long-Range Surface Phonon-Polaritons
(68.30 and 71.36; S7.11; S7.12; S9; S9.12)
(b) Vol. **137**, No. 1, p. 269—290 (1986).

N. Dutt, A. J. Kaur, and J. Shanker
Effective Potentials and Crystalline State Properties of Some Fluorite-Structure Crystals
(61.50 and 64.30; 62.20; 63.90; 65.50; 77.20; S9.12)
(b) Vol. **137**, No. 2, p. 459—468 (1986).

Uma Subramanian and M. L. Mukherjee
Thermoluminescence Study of CaF_2:Nd Single Crystals from LNT to 400 °C
(78.60; 71.55; S9.12)
(b) Vol. **137**, No. 2, p. 675—681 (1986).

A. Batana and M. R. Soriano
Low Temperature Thermal Expansion of Cubic Ionic Crystals
(65.70; S9.11; S9.12)
(a) Vol. **98**, No. 1, p. K15—K18 (1986).

S9.15. Mixed crystals

Original Papers and Short Notes

R. FLAGMEYER and B. SCHUMANN
Rutherford Backscattering/Channeling Analyses of Lattice-Matched $Ca_{0.43}Sr_{0.57}F_2$ Films on GaAs
(68.55; 61.80; 68.48; S9.15)
(a) Vol. **93**, No. 2, p. K117—K121 (1986).

A. K. GIRI and G. B. MITRA
Theoretical Prediction of Elastic Constants of Mixed Systems
(62.20; S7.15; S9.15)
(b) Vol. **134**, No. 1, p. K11—K16 (1986).

G. L. SHARMA, S. D. SHARMA, and V. K. AGRAWAL
Polytypism in Mixed Crystals of $CdBr_{2-x}I_x$
(61.60; S9.15)
(a) Vol. **95**, No. 1, p. 57—61 (1986).

M. SIESKIND, M. AYADI, and G. ZACHMANN
Infrared Lattice Vibration, Dielectric Dispersion, and Lattice Dynamics of BaFCl
(63.20; and 78.30; S9.15)
(b) Vol. **136**, No. 2, p. 489—495 (1986).

J. SHANKER and S. C. SHARMA
Evaluation on Electronic Dielectric Constant and Fractional Ionic Character of Mixed Alkali Halide Crystals
(64.70 and 77.20; S9.15)
(b) Vol. **136**, No. 2, p. 505—511 (1986).

K. INABE and N. TAKEUCHI
On the X-Ray Production of F-Centers in BaFCl Doped with Mono- or Trivalent Impurities
(61.70; 78.50; S9.15)
(b) Vol. **136**, No. 2, p. K97—K99 (1986).

A. L. GOLOVIN and U. PIETSCH
X-Ray Investigations of Submicrometer Layer Heterostructures
(61.10; 68.55; S7.12; S9.15)
(a) Vol. **96**, No. 2, p. K111—K115 (1986).

S9.16. Higher compounds

Original Papers and Short Notes

T. NATTERMANN
Dielectric Tails in K_2ZnCl_4: No Evidence for an Intrinsic Chaotic State
(77.20; 71.45; S9.16)
(b) Vol. **133**, No. 1, p. 65—70 (1986).

G. SORGE, H. MAACK, and L. A. SHUVALOV
Low-Frequency Dielectric Properties of Rb_2ZnCl_4
(77.80; 77.20; S9.16)
(a) Vol. **93**, No. 1, p. 315—320 (1986).

M. HIDAKA, H. FUJII, B. J. GARRARD, and B. M. WANKLYN
Structural Phase Transitions in the Layer Compound $CsFeF_4$
(64.70; S9.16)
(a) Vol. **95**, No. 2, p. 413—421 (1986).

P. VANĚK, B. BŘEZINA, M. HAVRÁNKOVÁ, and J. BIROŠ
Room-Temperature Phase Transition in Rb_2CoCl_4. Heat Capacity Measurements
(64.70; 65.40; S9.16)
(a) Vol. **95**, No. 2, p. K101—K103 (1986).

B. T. DESHMUKH, S. V. BODADAE, and S. V. MOHARIL
Thermoluminescence in K_2ZnCl_4
(78.60; S9.16)
(a) Vol. **95**, No. 2, p. K145—K148 (1986).

M. V. RAMANA REDDY and K. NARASIMHA REDDY
Characterization and Electrical Behaviour of $NaYF_4$ Thin Films
(73.60; S9.16)
(a) Vol. **95**, No. 2, p. K193—K198 (1986).

A. BRILINGAS, R. L. DAVIDOVICH, J. GRIGAS, S. LAPINSKAS, M. A. MEDKOV, V. SAMULIONIS, and V. SKRITSKII
Relaxational Dynamics of the Improper Ferroelectric $K_3BiCl_6 \cdot 2\,KCl \cdot KH_3F_4$
(77.80; 62.65; S9.16)
(a) Vol. **96**, No. 1, p. 101—109 (1986).

TH. GRIEB, K. WACKER, E. NAUMANN, W. TREUTMANN, G. HEGER, and R. GEICK
Magnetic Properties of $Rb_2Mn_xCr_{1-x}Cl_4$ Mixed Crystals
(75.25 and 75.60; 75.30; S9.16)
(b) Vol. **136**, No. 2, p. 585—591 (1986).

P. STUDZINSKI and J.-M. SPAETH
ENDOR Investigation of Cr^{3+} Centres in the Tetragonal Phase of $RbCdF_3$
(76.70; 64.70; 71.70; S9.16)
(b) Vol. **136**, No. 2, p. 735—742 (1986).

M. Hidaka, H. Fujii, B. J. Garrard, and B. M. Wanklyn
Structural Phase Transitions in the Layer Compound $CsVF_4$
(64.70; S9.16)
(a) Vol. **96**, No. 2, p. 415—423 (1986).

H. Grimm and J. Slak
Observation of Gradual Localisation of Protons in the Mixed System $Rb_{1-x}(NH_4)_xH_2(P, As)O_4$ by Incoherent Neutron Scattering
(75.25; 77.80; S9.16)
(b) Vol. **137**, No. 1, p. K37—K41 (1986).

J. Barnaś
Non-Reciprocal Character of Wave Propagation in Magnetic Systems with Spontaneous Electric Polarization
(71.36 and 75.30; 78.20; S9.16)
(b) Vol. **137**, No. 1, p. K43—K47 (1986).

A. A. Kaminskii
Stimulated Emission Spectroscopy of Ln^{3+} Ions in Tetragonal $LiLuF_4$
(78.45; 71.70; 78.55; S9.16)
(a) Vol. **97**, No. 1, p. K53—K58 (1986).

E. V. Kholopov, A. M. Panich, N. K. Moroz, and Yu. G. Kriger
One-Dimensional Mechanism of Antiferroelectric Ordering in a Non-Polar Crystal
(64.70 and 77.80; 65.40; 76.30; S9.16)
(b) Vol. **137**, No. 2, p. K121—K125 (1986).

L. Misiak, P. Mikołajczak, and M. Subotowicz
Lithium Rare-Earth Fluoride Crystal Growth and Thermal Variation of Lattice Constants
(61.50; 65.70; S9.16)
(a) Vol. **97**, No. 2, p. 353—359 (1986).

M. Hidaka, H. Akiyama, and B. M. Wanklyn
Structural Phase Transitions in the Layer Compound $RbFeF_4$
(64.70; S9.16)
(a) Vol. **97**, No. 2, p. 387—395 (1986).

A. A. Kaminskii and S. E. Sarkisov
Stimulated-Emission Spectroscopy of Pr^{3+} Ions in Monoclinic BaY_2F_8 Fluoride
(71.55; 78.45; 78.55; S9.16)
(a) Vol. **97**, No. 2, p. K163—K168 (1986).

A. Martin, F. J. López, E. Diéguez, F. Agulló-López, and H. Arend
Luminescence Properties of Rb_2ZnCl_4 and K_2ZnCl_4 Single Crystals
(78.55 and 78.60; 71.35; S9.16)
(b) Vol. **138**, No. 1, p. 245—252 (1986).

S10. Oxides

Review Article

K. V. S. Rama Rao and S. Ramaprabhu
Nuclear Quadrupole Resonance Studies on Inorganic Solids
(76.60; S8; S9; S10; S11)
(a) Vol. **93**, No. 1, p. 17—27 (1986).

Original Papers and Short Notes

A. K. Petrosyan and A. A. Mirzakhanyan
Zero-Field Splitting and *g*-Values of d^8 Ions in a Trigonal Crystal Field
(71.70; S10; S10.1)
(b) Vol. **133**, No. 1, p. 315—319 (1986).

V. E. Borisenko and V. P. Parkhutik
RBS Study of Transient Thermal and Anodic Oxidation of Tantalum Films
(68.55; S10)
(a) Vol. **93**, No. 1, p. 123—130 (1986).

H. Lass, J. Pebler, M. Pekała, and A. Polaczek
Magnetic and Mössbauer Effect Study of the Cluster Distribution Function in the Spin-Glass Phase θ-$Me_{0.35}V_2O_5$ (Me = Fe, Al)
(75.30; 76.80; S10)
(a) Vol. **93**, No. 1, p. 233—240 (1986).

M. Ben Salem, B. Yangui, G. Schiffmacher, and C. Boulesteix
Erratum to: Twinning of the Hexagonal (A) Structure of Rare Earth Sesquioxides
(61.60; S10)
(a) Vol. **93**, No. 1, p. 397 (1986).

Y. Shibata, T. Suemoto, and M. Ishigame
Raman Scattering Studies of Mobile Ions in Superionic Conductor Hollandites
(78.30; S10)
(b) Vol. **134**, No. 1, p. 71—79 (1986).

J. Roset, A. Fernández, X. Obradors, and J. Tejada
Mössbauer Emission Studies of ^{57}Co-Doped ZnO, Cu_2O, and $TiCoO_3$
(76.80; S10; S10.15)
(b) Vol. **134**, No. 1, p. 297—302 (1986).

F. A. Benko and F. P. Koffyberg
The Optical Interband Transitions of the Semiconductor $CuGaO_2$
(71.25; 72.40; S10)
(a) Vol. **94**, No. 1, p. 231—234 (1986).

B. C. Grabmaier and R. Oberschmid
Properties of Pure and Doped $Bi_{12}GeO_{20}$ and $Bi_{12}SiO_{20}$ Crystals
(61.70 and 71.55; 72.40; 78.50; 78.55; S10)
(a) Vol. **96**, No. 1, p. 199—210 (1986).

S10.1. Simple oxides

Review Article

J. SHANKER and W. N. BHENDE
Higher-Order Elastic Constants and Thermoelastic Properties of Ionic Solids
(62.20 and 63.90; 64.30; 65.70; S7.1; S9.11; S9.12; S10.1)
(b) Vol. **136**, No. 1, p. 11—30 (1986).

Original Papers and Short Notes

M. KUMARI and N. DASS
On the Pressure Dependence of Grüneisen Parameter in Solids
(63.90; 65.50; S2; S9.11; S9.12; S10.1)
(b) Vol. **133**, No. 1, p. 101—110 (1986).

A. K. PETROSYAN and A. A. MIRZAKHANYAN
Zero-Field Splitting and *g*-Values of d^8 Ions in a Trigonal Crystal Field
(71.70; S10; S10.1)
(b) Vol. **133**, No. 1, p. 315—319 (1986).

H.-J. FITTING and A. VON CZARNOWSKI
Electron Breakdown in Polar Insulating and Semiconducting Layers
(73.60; 71.38; 79.70; S8.11; S10.1)
(a) Vol. **93**, No. 1, p. 385—396 (1986).

G. BOTTONI, D. CANDOLFO, A. CECCHETTI, F. MASOLI, E. MELLO CERESA, and A. PAPOIAN
Magnetic Parameters and Crystal Lattice Modifications of Doped CrO_2 Particles Subjected to Thermal Treatment
(61.70 and 75.50; 61.10; S10.1)
(a) Vol. **93**, No. 1, p. K41—K45 (1986).

A. A. AKHMEDOV, S. KH. KHALILOV, A. G. KYAZYMZADE, and YA. A. BAIRAMOV
Prolonged Relaxation and Residual Conductivity in Sn_2O–GaSe Heterotransitions
(73.40; S8; S10.1)
(a) Vol. **93**, No. 1, p. K79—K83 (1986).

S. V. VLASOV and O. V. FARBEROVICH
APW Research of Europium Monochalcogenides. II. Magnetic Form Factor and Insulator–Metal Transition in EuO under Pressure
(71.10; 71.25; 71.30; 75.25; S10.1)
(b) Vol. **133**, No. 2, p. 601—609 (1986).

V. A. RADTSIG and A. A. BOBYSHEV
Twofold Coordinated Si and Ge Atoms and "Hydrogen" Paramagnetic Centers in Amorphous SiO_2 and GeO_2
(71.55; 61.70; 76.30; S10.1)
(b) Vol. **133**, No. 2, p. 621—627 (1986).

R. Butz and H. Wagner
Reactions at the Ti/SiO_2 Interface Studied by EELS
(68.48; 68.30; 61.80; S1.2; S10.1)
(a) Vol. **94**, No. 1, p. 71—76 (1986).

K. N. Jog, S. P. Sanyal, and R. K. Singh
Effect of Three-Body Interactions on the High-Pressure Behaviour of MgO and CaO Crystals
(62.20 and 64.70; S10.1)
(a) Vol. **94**, No. 1, p. 107—114 (1986).

F. Berg, H. Pagnia, and N. Sotnik
Influence of Ion Bombardement on the Electrical Behaviour of Discontinuous Metal Film Diodes
(73.40; 61.80; S10.1)
(a) Vol. **94**, No. 1, p. K89—K92 (1986).

C. Boulesteix, Zenchuan Kang, and M. Lottiaux
HREM Study of Crystalline and Amorphous Regions of a TiO_2 Thin Layer. Modification of the Lattice Parameter inside Small Clusters
(68.55; 61.16; S10.1)
(a) Vol. **94**, No. 2, p. 499—506 (1986).

I. P. Mikhailovskii, P. B. Potapov, and A. E. Epov
Sign of the Charge Accumulated in Thermal SiO_2 Films of Silicon MIS Structures under High Electric Field Condition
(77.55; S10.1)
(a) Vol. **94**, No. 2, p. 679—685 (1986).

M. Matlak and K. Skrobiś
Curie Temperature of a Doped Ferromagnetic Semiconductor
(75.10 and 75.50; S8; S10.1)
(b) Vol. **135**, No. 1, p. 155—161 (1986).

M. Mertig, G. Pompe, and E. Hegenbarth
Specific Heat of Nb-Doped RuO_2 Single Crystals
(65.40 and 71.20; S10.1)
(b) Vol. **135**, No. 1, p. 335—342 (1986).

V. O. Sokolov and V. B. Sulimov
Semiempirical Calculation of Oxygen Vacancy in Vitreous Silica
(71.55; S10.1)
(b) Vol. **135**, No. 1, p. 369—377 (1986).

J. Fontcuberta, J. A. Cusidó, and J. Tejada
On the Electric Field Effect in the Mössbauer Emission Studies
(76.80; 71.55; S10.1)
(b) Vol. **135**, No. 1, p. K27—K31 (1986).

R. A. Collins, S. C. Edwards, F. J. Johnson, A. G. Jones, and L. D. McMahon
Time-Dependent Reverse Negative Differential Resistance in Thin-Film MIM Devices
(73.40; S9; S10.1)
(a) Vol. **95**, No. 1, p. 353—359 (1986).

P. W. Deutsch and E. S. Banas
Computational Simulation of Ice Ih X-Ray Powder Spectrum at 20 K
(61.10; S10.1)
(b) Vol. **136**, No. 1, p. K1—K3 (1986).

V. P. Parkhutik, V. A. Labunov, and G. G. Chigir
Kinetics and Mechanism of Transient Thermal Oxidation of Silicon
(68.55; 73.20; 73.40; 79.60; S5.11; S10.1)
(a) Vol. **96**, No. 1, p. 11—18 (1986).

N. V. Rumak and V. V. Khatko
Stimulated Crystallization of Thermal Silica Films. II. Treatment with Metal Particles and/or Inert Gas Ions
(68.55; 68.48; S10.1)
(a) Vol. **96**, No. 1, p. 19—24 (1986).

L. N. Skuja and W. Entzian
Cathodoluminescence of Intrinsic Defects in Glassy SiO_2, Thermal SiO_2 Films, and α-Quartz
(78.60; S10.1)
(a) Vol. **96**, No. 1, p. 191—198 (1986).

S. Alexandrova and A. Szekeres
Low-Temperature RF Plasma Treatment of Si–SiO_2 Structures as a Substitution for High-Temperature Anneals
(73.40; S5.11; S10.1)
(a) Vol. **96**, No. 1, p. 363—366 (1986).

O. A. Nabutovskaya, E. V. Kholopov, and I. E. Paukov
Nonpower-Law Behaviour in the Heat Capacity of CoO in the Vicinity of Strong Fluctuations
(75.40; S10.1)
(a) Vol. **96**, No. 1, p. K57—K60 (1986).

A. K. M. A. Islam, K. Islam, M. N. Islam, and G. Mortuza
Polarizabilities, Energies, and Pressure Derivatives of Bulk Moduli of Some Rutile-Type Oxide Crystals
(61.50 and 62.20; S10.1)
(b) Vol. **136**, No. 2, p. 443—451 (1986).

P. Van den Keybus and W. Grevendonk
Comparison of Optical Activity and Faraday Rotation in Crystalline SiO_2
(78.20; S10.1)
(b) Vol. **136**, No. 2, p. 651—659 (1986).

R. B. Bossoli and L. E. Halliburton
^{27}Al Hyperfine and Quadrupole Interactions for the $[AlO_4]^\circ$ Center in Quartz
(71.70 and 76.70; 61.70; S10.1)
(b) Vol. **136**, No. 2, p. 709—714 (1986).

V. Bethkenhagen and K. Hübner
Cluster-Bethe-Lattice Approach to the Electronic Structure of Intrinsic Defects in v-SiO_2
(71.55; 71.20; 78.50; 78.55; S10.1)
(b) Vol. **136**, No. 2, p. 721—734 (1986).

U. Hoppe
Correction of the Scattering Cross-Section of Neutron Diffraction Measurements for Inelastic Scattering in the Case of Amorphous Silica
(61.12; S10.1)
(b) Vol. **137**, No. 1, p. K5—K8 (1986).

M. Georgiev and A. Manov
Non-Radiative Charge Transfer Rate at the E_1' Center in α-Quartz
(71.55; S10.1)
(b) Vol. **137**, No. 1, p. K83—K87 (1986).

I. M. Chapnik
On the Relation between the Energy Gap Data of Semiconducting Compounds and the Position of the Components of the Compounds in the Periodic Table
(72.80; S5; S6; S7; S8; S10.1)
(b) Vol. **137**, No. 1, p. K95—K100 (1986).

A. Scharmann and G. Schmirler
Influence of Pyroelectric Properties on the Thermally Stimulated Exoelectron Emission (TSEE) of BeO
(77.70 and 79.75; S10.1)
(a) Vol. **97**, No. 1, p. 265—274 (1986).

M. Ben Salem, B. Yangui, G. Schiffmacher, and C. Boulesteix
Erratum to: Twinning of the Hexagonal (A) Structure of Rare Earth Sesquioxides
(61.70; S10.1)
(a) Vol. **97**, No. 1, p. 331 (1986).

N. N. Koren, S. P. Zhukov, and V. V. Kindyak
On the Layer Structure of Piezoelectric ZnO Films
(68.55; S10.1)
(a) Vol. **97**, No. 1, p. K1—K3 (1986).

H. Haberlandt
Determination of Titanium CNDO Parameters and Application in Cluster Calculations of Different Titanium Compounds
(71.25; S1.2; S10.1)
(b) Vol. **137**, No. 2, p. 581—589 (1986).

V. O. Sokolov and V. B. Sulimov
Semiempirical Calculation of Fluorine Substitutional Impurity in Vitreous Silicon Dioxide
(61.70 and 71.55; S10.1)
(b) Vol. **137**, No. 2, p. K167—K170 (1986).

P. Yang, S. S. Jiang, H. N. Zhou, and D. Feng
The X-Ray Dynamical Diffraction Contrast Images of Brazil Twin Boundaries in Natural Quartz
(61.10 and 61.70; S10.1)
(a) Vol. **97**, No. 2, p. 411—420 (1986).

N. Amin and S. Arajs
Magnetic Behaviour of Nearly Spherical Fe_3O_4 Particles in Low Magnetic Fields at Elevated Temperatures
(75.60; S10.1)
(a) Vol. **97**, No. 2, p. 479—484 (1986).

P. Bonasewicz, W. Hirschwald, and G. Neumann
The Investigation of the Pressure and Temperature Dependence of the Electrical Conductivity of Thin Zinc Oxide Films with High Resistances
(73.60; S10.1)
(a) Vol. **97**, No. 2, p. 593—599 (1986).

S. García and E. Altshuler
Reactivity Study of Ferric Oxides from Different Origins
(61.50; S10.1)
(a) Vol. **97**, No. 2, p. K119—K121 (1986).

J. Llopis, J. Piqueras, J. A. García, and A. Remón
Red Cathodoluminescence from Deformed MgO Crystals
(78.60; S10.1)
(a) Vol. **97**, No. 2, p. K159—K162 (1986).

J. Hugel and C. Carabatos
Band Structure and Optical Properties of CoO
(71.25; 78.20; S10.1)
(b) Vol. **138**, No. 1, p. K43—K47 (1986).

M. Sanchez, A. Dominguez-Rodriguez, R. Marquez, and J. Castaing
Creep of CoO Single Crystals. Substructure and Stress Distribution
(62.20; S10.1)
(a) Vol. **98**, No. 1, p. 203—208 (1986).

V. M. Efimov, D. G. Esaev, E. E. Meerson, L. M. Logvinskii, and G. N. Perov
Charge Transport through Layers of Thermally Nitrided SiO_2
(77.55; S10.1)
(a) Vol. **98**, No. 1, p. 319—327 (1986).

J. F. Boas
Electron Spin Resonance Studies of Light Sensitive Centers with Tetragonal Symmetry in CaO
(61.70 and 76.30; 71.55; S10.1)
(b) Vol. **138**, No. 2, p. 501—507 (1986).

R. Manaila and M. Zaharescu
Local Order in High Surface Area Amorphous Silica
(61.40 and 68.20; S10.1)
(a) Vol. **98**, No. 2, p. 377—382 (1986).

B. P. Rai
The Optical Absorption Edge of Thin GeO_2 Films Doped with Copper
(78.65; S10.1)
(a) Vol. **98**, No. 2, p. K171—K174 (1986).

O. K. Malinovskaya, A. F. Plotnikov, and V. A. Tolokonnikov
Determination of Energy Barriers for the Photocharging of Trapping Sites Contained in ZnSe Layer of Ni–ZnSe–SiO_2–Si Structure by Electrons and Holes
(73.40; S5.11; S8.12; S10.1)
(a) Vol. 98, No. 2, p. K191—K195 (1986).

S10.15. Mixed crystals

Original Papers and Short Notes

Sh. M. Efendiev, N. G. Darvishov, and F. K. Yusifov
Photoluminescence and Excitation Spectra in $Sr_xCa_{1-x}MoO_4$ Solid Solutions
(78.55; S10.15)
(a) Vol. **94**, No. 2, p. K113—K117 (1986).

J. Bhattacharyya, S. Chaudhuri, and A. K. Pal
Studies of the Optical Properties and the Burstein-Moss Shift in Indium Tin Oxide Films
(78.65; 68.55; S10.15)
(a) Vol. **95**, No. 1, p. 239—248 (1986).

Sh. M. Efendiev, V. E. Bagiev, and R. A. Aliev
Photoelectrical Fatigue Effect in $Bi_{12}TiO_{20}$ Crystals
(72.40; 71.55; S10.15)
(a) Vol. **95**, No. 2, p. 709—716 (1986).

Y. Miyamoto, T. Yamamoto, M. Koizumi, and M. Shimada
Magnetic and Electrical Properties of Cubic Perovskite $Sr(Fe_{1-x}Ni_x)O_3$ ($0 \leqq x \leqq 0.4$) Synthesized under High Oxygen Pressure
(72.80 and 75.50; S10.15)
(a) Vol. **95**, No. 2, p. K187—K191 (1986).

E. Culea, C. Gheorghiu, and Al. Nicula
Electrical Conductivity of Vitreous $75\% V_2O_5$–$25\% (As_2O_3 \cdot B_2O_3)$
(72.80; S10.15)
(a) Vol. **96**, No. 1, p. K85—K88 (1986).

B. Poumellec, J. F. Marucco, and B. Touzelin
X K Edge Absorption of Titanium and L Edges of Tantalum in $(Ti, Ta)O_2$ Rutile Solid Solutions
(78.70; 71.35; 71.70; S10.15)
(b) Vol. **137**, No. 2, p. 519—531 (1986).

C. A. Hogarth and Z. T. Al-Dhhan
Optical Absorption in Thin Films of Cerium Dioxide and Cerium Dioxide Containing Silicon Monoxide
(78.65; S10.15)
(b) Vol. **137**, No. 2, p. K157—K160 (1986).

H. Gruber, E. Krautz, H. P. Fritzer, K. Gatterer, and A. Popitsch
Changes of Electrical Conductivity, Magnetic Susceptibility, and IR Spectra in the Ternary System $Mo_{n-x}W_xO_{3n-1}$
(72.80; 75.50; 78.30; S10.15)
(a) Vol. **98**, No. 1, p. 297—304 (1986).

D. Bahadur
Synthesis and Characterization of NASICON Based Glass
(66.30; 61.40; S10.15)
(a) Vol. **98**, No. 1, p. K23—K26 (1986).

S11. Other higher compounds

Review Article

K. V. S. Rama Rao and S. Ramaprabhu
Nuclear Quadrupole Resonance Studies on Inorganic Solids
(76.60; S8; S9; S10; S11)
(a) Vol. **93**, No. 1, p. 17—27 (1986).

Original Papers and Short Notes

D. Wolf and S. Kemmler-Sack
The Luminescence of Nd^{3+} in the Polymorphic Perovskites $Ba_6B_2W_3O_{18}$ (B = Gd, Lu, Y). A Case with a Low-Lying Charge Transfer Transition
(78.55; 71.55; 78.40; S11)
(a) Vol. **93**, No. 1, p. 271—275 (1986).

A. S. Litovchenko, V. A. Chernenko, I. V. Matyash, and O. D. Ishutina
Influence of Pressure on Dehydroxylation Effect in Muscovite
(64.75; 76.60; 77.20; S11)
(a) Vol. **93**, No. 1, p. K9—K13 (1986).

E. Trif and Al. Nicula
EPR of Nd^{3+} Ions in Y-Type Zeolites
(76.30; S11)
(b) Vol. **133**, No. 2, p. 683—686 (1986).

R. Jakubas, Z. Czapla, Z. Galewski, L. Sobczyk, O. J. Żogał, and T. Lis
Structure and Phase Transition in $(CH_3NH_3)_3Sb_2Cl_9$
(61.50 and 64.70; 76.60; 77.20; S11)
(a) Vol. **93**, No. 2, p. 449—455 (1986).

T. N. Mamontova, Kh. K. Mukanov, G. Z. Vinogradova, N. G. Maisashvili, and I. Z. Babievskaya
A Study of the Optical Properties and Photoluminescence of Glasses in the Ge–P–S System
(78.30 and 78.55; S11)
(a) Vol. **93**, No. 2, p. 635—644 (1986).

A. S. Litovchenko, V. E. Yachemenev, and V. A. Vilkovskii
Temperature Dependence of the Heat Capacity of Ferriferrous Micas
(65.40; S11)
(a) Vol. **93**, No. 2, p. K133—K135 (1986).

K. K. P. Srivastava and S. N. Choudhary
Orbit–Lattice Interaction and Mössbauer Quadrupole Splitting of Fe(II) in $(NMe_4)_2FeCl_4$
(71.70; 63.20; S11)
(b) Vol. **134**, No. 1, p. 289—296 (1986).

F. SUDA, J. HATANO, S. CHIKAMATSU, and H. FUTAMA
Bulging-Out Effect of the Domain Switching in Ferroelectric Guanidinium Aluminum Sulfate Hexahydrate (GASH)
(77.80; S11)
(a) Vol. **95**, No. 2, p. 505—509 (1986).

K. HEMACHANDRAN and A. R. CHETAL
X-Ray K-Absorption Study of Copper in Malachite Mineral
(78.70; S11)
(b) Vol. **136**, No. 1, p. 181—185 (1986).

R. K. SINGH, N. KUMAR, and G. B. VARSHNEY
Multipole Interaction Effects on Cohesive and Anharmonic Properties of Alkali Cyanide Crystals
(62.20; S11)
(b) Vol. **136**, No. 2, p. 457—465 (1986).

A. TRESSAUD, S. KHAÏROUN, L. RABARDEL, T. KOBAYASHI, T. MATSUO, and H. SUGA
Phase Transitions in Ammonium Hexafluorometallates(III)
(64.70; 65.40; S11)
(a) Vol. **96**, No. 2, p. 407—414 (1986).

M. SHIOJIRI, T. MAEDA, T. ISSHIKI, Y. HIROTA, S. SEKIMOTO, Y. IKEDA, and K. IWAUCHI
Crystal Structures and Magnetic Properties of $BaTi_{2-x}Sn_xFe_4O_{11}$
(61.60 and 75.50; S11)
(a) Vol. **97**, No. 1, p. 231—241 (1986).

R. K. GARTIA and A. B. AHMED
Low-Temperature Thermally Stimulated Luminescence in $KAlSi_3O_8$
(78.60; 71.55; S11)
(a) Vol. **97**, No. 1, p. 257—264 (1986).

H. V. VARMA and J. VARMA
Mössbauer Effect Study in Natural Staurolite
(76.80; S11)
(a) Vol. **97**, No. 1, p. 275—278 (1986).

V. N. VASYUKOV
Dependence of the Low Temperature EPR Spectrum of Cu^{2+} Ion on the Microwave Field Frequency and Temperature
(76.30; 71.70; S11)
(b) Vol. **137**, No. 2, p. 623—631 (1986).

A. A. KAMINSKII, K. KURBANOV, and A. G. PETROSYAN
Spectral Composition and Kinetics of 2 μm Stimulated Emission of Ho^{3+} Ions in Sensitized $Y_3Al_5O_{12}$ and $Lu_3Al_5O_{12}$ Single Crystals
(78.45; 78.55; S11)
(a) Vol. **98**, No. 1, p. K57—K62 (1986).

S11.1. Salts of oxy-acids

Original Papers and Short Notes

V. P. SETH, A. YADAV, and R. S. BANSAL
Solution of the Spin-Hamiltonian with Orthorhombic HF and g-Tensors ($S = \frac{1}{2}$). II. Application to VO^{2+} Centres in Oxide Glasses
(76.30; 71.70; S11.1)
(b) Vol. **133**, No. 1, p. 297—300 (1986).

M. KRUPSKI, A. A. MIRZAKHANYAN, W. HILCZER, and E. G. SHAROYAN
Effect of Pressure and Temperature on the EPR Spectrum of Cr^{3+} Ions in α-$LiIO_3$ Single Crystals
(76.30; 71.70; S11.1)
(b) Vol. **133**, No. 1, p. 355—361 (1986).

V. K. YANOVSKII and V. I. VORONKOVA
Polymorphism and Properties of Bi_2WO_6 and Bi_2MoO_6
(61.50; 64.70; 77.80; S11.1)
(a) Vol. **93**, No. 1, p. 57—66 (1986).

H. VOGT, R. ULM, P. WÜRFEL, and W. RUPPEL
Anomalous Charge Storage in $NaNO_2$ Layers
(77.50; 66.90; S11.1)
(a) Vol. **93**, No. 1, p. 299—308 (1986).

P. DORDOR, S. JOIRET, J. P. DOUMERC, J. C. LAUNAY, J. CLAVERIE, and P. HAGENMÜLLER
Electrical Properties of Lead-Doped $LaCoO_3$ Single Crystals
(72.80; 72.60; S11.1)
(a) Vol. **93**, No. 1, p. 321—328 (1986).

M. DERNO, G. HERMS, and H. STEIL
X-Ray Study on the Boron Coordination in Sodium Borate Glasses at Higher Temperatures
(61.40; 61.10; S11.1)
(a) Vol. **93**, No. 1, p. K5—K8 (1986).

R. K. GARTIA, R. SINGHA, and T. SHANKER RAO
Thermoluminescence of Topaz X-Irradiated at 80 K
(78.60; S11.1)
(a) Vol. **93**, No. 1, p. K59—K63 (1986).

T. Y. AYYUB and W. BOGUSZ
On Electrical Conductivity of $Li_{16-2x}Zn_x(SiO_4)_4$
(72.80; S11.1)
(a) Vol. **93**, No. 1, p. K97—K100 (1986).

A. A. BOKOV, I. P. RAYEVSKII, V. G. SMOTRAKOV, and O. I. PROKOPALO
Kinetics of Compositional Ordering in $Pb_2B'B''O_6$ Crystals
(77.80; 61.40; S11.1)
(a) Vol. **93**, No. 2, p. 411—417 (1986).

N. M. Plakida
Superionic Phase Transitions in Hydrogen Bonded Crystals
(64.60; 62.20; S11.1)
(b) Vol. **135**, No. 1, p. 133—139 (1986).

A. M. El-Dib and H. F. Hassan
Energy Gap, Effective Mass, and Critical Phenomena in $NaNO_2$ Crystal
(71.25 and 77.80; 65.70; S11.1)
(b) Vol. **135**, No. 1, p. 253—259 (1986).

P. Anuradha and I. V. K. Bhagavan Raju
Study of Plastic Flow in $CaCO_3$ and $NaNO_3$ Crystals under Concentrated Loads
(62.20; S11.1)
(a) Vol. **95**, No. 1, p. 113—119 (1986).

T. Hammad, F. El-Kabbany, and Y. Badr
Ferroelectric and Dielectric Properties of Thin $NaNO_2$ Layers
(77.80; S11.1)
(a) Vol. **95**, No. 1, p. 121—125 (1986).

F. El-Kabbany, Y. Badr, G. Said, S. Taha, and S. Mahrous
A Study of the Thermal Hysteresis in $AgNO_3$
(64.70; S11.1)
(a) Vol. **95**, No. 1, p. 127—134 (1986).

T. Juhász, J. S. Bakos, and Cs. Kuti
Direct Lifetime Measurement of the A_{1g} Phonon Mode in Calcite
(63.20; 78.30; S11.1)
(b) Vol. **135**, No. 2, p. K99—K102 (1986).

R. Poprawski and W. Salejda
The Phenomenological Model of Phase Transitions in NH_4HSeO_4 Crystals
(77.80; 65.50; 77.70; S11.1)
(a) Vol. **95**, No. 2, p. 459—465 (1986).

F. El-Kabbany, G. Said, S. Mahrous, and S. Taha
Comparative Electrical Study of Thermal Hysteresis in KNO_3, $AgNO_3$, and $NaNO_3$
(64.60; 77.80; S11.1)
(a) Vol. **95**, No. 2, p. 495—503 (1986).

S. E. Sarkisov and A. A. Kaminskii
Luminescence of Nd^{3+} Ions in Semiconductor $Bi_{12}GeO_{20}$ Crystals
(71.55 and 78.55; 61.60; 78.20; 78.50; S11.1)
(a) Vol. **95**, No. 2, p. 641—649 (1986).

A. M. El-Dib and H. F. Hassan
The Electronic Energy Bands of $AgNO_2$ Crystal at Room Temperature
(71.25; S11.1)
(b) Vol. **136**, No. 1, p. 159—166 (1986).

F. El-Kabbany, Y. Badr, G. Said, and S. Taha
Ferroelectricity Associated with the Metastable Phase III of $AgNO_3$
(77.80; 64.70; S11.1)
(a) Vol. **98**, No. 1, p. 117—125 (1986).

B. T. Deshmukh, S. V. Bodade, and S. V. Moharil
Thermoluminescence of $K_2Cd_2(SO_4)_3$
(78.60; 64.70; S11.1)
(a) Vol. **98**, No. 1, p. 239—246 (1986).

D. Čiplys, P. Hegedus, and R. Rimeika
Contributions of Elastic Strain and Electric Field to Acousto-Optic Diffraction in $LiNbO_3$:Ti
(78.20; S11.1)
(a) Vol. **98**, No. 1, p. K11—K14 (1986).

Z. Czapla, R. Jakubas, E. Narewski, and L. Sobczyk
Dielectric Relaxation in ND_4DSeO_4 (d-AHSe)
(77.40; S11.1)
(a) Vol. **98**, No. 1, p. K87—K91 (1986).

S. Gharbage, J.-C. Bissey, and Y. Servant
Electron Paramagnetic Resonance (EPR) and Related Magneto-Microwave Birefringence (MMB) Powder Spectra of Copper Nitrate $Cu(NO_3)_2 \cdot 2.5\ H_2O$
(76.30; 75.50; 76.90; S11.1)
(b) Vol. **138**, No. 2, p. 655—659 (1986).

Fat Duen Ho
Further Calculations of the F^+-Center Wave Function for Lithium Niobate
(71.55; S11.1)
(b) Vol. **138**, No. 2, p. 693—697 (1986).

K. Kurbanov, E. S. Dosmagambetov, R. Ehlert, D. Schultze, and A. A. Kaminskii
Crystal Growth and Spectroscopic Properties of Nd^{3+} Ions in Potassium Bismuth Niobate
(78.55; S10.15; S11.1)
(a) Vol. **98**, No. 2, p. K175—K177 (1986).

S11.2. Spinels, garnets, ferrites

Original Papers and Short Notes

A. Collomb, B. Lambert-Andron, J. X. Boucherle, and D. Samaras
Crystal Structure and Cobalt Location in the W-Type Hexagonal Ferrite [Ba]Co_2–W
(61.60; S11.2)
(a) Vol. **96**, No. 2, p. 385—395 (1986).

J. Abellán and M. Ortuño
The Verwey Transition in Polycrystalline Cobalt–Iron Ferrites
(75.30; 75.50; S11.2)
(a) Vol. **96**, No. 2, p. 581—586 (1986).

J. D. Stephenson and B. Strocka
White Synchroton Radiation Topography of $Ba_{1.03}Fe_{11.97}O_{19}$ and $Ba_{1.06}Fe_{10.92}Mn_{1.02}O_{19}$ Single Crystals
(61.10 and 75.50; 61.70; S11.2)
(a) Vol. **96**, No. 2, p. 597—602 (1986).

A. F. Rakov
Investigation of Colour Centres in γ-Irradiated $Y_3Al_5O_{12}$ Crystals at Room Temperature
(61.70; 78.50; S11.2)
(a) Vol. **96**, No. 2, p. K169—K173 (1986).

V. M. Talanov and G. V. Bezrukov
Thermodynamic Model of Interacting Multilevel Systems with a Single Order Parameter. II. Phase Diagrams
(64.60; S11.2)
(a) Vol. **97**, No. 1, p. 111—119 (1986).

H. Hemme, H. Dötsch, and J. Middelberg
Phase Matching of Optical Modes in Orthorhombic Garnet Films
(78.65; S11.2)
(a) Vol. **97**, No. 1, p. 249—256 (1986).

S. Ligenza
On the Dynamical Nature of Superexchange Interaction in YIG Studied by Neutron Spectroscopy
(72.25; S11.2)
(b) Vol. **137**, No. 2, p. K131—K135 (1986).

L. F. Magaña, M. A. Escobar, and R. Valenzuela
Effect of the Grain Size Distribution on the Ferromagnetic Hysteresis Loop
(75.60; S11.2)
(a) Vol. **97**, No. 2, p. 495—500 (1986).

A. Collomb, O. Abdelkader, A. Isalgué and J. P. Mignot
Magnetic Properties of Hexagonal Ferrites: $BaMg_2$–W and $BaCo_2$–W
(75.50; S11.2)
(a) Vol. **97**, No. 2, p. 511—519 (1986).

S12. Organic compounds

Original Papers and Short Notes

V. K. Jindal and J. Kalus
Calculation of Thermal Expansion and Phonon Frequency Shift in Deuterated Naphthalene
(63.20 and 65.70; 62.20; S12)
(b) Vol. **133**, No. 1, p. 89—99 (1986).

S. K. Hoffmann, J. Goslar, and L. S. Szczepaniak
Low-Symmetry Doped Cu(II) Complexes in $(gly)_2CaCl_2 \cdot 4\ H_2O$ Single Crystals: EPR Studies
(76.30; 71.70; S12)
(b) Vol. **133**, No. 1, p. 321—328 (1986).

B. Kumar
Carrier Drift Mobilities in Some Organic Solids. A Search for Organic Photovoltaic Materials. I. Charge Transport Calculations in 2-2′ bipyridine and pyridazino [4,5-d] pridazine
(72.40 and 72.80; 71.25; S12)
(b) Vol. **133**, No. 1, p. 395—402 (1986).

J. W. Clark, A. J. Leadbetter, M. Godlewska, and S. Urban
The Phase Transitions in 2,2-Dinitropropane
(64.70; S12)
(a) Vol. **93**, No. 1, p. 87—91 (1986).

K. R. K. Rao, J. V. Rao, P. Venkatacharyulu, and V. Baliah
Phase Transition Studies of the Liquid Crystals Heptyloxybenzylidene Phenylazoaniline
(64.70; 65.70; S12)
(a) Vol. **93**, No. 1, p. 93—97 (1986).

T. Kawakubo
Color Centers in Irradiated Para-Dichlorobenzene Crystals
(61.70; 76.30; 78.50; S12)
(a) Vol. **93**, No. 1, p. 177—184 (1986).

G. S. Mingaleev, A. P. Tyutnev, B. P. Gerasimov, and I. A. Kulchitskaya
Numerical Analysis of the Transient Radiation-Induced Conductivity in the Framework of the Rose-Fowler-Vaisberg Formalism
(72.40; 71.55; S12)
(a) Vol. **93**, No. 1, p. 251—262 (1986).

M. G. Kaplunov, E. B. Yagubskii, L. P. Rosenberg, and Yu. G. Borodko
Erratum to: Optical Properties of the Two Crystal Modifications of the Organic Conductor $(BEDT\text{-}TTF)_2I_3$
(78.30; S12)
(a) Vol. **93**, No. 1, p. 399 (1986).

A. A. Aly, I. H. Ibrahim, A. M. Abdeen, and M. M. Abdel-Gawad
Variances in Molecular Magnetic Susceptibilities of Monoclinic Crystals
(75.30; S12)
(b) Vol. **134**, No. 1, p. 119—124 (1986).

A. J. Pal and T. N. Misra
Delayed Luminescence of Organic Mixed Crystals. Guest Concentration Effect in the p-Terphenyl Host–Pyrene Guest System
(78.55; 71.35; S12)
(b) Vol. **134**, No. 1, p. 243—249 (1986).

R. Coelho, L. Levy, and D. Sarrail
On the Natural Decay of Corona Charged Insulating Sheets
(73.25 and 77.50; 77.30; S12)
(a) Vol. **94**, No. 1, p. 289—298 (1986).

T. Barbaszewski, M. Drwiega, E. Lipinska, and F. Starzyk
Iodine Implantation in Poly-N-Vinylcarbazole — an Organic Semiconductor
(61.70; 61.80; 78.65; S12)
(a) Vol. **94**, No. 1, p. 419—422 (1986).

K.-P. Meyer
On the Orientation of the Optic Plane in TGS
(77.80; S12)
(a) Vol. **94**, No. 1, p. K1—K3 (1986).

M. K. Kerimov and B. A. Suleimanov
Effect of Electric Field Strength on the Intensity Distribution of Electrostimulated Thermoluminescence of Polyethylene
(78.60; S12)
(a) Vol. **94**, No. 1, p. K51—K54 (1986).

S. N. Drozhdin, L. N. Kamysheva, and Z. A. Liberman
Investigation of the Internal Field of Doped TGS Crystals
(77.80; S12)
(a) Vol. **94**, No. 1, p. K69—K73 (1986).

A. N. Botvich, V. F. Shabanov, A. N. Vtyurin, and T. A. Pozdnyakova
Photon–Phonon Interaction Influence on the External Vibration Spectra of Molecular Crystals
(63.20 and 71.36; 78.30; S12)
(b) Vol. **134**, No. 2, p. 515—522 (1986).

S. Bedenk, J. Lužnik, J. Pirnat, Z. Trontelj, and W. Windsch
Bromine Nuclear Quadrupole Resonance in Tri-Sarcosine Calcium Bromide
(76.60; S12)
(a) Vol. **95**, No. 1, p. K33—K35 (1986).

J. Dolinšek, D. Hanžel, R. Blinc, and M. Koren
^{13}C and Proton NMR of Molecular Motion in $[N(CH_3)_4]_2ZnCl_4$
(76.60; S12)
(a) Vol. **95**, No. 1, p. K53—K56 (1986).

W. KUCZYŃSKI
Refractometric Measurements in Cholesteric Blue Phases
(64.70 and 78.20; S12)
(a) Vol. **96**, No. 2, p. K127—K131 (1986).

A. A. ALY, M. GAAFAR, N. S. ALY, and M. M. ABDEL-GAWAD
Variances in Molecular Refractivities and Anisotropies in Monoclinic Crystals
(78.20; S12)
(a) Vol. **96**, No. 2, p. K181—K186 (1986).

P. REINEKER, R. G. WINKLER, and H. PORT
Temperature Dependence of Resonance Splittings in Molecular Pairs. Comparison of Optical Experiments and Theoretical Results for the Phenazine Triplet State
(63.20 and 71.35; S12)
(b) Vol. **137**, No. 1, p. 137—147 (1986).

S. HEDEWY and S. K. HOFFMANN
Electron Paramagnetic Resonance of Ferric Acetyl Acetonate
(76.30; S12)
(a) Vol. **97**, No. 1, p. 129—133 (1986).

L. N. KAMYSHEVA, S. N. DROZHDIN, and O. M. SERDYUK
The Influence of Defects on the Processes of Polarization Relaxation in TGS Crystals
(61.70 and 77.80; S12)
(a) Vol. **97**, No. 1, p. K29—K34 (1986).

S. L. DRECHSLER and M. BOBETH
Momentum-Dependent Dielectric Response of Strongly Heterogeneous Materials with Anisotropic Components
(77.20; S12)
(b) Vol. **137**, No. 2, p. 555—567 (1986).

A. M. BIRADAR, S. S. BAWA, S. B. SAMANTA, and S. CHANDRA
Measurements of a Free Value of Helical Pitch in a Thin Sample of Ferroelectric Liquid Crystal
(61.30 and 77.80; S12)
(a) Vol. **97**, No. 2, p. 427—430 (1986).

G. KAMPFRATH, D. DUSCHL, and C. HAMANN
Doped Polymer Films Prepared by Glow Discharge Polymerization
(61.40 and 73.60; 61.65; S12)
(a) Vol. **97**, No. 2, p. 619—626 (1986).

L. KOMITOV, G. HAUCK, and H. D. KOSWIG
Initially Splayed Nematic Liquid Crystal Layer under an Electric Field. Induction of π-Twist Metastable State
(61.30; 77.90; 78.65; S12)
(a) Vol. **97**, No. 2, p. 645—655 (1986).

S12.1. Quasi-one-dimensional systems

Original Papers and Short Notes

KH. I. PUSHARKOV and M. T. PRIMATAROVA
Solitary Excitations in Molecular Chains in the Presence of Vibronic States
(71.38; 71.35; S12.1)
(b) Vol. **133**, No. 1, p. 253—265 (1986).

V. V. KRASNOGOLOVETS and P. M. TOMCHUK
Erratum to: The Coulomb Correlations and the Electric Field Local Heterogeneities on the Proton Conductivity of Hydrogen Bonded Chains
(71.38 and 77.30; S12.1)
(b) Vol. **133**, No. 1, p. 431 (1986).

M. A. YURISHCHEV and A. M. STERLIN
Three-Chain Ising Strip in a Nonuniform Field
(75.10; S12.1)
(b) Vol. **133**, No. 2, p. 539—546 (1986).

G. VÖLKEL and W. BRUNNER
Electron Spin–Lattice Relaxation of Trapped EPR Active Centres in O_2-Contaminated Polyacetylene
(76.30; S12.1)
(a) Vol. **94**, No. 2, p. 673—678 (1986).

W. SCHILLER, E. CHILLA, and V. AKSENOV
The Influence of Anharmonicity on the Electrical Resistivity in TTF–TCNQ
(72.10; S12.1)
(b) Vol. **135**, No. 2, p. 837—848 (1986).

A. BARTL and A. STASKO
ENDOR Studies on Conjugated Polymers
(76.30 and 76.70; S12.1)
(a) Vol. **95**, No. 2, p. 659—663 (1986).

C.-M. LIEGENER, G. BICZÓ, and A. GODZIK
On the Concept of Intermediate One-Particle States in Finite Chains. A Self-Consistent Field Study
(71.50; 73.20; S12.1)
(b) Vol. **136**, No. 1, p. 281—289 (1986).

LJ. RISTOVSKI, G. S. DAVIDOVIĆ-RISTOVSKI, and V. RISTIĆ
Bisolitons in a Molecular Polymer Chain
(71.45; S12.1)
(b) Vol. **136**, No. 2, p. 615—622 (1986).

S. ICHINOSE
Bound Domain Wall Pairs in Wolffram's Red Salt
(75.10 and 75.60; S12.1)
(b) Vol. **137**, No. 2, p. 479—485 (1986).

Author Index

Attention

To find a paper of your interest please choose first the special entry of the subject index and then use the volume number and page.

Substance Classification*)

S1. Transition metals and their alloys compounds (for oxides see S10)
- S1.1. Fe
- S1.2. Other 3d metals
- S1.3. Noble metals
- S1.4. Lanthanides
- S1.5. Actinides
- S1.61. Carbides, silicides, germanides
- S1.62. Nitrides, phosphides, arsenides
- S1.63. Sulfides, selenides, tellurides

S2. Alkali metals and their alloys and compounds (for alkali halides, see S9.11)

S3. Alkaline earth metals and their alloys and compounds (for alkaline earth halides, see S9.12)

S4. Other metals and their alloys

S5. Nonmetallic elements
- S5.11. Silicon
- S5.12. Germanium
- S5.2. Solidified Gases

S6. Group IV compounds (for transition metal compounds, see S1.61)

S7. Group V compounds (for transition metal compounds, see S1.62)
- S7.1. III–V compounds
- S7.11. Phosphides
- S7.12. Arsenides
- S7.13. Antimonides
- S7.15. Mixed crystals
- S7.16. Higher compounds

S8. Chalcogenides (for transition metal compounds, see S1.63)
- S8.1. II–VI compounds
- S8.11. Sulfides
- S8.12. Selenides
- S8.13. Tellurides
- S8.15. Mixed crystals
- S8.16. Higher compounds

S9. Halides
- S9.11. Alkali halides
- S9.12. Alkaline earth halides
- S9.15. Mixed crystals
- S9.16. Higher compounds

S10. Oxides
- S10.1. Simple oxides
- S10.15. Mixed crystals

S11. Other higher compounds
- S11.1. Salts of oxy-acids
- S11.2. Spinels, garnets, ferrites

S12. Organic compounds
- S12.1. Quasi-one-dimensional systems

For comments on the use of the classification scheme see p. 10.

---

*) Outside the ICSTI Classification for Physics.

Manuscripts and letters for physica status solidi (b) — basic research and physica status solidi (a) — applied research

should be sent to

Leipziger Straße 3—4, P. B. 1233, DDR-1086 Berlin

or

Kunst und Wissen, Erich Bieber OHG, Postfach 330104, D-1000 Berlin 33.

The object of "physica status solidi (b)" and "physica status solidi (a)" is to publish rapidly important and topical results concerning fundamental research in the field of solid state physics and aspects of application and experimental techniques of fundamental solid state research, respectively. The issues of "physica status solidi (b)" and "physica status solidi (a)" will be published on the 1st and the 16th day of each month, respectively. Either series will contain the following categories:

1. Review articles written by arrangement with the editors. They will be remunerated.
2. Original papers selected from those which have not been submitted for publication elsewhere, after being reviewed. (The latest date for acceptance of contributions is the 8th and the 23rd day of each month for "physica status solidi (b)" and "physica status solidi (a)", respectively, 50 days before the delivery of the journal.)
3. Short notes published in the off-set part of the journal. They are, if necessary after being reviewed, selected by editorial staff. (The latest date for acceptance of manuscripts for this part is the 12th and the 27th day of each month for "physica status solidi (b)" and "physica status solidi (a)", respectively, 16 days before the delivery of the journal.)
4. Pre-printed titles of papers to be published in the next issue of both series.

Editorial Information

Manuscripts should be submitted preferentially in English. Besides, papers in German, French, and Russian are admitted. The final decision on the incorporation of the papers in the series (b) or (a) is taken by the editorial staff after consultation of the referee. Papers on liquids are only published in physica status solidi if the problems treated are also relevant for the solid state. Authors are asked to observe the following rules:

1. Manuscripts have to be written as consisely as possible. At most, 20 and 4 pages (28 lines per page; abstracts, tables, references, figures, and captions of figures included, where 3 figures are counted as about one page) are admitted for original papers and short notes, respectively. All manuscripts should be typewritten on one side only. Two copies are required. Badly arranged manuscripts are returned.
2. The author's name, initials, and institution should be given.
3. One abstract of about ten lines in English and one in German, French, or Russian are required for original papers.
4. Because of lack of space conclusions (if any) must not have the character of a summary (i.e. substantially a repetition of the contents of the abstract).
5. Formulae should not be written too small and not with pencil. Separate lines for formulae are desirable. Indistinct writing will lead to a return of the manuscript.
6. Drawings should be enclosed on separate sheets of paper, not exceeding the maximum and minimum sizes of 20×30 and 10×15 cm^2, respectively. Photographs should be submitted, on white paper, only for review articles and original papers. Duplicates are required.
7. Captions for the figures should be submitted on separate sheets of paper, and special instructions with regard to the size of illustrations should be given on these.
8. References should be numbered in order of appearance and should be given, on a separate sheet, in the form used in physica status solidi.
9. SI-units should be used.
10. Each paper must contain a proposal for the subject classification according to the classification scheme appearing in the January issues of each annual set.

The authors will be sent galley proofs of review articles and original papers. Proof reading of short notes and pre-printed titles will be done by the editorial staff.

For each original paper and short note a total of 75 reprints is available free of charge. The number of reprints free of charge per review article is 25. Requests for additional reprints (against payment) should be given together with the submitted manuscripts. **The deadline for orders of additional reprints is the 15th and the 30th of each month** for "physica status solidi (b)" and "physica status solidi (a)", respectively, two weeks before the delivery of the journal.

Bei Fragen zur Produktsicherheit wenden Sie sich bitte an:
If you have any questions regarding product safety,
please contact:

Walter de Gruyter GmbH
Genthiner Straße 13
10785 Berlin
productsafety@degruyterbrill.com